# OPPORTUNISTIC INFECTIONS:
## *Toxoplasma, Sarcocystis,* and Microsporidia

# World Class Parasites

## VOLUME 9

Volumes in the World Class Parasites book series are written for researchers, students and scholars who enjoy reading about excellent research on problems of global significance. Each volume focuses on a parasite, or group of parasites, that has a major impact on human health, or agricultural productivity, and against which we have no satisfactory defense. The volumes are intended to supplement more formal texts that cover taxonomy, life cycles, morphology, vector distribution, symptoms and treatment. They integrate vector, pathogen and host biology and celebrate the diversity of approach that comprises modern parasitological research.

*Series Editors*
Samuel J. Black, *University of Massachusetts, Amherst, MA, U.S.A.*
J. Richard Seed, *University of North Carolina, Chapel Hill, NC, U.S.A.*

# OPPORTUNISTIC INFECTIONS:
## *Toxoplasma, Sarcocystis,* and
# Microsporidia

*edited by*

## David S. Lindsay
*Virginia-Maryland Regional College of Veterinary Medicine*
*Blacksburg, Virginia*

*and*

## Louis M. Weiss
*Albert Einstein College of Medicine*
*Bronx, New York*

**KLUWER ACADEMIC PUBLISHERS**
**Boston / Dordrecht / New York / London**

**Distributors for North, Central and South America:**
Kluwer Academic Publishers
101 Philip Drive
Assinippi Park
Norwell, Massachusetts 02061 USA
Telephone (781) 871-6600
Fax (781) 681-9045
E-Mail: kluwer@wkap.com

**Distributors for all other countries:**
Kluwer Academic Publishers Group
Post Office Box 322
3300 AH Dordrecht, THE NETHERLANDS
Telephone 31 786 576 000
Fax 31 786 576 254
E-Mail: services@wkap.nl

  Electronic Services <http://www.wkap.nl>

**Library of Congress Cataloging-in-Publication Data**

A C.I.P. Catalogue record for this book is available
from the Library of Congress.

Opportunistic Infections: Toxoplasma, Sarcocystis, and Microsporidia
edited by David S. Lindsay and Louis M. Weiss
ISBN 1-4020-7814-5
e-Book ISBN 1-4020-7846-3

RC
186
.T75
066
2004

*The Publisher offers discounts on this book for course use and bulk purchases. For further information, send email to <Joseph.Burns@wkap.com>.*

# TABLE OF CONTENTS

**Preface**............................................................................vii

**Chapter 1. Biology of *Toxoplasma gondii* in cats and other animals.**
J.P. Dubey and David S. Lindsay..................................................1

**Chapter 2. Proteases as potential targets for blocking *Toxoplasma gondii* invasion and replication.** Vern B. Carruthers...............................21

**Chapter 3. Targeting the *Toxoplasma gondii* apicoplast for chemotherapy.** Sunny C. Yung and Naomi Lang-Unnasch....................39

**Chapter 4. Factors determining resistance and susceptibility to infection with *Toxoplasma gondii*.** Yasuhiro Suzuki....................................51

**Chapter 5. Immune responses to *Toxoplasma gondii* in the central nervous system.** Sandra Halonen....................................................67

**Chapter 6. Developmental stage conversion: Insights and possibilities**
Kami Kim and Louis Weiss ..........................................................89

**Chapter 7. *Sarcocystis* of humans**. Ronald Fayer...........................111

**Chapter 8. Zoonotic microsporidia from animals and arthropods with a discussion of human infections.** K.F. Snowden...............................123

**Chapter 9. Insights into the immune responses to microsporidia.**
Imtiaz A. Khan and Elizabeth S. Didier........................................135

**Chapter 10. Chemotherapy of microsporidiosis: Benzimidazoles, fumagillin and polyamine analogues.** Cyrus J. Bacchi and Louis M. Weiss.................................................................................159

**Chapter 11. Phylogenetics: Taxonomy and the microsporidia as derived fungi.** Charles R. Vossbrinck, Theodore G. Andreadis and Louis M. Weiss.................................................................................189

**Chapter 12. The Microsporidia genome: Living with minimal genes as an intracellular eukaryote.** C.P. Vivarès and G. Méténier.................215

**Index**...............................................................................243

# Preface

Protozoan parasites compromise an extremely diverse group of eukaryotic life forms. The Phyla Apicomplexa and Microsporidia are obligatory parasites. All Microsporidian parasites are intracellular in host cells and most Apicomplexan parasites live inside of host cells. These two Phyla of protists contain some of the most interesting and diverse groups of parasites infecting vertebrates and invertebrates.

This book will focus on two important Genera of Apicomplexan parasites, *Toxoplasma gondii* and *Sarcocystis* species, and the medically important members of the Phylum Microsporida. We have been fortunate in obtaining excellent contributions from many experts in the field.

*Toxoplasma gondii* has long been recognized as an important parasite of humans and their domestic animals. The devastating effects of congenital toxoplasmosis were well known long before the life cycle of the parasite was fully understood. When the AIDS epidemic began in the early 1980's *T. gondii* gained more prominence as a severe opportunistic parasite and AIDS defining infection. About 1.5 million cases human toxoplasmosis occur in the United States each year. This volume begins with a discussion of the *T. gondii* life cycle and the impact of this parasite on humans and animals and potential sources of human infection by J.P. Dubey and David S. Lindsay. *Toxoplasma gondii* has an actively replicating stage the tachyzoite and a tissue cyst stage that contains latent stages called bradyzoites. Vern Carruthers examines how these parasites invade the host cells to set up house keeping. He emphasizes the importance of proteases in *T. gondii* invasion and discusses proteases as potential targets for blocking *T. gondii* cell invasion and replication. *Toxoplasma gondii* and other Apicomplexan parasites contain a plastid-like organelle the apicoplast. This is an important potential drug target and Sunny C. Young and Naomi Lang-Unash provide current information into the function of the apicoplast and on the possibility of using metabolic functions of this organelle as a drug target. Yasuhiro Suzuki discusses host and parasite genetic and other factors that influence the pathogenesis of *T. gondii* infection. Toxoplasmic encephalitis and the host's response to infection in the central nervous system are examined by Sandra Halonen. The factors that govern the conversion of tachyzoites to bradyzoites and bradyzoites to tachyzoites in host tissues are discussed by Kami Kim and Louis M. Weiss.

Humans are definitive and intermediate hosts for *Sarcocystis* species. This parasite must have two hosts in its life cycle. *Sarcocystis hominus* and *S. suihominus* are the two species of this parasite that use humans as a definitive host. Intestinal disease similar to intestinal isosoporiasis often occurs in human intestinal *Sarcocystis* infections. Ronald Fayer describes the biology *Sarcocystis* infections in humans and addresses the unusual occurrence of this parasite in the muscles of humans.

The Microsporidia have become important opportunistic parasites due to the AIDS epidemic. These protists are small (2 μm) and are often difficult to diagnose. Little is known about the sources of the microsporidian infection seen in humans and Karen F. Snowden discusses the recent findings on zoonotic microsporidia and their transmission to humans. Research into the immune response to microsporidiosis has advanced our knowledge at a rapid pace since the importance of these organisms became fully appreciated in the 1980's. Current knowledge of the cell mediated and other immune responses in microsporidiosis are presented by Imtiaz A. Khan and Elizabeth S. Dider. New insights gained in the metabolic pathways and polyamine metabolism in microsporidian parasites are discussed by Cyrus J. Bacchi and Louis M. Weiss. The utility of these pathways as potential new drug targets for these parasites is also discussed. Charles R. Vossbrinck, Theodore G. Andreadis and Louis M. Weiss introduce and discuss the phylogeny of the Microsporidia and evidence of the relationship of Microsporidia to the fungi. Our understanding of the genomes of parasites is increasing as more genome projects on pathogenic protists are completed. Microsporidia have the smallest genomes yet identified in eukaryotes (less than 3 Mbp). The mechanism by which these eukaryotes have developed and function with such small genomes has relevance to many areas of biology. The information on the completed *Encephalitozoon cuniculi* genome and the impact of these findings on future research are discussed by Christian P. Vivarès and Guy Méténier.

We express our thanks to our wonderful group of experts who provide the concise and current material for this volume. We thank Terry Lawrence for the cover artwork. We are grateful to Carly N. Wetch for creating the index and proof reading the text.

David S. Lindsay and Louis M. Weiss, December 2003

# BIOLOGY OF *TOXOPLASMA GONDII* IN CATS AND OTHER ANIMALS

J.P. Dubey[1] and D.S. Lindsay[2]

[1]*Animal Parasitic Diseases Laboratory, Animal and Natural Resources Institute, Agricultural Research Service, U.S. Department of Agriculture, Building 1001, BARC-East, 10300 Baltimore Avenue, Beltsville, Maryland 20705-2350, USA* [2]*Center for Molecular Medicine and Infectious Diseases, Department of Biomedical Sciences and Pathobiology, Virginia-Maryland Regional College of Veterinary Medicine, Virginia Tech, 1410 Prices Fork Road, Blacksburg, Virginia 24061-0342, USA*

## ABSTRACT

*Toxoplasma gondii* is a protozoan parasite of warm-blooded animals including humans. It has a worldwide distribution. Cats, including all felines, are its definitive hosts and excrete environmentally-resistant oocysts in their feces. Hosts become infected by ingesting food or drink contaminated with oocysts or by ingesting undercooked meat infected with *T. gondii*. It causes mental retardation and loss of vision in congenitally-infected children and abortion in livestock. This chapter describes life cycle, epidemiology, diagnosis, and prevention of *T. gondii*.

**Key words**: *Toxoplasma gondii*, cat, oocyst, tachyzoite, bradyzoite, tissue cyst

## INTRODUCTION

Infection with the protozoan parasite *Toxoplasma gondii* is one of the most common parasitic infections of man and other warm-blooded animals (Dubey and Beattie, 1988). It has been found worldwide from Alaska to Australia. Nearly one-third of humanity has been exposed to this parasite (Dubey and Beattie, 1988; Tenter et al., 2000). In most adults it does not cause serious illness, but it can cause blindness and mental retardation in congenitally infected children and devastating disease in those with depressed immunity. Domestic cats and other felids are the only known definitive hosts. This review is devoted to examining the biology of *T. gondii* in the definitive host and common intermediate hosts.

## LIFE CYCLE

*Toxoplasma gondii* is a coccidian parasite with cats as the definitive host, and warm-blooded animals as intermediate hosts (Frenkel et al., 1970) (Figure 1). It is among the most important of parasites of animals. There is only 1 species of *Toxoplasma, T. gondii.* Coccidia in general have complicated life cycles. Most coccidia are host-specific, and transmitted by a fecal-oral cycle. *Toxoplasma gondii* has adapted to use transmission by the fecal-oral cycle, by carnivorism, and transplacentally.

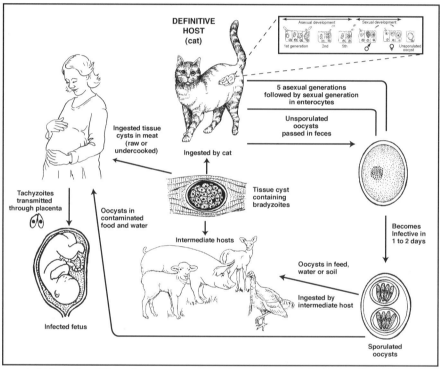

*Figure 1. Life cycle of Toxoplasma gondii.*

There are 3 infectious stages of *T. gondii*: the tachyzoites (in groups), the bradyzoites (in tissue cysts), and the sporozoites (in oocysts), as shown in Figures 1 and 2.

The tachyzoite is often crescent-shaped and is approximately the size (2 x 6 μm) of a red blood cell (Figure 2A). Its anterior end is pointed and its posterior end is round. It has a pellicle (outer covering), several organelles including subpellicular microtubules, mitochondrium, endoplasmic reticulum, a Golgi apparatus, an apicoplast, ribosomes, rough-surfaced endoplasmic reticulum, a micropore, and a well-defined nucleus. The nucleus is usually situated toward the posterior end or in the central area of the cell.

The tachyzoite enters the host cell by active penetration of the host cell membrane and can tilt, extend, and retract as it searches for a host cell. After entering the host cell, the tachyzoite becomes ovoid in shape and becomes surrounded by a parasitophorous vacuole. *Toxoplasma gondii* in a parasitophorous vacuole is protected from host defense mechanisms. The tachyzoite multiplies asexually within the host cell by repeated divisions in which 2 progeny form within the parent parasite, consuming it (Figure 1A). Tachyzoites continue to divide until the host cell is filled with parasites.

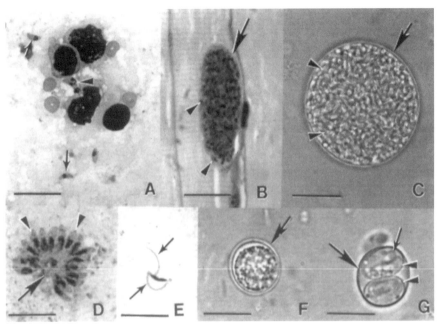

*Figure 2. Stages of Toxoplasma gondii. Scale bar in A-D= 20 μm, in E-G = 10 μm. A. Tachyzoites in smear of lung. Note crescent-shaped individual tachyzoites (arrows), and dividing tachyzoites (arrowheads). Giemsa stain. B. Tissue cysts in section of muscle. The tissue cyst wall is very thin (arrow) and encloses many tiny bradyzoites (arrowheads). Hematoxylin and eosin stain. C. Tissue cyst separated from host tissue by homogenization of infected brain. Note tissue cyst wall (arrow) and hundreds of bradyzoites (arrowheads). Unstained. D. Schizont (arrow) with several merozoites (arrowheads) separating from the main mass. Impression smear of infected cat intestine. Giemsa stain. E. A male gamete with two flagella (arrows). Impression smear of infected cat intestine. Giemsa stain. F. Unsporulated oocyst in fecal float.. Unstained. Note double layered oocyst wall (arrow) enclosing a central undivided mass. G. Sporulated oocyst with a thin oocyst wall (large arrow), 2 sporocysts (arrowheads). Each sporocyst has 4 sporozoites (small arrows). Unstained.*

After a few divisions, *T. gondii* form another stage called tissue cysts. Tissue cysts grow and remain intracellular. Tissue cysts vary in size from 5-70 Fm (Figures 2B,C). Although tissue cysts may develop in visceral organs, including lungs, liver, and kidneys, they are more prevalent in muscular and neural tissues (Figure 2B), including the brain (Figure 2C), eye, skeletal, and cardiac muscle. Intact tissue cysts probably do not cause any harm and can persist for the life of the host.

The tissue cyst wall is elastic, thin ($< 0.5$ μm), and may enclose hundreds of the crescent-shaped slender *T. gondii* stage known as bradyzoites (Figure 2C). The bradyzoites are approximately 7 x 1.5 μm. Bradyzoites differ structurally only slightly from tachyzoites. They have a nucleus situated toward the posterior end whereas the nucleus in tachyzoites is more centrally located. Bradyzoites are more slender than are tachyzoites and they are less susceptible to destruction by proteolytic enzymes than are tachyzoites.

All coccidian parasites have a resistant stage in their life cycle, called an oocyst. Oocysts of *T. gondii* are formed only in cats, not only domestic cats, but probably all felines (Figures 3, 4). Cats shed oocysts after ingesting any of the 3 infectious stages of *T. gondii*, i.e., tachyzoites, bradyzoites, and sporozoites (Dubey and Frenkel, 1972, 1976; Freyre et al., 1989; Dubey, 1996, 2002b). Less than 50% of cats shed oocysts after ingesting tachyzoites or oocysts whereas nearly all cats shed oocysts after ingesting tissue cysts (Dubey and Frenkel, 1976).

After the ingestion of tissue cysts by cats, the tissue cyst wall is dissolved by the proteolytic enzymes in the stomach and small intestine. The released bradyzoites penetrate the epithelial cells of the small intestine and initiate development of numerous generations of asexual and sexual cycles of *T. gondii* (Dubey and Frenkel, 1972). *Toxoplasma gondii* multiplies profusely in intestinal epithelial cells of cats (entero-epithelial cycle) and these stages are known as schizonts (Figure 2D). Organisms (merozoites) released from schizonts form male and female gametes. The male gamete has two flagella (Figure 2E) and it swims to and enters the female gamete. After the female gamete is fertilized by the male gamete (Figure 2E), oocyst wall formation begins around the fertilized gamete. When oocysts are mature, they are discharged into the intestinal lumen by the rupture of intestinal epithelial cells.

In freshly passed feces, oocysts are unsporulated (non-infective). Unsporulated oocysts are subspherical to spherical and are 10 x 12 μm in diameter (Figure 2F). They sporulate (become infectious) outside the cat within 1 to 5 days depending upon aeration and temperature. Sporulated oocysts contain 2 ellipsoidal sporocysts (Figure 2G). Each sporocyst contains 4 sporozoites. The sporozoites are 2 x 6 to 8 μm in size.

As the entero-epithelial cycle progresses, bradyzoites penetrate the lamina propria of the feline intestine and multiply as tachyzoites. Within a few hours after infection of cats, *T. gondii* may disseminate to extra-intestinal tissues. *Toxoplasma gondii* persists in intestinal and extra-intestinal tissues of cats for at least several months, and possibly for the life of the cat.

Hosts can acquire *T. gondii* infection by ingesting tissues of infected animals or ingesting food or drink contaminated with sporulated oocysts, or transplacentally. After ingestion, bradyzoites released from tissue cysts or sporozoites from oocysts penetrate intestinal tissues, transform to tachyzoites, multiply locally, and are disseminated in the body via blood or lymph. After a few multiplication cycles, tachyzoites encyst in many tissues. *Toxoplasma gondii* infection during pregnancy can lead to infection of the fetus. Congenital toxoplasmosis in humans, sheep, and goats can cause devastation of the fetus.

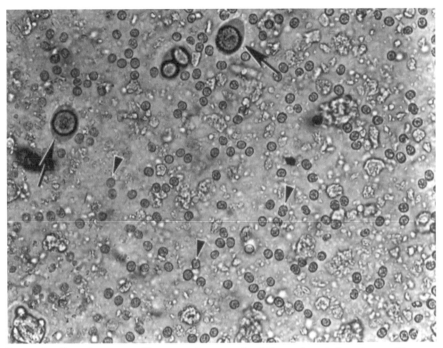

*Figure 3. Toxoplasma gondii oocysts in sugar fecal float of an infected cat. Note many spherical T. gondii oocysts (arrowheads). Also note oocysts of Isospora felis (arrows) which are often present in cat feces. Isospora felis is the most common coccidium of cats. Its oocysts are about 4 times the size of T. gondii oocysts and they often in different focus than T. gondii oocysts.*

## HOST-PARASITE RELATIONSHIP

*Toxoplasma gondii* can multiply in virtually any cell in the body. How *T. gondii* is destroyed in immune cells is not completely known. All extracellular forms of the parasite are directly affected by antibody but intracellular forms are not. It is believed that cellular factors, including lymphocytes and lymphokines, are more important than humoral factors in immune mediated destruction of *T. gondii.*

Immunity does not eradicate infection. *Toxoplasma gondii* tissue cysts persist several years after acute infection. The fate of tissue cysts is not fully known. It has been proposed that tissue cysts may at times rupture during the life of the host. The released bradyzoites may be destroyed by the host's immune responses. The reaction may cause local necrosis accompanied by inflammation. Hypersensitivity plays a major role in such reactions (Frenkel, 1973). After such events, inflammation usually again subsides with no local renewed multiplication of *T. gondii* in the tissue, though, occasionally there may be formation of new tissue cysts.

In immunosuppressed patients, such as those given large doses of immunosuppressive agents in preparation for organ transplants and in those with AIDS, rupture of a tissue cyst may result in transformation of bradyzoites into tachyzoites and renewed multiplication. The immunosuppressed host may die from toxoplasmosis unless treated.

Pathogenicity of *T. gondii* is determined by the virulence of the strain and the susceptibility of the host species. *Toxoplasma gondii* strains may vary in their pathogenicity in a given host. Certain strains of mice are more susceptible than others and the severity of infection in individual mice within the same strain may vary. Certain species are genetically resistant to clinical toxoplasmosis. For example, adult rats do not become ill, while young rats can die of toxoplasmosis. Mice of any age are susceptible to clinical *T. gondii* infection. Adult dogs, like adult rats, are resistant, whereas puppies are fully susceptible to clinical toxoplasmosis. Cattle and horses are among the hosts more resistant to clinical toxoplasmosis, whereas certain marsupials and New World monkeys are the most susceptible to *T. gondii* infection (Dubey and Beattie, 1988). Nothing is known concerning genetically-determined susceptibility to clinical toxoplasmosis in higher mammals, including humans.

## INFECTION IN CATS AND OTHER ANIMALS

*Toxoplasma gondii* is capable of causing severe disease in many species of animals other than humans. This subject is too broad to review here in detail and has been reviewed by Dubey and Beattie (1988), and

Tenter et al. (2000). Among livestock, toxoplasmosis causes great losses in sheep and goats. *Toxoplasma gondii* may cause embryonic death and resorption, fetal death and mummification, abortion, stillbirth and neonatal death in these animals. Disease is more severe in goats than in sheep. Outbreaks of toxoplasmosis in pigs have been reported from several countries, especially Japan. Mortality in young pigs is more common than mortality in adult pigs. Pneumonia, myocarditis, encephalitis and placental necrosis have been reported to occur in infected pigs. Sporadic and widespread outbreaks of toxoplasmosis occur in rabbits, mink, birds and other domesticated and wild animals. Among

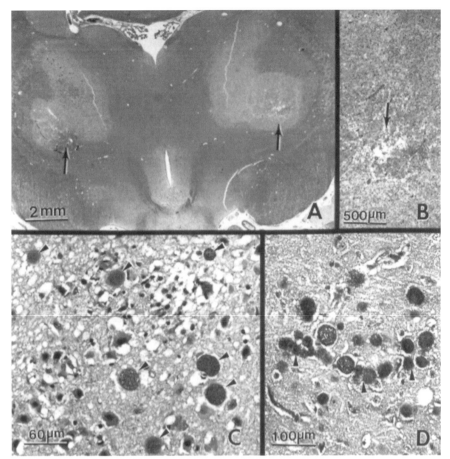

*Figure 4. Sections of brain of cats naturally-infected with T. gondii. (A) Thalamus of a thiamine deficient cat showing characteristic bilateral necrosis (arrows) of lateral geninuclate bodies. (B) Higher magnification of 4A showing necrosis of neuropil. Tachyzoites of T. gondii are present but not visible at this magnification. (C) and (D) Numerous tissue cysts (arrowheads) in cerebrum of a cat associated with focal lesion in C and without inflammation in D.*

zoo animals, toxoplasmosis causes severe, often fatal disease in Australian marsupials, New World monkeys, Pallas cats, and canaries (*Serinus canarius*) (Dubey and Beattie, 1988; Dubey and Odening, 2001; Dubey, 2002a).

Among companion animals, fatal toxoplasmosis in dogs is mostly secondary to immunosuppression by concurrent distemper virus infection, and clinical canine toxoplasmosis is now rare in dogs that are vaccinated against distemper virus (Dubey et al., 1989). Although cats of any age can die of toxoplasmosis, it is more frequent in kittens and those with depressed immunity (Dubey and Carpenter, 1993a,b). Among 100 clinical cases of feline toxoplasmosis diagnosed in one hospital, 36 were considered to have generalized toxoplasmosis, 26 predominantly pulmorary lesions, 16 abdominal, 2 hepatic, 1 pancreatic, 1 cardiac, 2 cutaneous, 7 neurologic, and 9 had neonatal toxoplasmosis. In 14 cats, concurrent microbial infections or other maladies were seen (Dubey and Carpenter, 1993b).

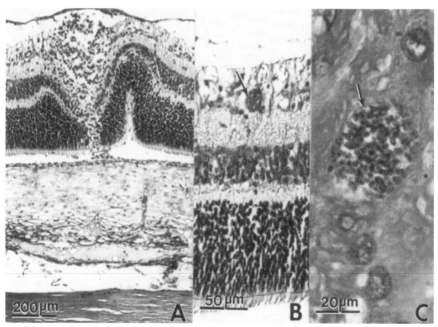

*Figure 5. Toxoplasma gondii-associated retinochoroditis in a 1-month-old naturally infected cat. H&E stain. (A) Focal infiltration of lymphocytes extending to all layers of retina. (B) A tissue cyst (arrow) in retina. (C) A tissue cyst (arrow) in choroid.*

*Toxoplasma gondii* can cause loss of vision in cats. Forty-one eyes from 27 of the cats were examined microscopically. Twenty-two of the

27 cats (81.5 %) had evidence of intraocular inflammation in one or both eyes. Multifocal iridocyclochoroiditis was the most common lesion and was seen in 18 (81.8 %) of the cats with ophthalmitis. The ciliary body was the most often severely affected portion of the uvea. Of the 22 cats with ocular toxoplasmosis, *T. gondii* was found in eyes of 10 (Figure 5). *Toxoplasma gondii* was found in the retina of 5 cats, the choroid of 2, the optic nerve of 1, the iris of 3, and the ciliary body of 4. *Toxoplasma gondii* was identified in 80 % of 55 brains, 70.0 % of 90 livers, 76.6 % of 86 lungs, 64.4 % of 45 pancreata, 62.7 % of 59 hearts, 45.8 % of 72 spleens, 41.5 % of 65 intestines, 17.7 % of 61 kidneys, and 60.0 % of 30 adrenal glands (Dubey and Carpenter, 1993b).

Subclinical *T. gondii* infections in cats and food animals are important epidemiologically and this is subject reviewed under the section of epidemiology.

## DIAGNOSIS

Diagnosis is made by biologic, serologic, or histologic methods or by some combination of the above. Clinical signs of toxoplasmosis are nonspecific and are not sufficiently characteristic for a definite diagnosis. Toxoplasmosis in fact mimics several other infectious diseases.

Detection of *T. gondii* antibody in patients may aid diagnosis. There are numerous serologic procedures available for detection of humoral antibodies; these include the Sabin-Feldman dye test, the indirect hemagglutination assay, the indirect fluorescent antibody assay (IFA), the direct agglutination test, the latex agglutination test, the enzyme-linked immunoabsorbent assay (ELISA), and the immunoabsorbent agglutination assay test (IAAT). The IFA, IAAT and ELISA have been modified to detect IgM antibodies. The IgM antibodies appear sooner after infection than the IgG antibodies and the IgM antibodies disappear faster than IgG antibodies after recovery (Remington et al., 1995).

*Toxoplasma gondii* can be isolated from patients by inoculation of laboratory animals and tissue cultures with secretions, excretions, body fluids, tissues taken by biopsy and tissues with macroscopic lesions taken postmortem. Using such specimens one may not only attempt isolation of *T. gondii*, but one may search for *T. gondii* microscopically or for toxoplasmal DNA by use of the polymerase chain reaction (Grover et al., 1990).

As just noted, diagnosis can be made by finding *T. gondii* in host tissue removed by biopsy or at necropsy. A rapid diagnosis may be made by microscopic examination of impression smears of lesions. After drying for 10 to 30 minutes, the smears are fixed in methyl alcohol and stained with one of the Romanowsky strains, the Giemsa stain being very satisfactory. Well preserved *T. gondii* are crescent-shaped (1A). In sections, the tachyzoites usually appear round to oval. Electron microscopy can aid diagnosis. *Toxoplasma gondii* tachyzoites are always located in vacuoles. Tissue cysts are usually spherical, lack septa and the cyst wall can be stained with a silver stain. The bradyzoites are strongly periodic acid Schiff (PAS) positive. The immunohistochemical staining of parasites with fluorescent or other types of labeled *T. gondii* antiserum can aid in diagnosis.

## EPIDEMIOLOGY

As noted earlier, toxoplasmosis may be acquired by ingestion of oocysts or by ingestion of tissue-inhabiting stages of the parasite. The contamination of the environment by oocysts is widespread as oocysts are shed by cats, not only the domestic cat but by other felines as well (Dubey and Beattie, 1988). Domestic cats are probably the major source of contamination as oocyst formation is greatest in the domestic cats, and they are extremely common. Widespread natural infection of the environment is possible since a cat may excrete millions of oocysts after ingesting as few as 1 bradyzoite or 1 tissue cyst and many tissue cysts may be present in one infected mouse (Dubey and Frenkel, 1972; Dubey, 2001). Sporulated oocysts survive for long periods under most ordinary environmental conditions. They can survive in moist soil, for example, for months and even years (Dubey and Beattie, 1988). Oocysts in soil do not always stay there as invertebrates like flies, cockroaches, dung beetles, and earthworms can mechanically spread these oocysts and even carry them onto food.

While a few cats may be shedding *T. gondii* oocysts (Table 1) at any given time (as few as 1%); the enormous numbers shed and their resistance to destruction assure widespread contamination. Under experimental conditions, infected cats can shed oocysts after reinoculation with tissue cysts (Dubey, 1995). Congenital infection can also occur in cats, and congenitally infected kittens can excrete oocysts, providing another source of oocysts for contamination. Infection rates in cats are determined by the rate of infection in local avian and rodent populations because cats are thought to become infected by eating these animals. The more oocysts in the environment the more likely prey animals would be infected and this in turn would increase the infection rate in cats. For example, *T. gondii* oocysts were found in 23.2 % of 237 cats in Costa Rica where infection in local rodents and birds was high compared with only 1% or less in cats in the United States (Table 1).

Infection in humans is probably most often the result of ingestion of tissue cysts contained in undercooked meat (Dubey and Beattie, 1988; Cook et al., 2000; Lopez et al., 2000) . *Toxoplasma gondii* infection is common in many animals used for food including, sheep, pigs, and rabbits.

Infection in cattle is less prevalent than is infection in sheep or pigs. *Toxoplasma gondii* in tissue cysts survive in food animals for years, and virtually all commercial cuts of meats can be infected (Dubey et al., 1986).

The prevalence of *T. gondii* in food animals may differ based on farm management practices, climatic conditions and different parts of the world (Dubey and Beattie, 1988; Tenter et al., 2000).

*Table 1. Selected reports of prevalence of Toxoplasma gondii oocysts in feces of naturally-infected cats*

| Country | No. of cats | No. positive | % prevalence | Reference |
|---|---|---|---|---|
| Australia | 74 | 1 | 1.3 | Jakob-Hoff and Dunsmore (1983) |
| Brazil | 185 | 1 | 0.5 | Nery-Guimarães and Lage (1973) |
| Costa Rica | 237 | 55 | 23.2 | Ruiz and Frenkel (1980b) |
| Egypt | 213 | 88 | 41.3 | Rifaat et al. (1976) |
| Germany | 502 | 5 | 0.9 | Janitschke and Kühn (1972) |
| Hungary | 200 | 2 | 1.0 | Varga 1982 |
| Italy | 250 | 1 | 0.4 | Pampiglione et al. (1973) |
| Japan | 446 | 4 | 0.8 | Ito et al. (1974) |
| Spain | 169 | 1 | 0.6 | Aparicio Garrido et al. (1972) |
| Uruguay | 138 | 1 | 0.7 | Freyre et al. (1981) |
| U.S.A | 1,604 | 12 | 0.7 | Wallace (1973) |
| | 1,000 | 7 | 0.7 | Dubey et al. (1977) |
| | 274 | 5 | 1.8 | Dubey et al. (1995) |

*Table 2. Prevalence of viable Toxoplasma gondii in pigs in the U.S.*

| Source | No. of animals | Location | Isolation | % positive | Reference |
|---|---|---|---|---|---|
| Slaughter house | 50 | Maryland | 12 | 24.0 | Jacobs et al. (1960) |
| Grocery store | 50 | California | 16 | 32.0 | Remington (1968) |
| Sows (slaughter house) | 1,000 | Iowa | 170 | 17.0 | Dubey et al. (1995b) |
| One farm market weight | 55 | Massachusetts | 51 | 92.7 | Dubey et al. (2002) |

*Toxoplasma gondii* has been isolated from pigs in the U.S. (Table 2). Its prevalence is declining drastically in pigs raised indoors in commercial enterprises (Table 3).

*Toxoplasma gondii* has not been isolated from edible beef in the U.S. (Dubey and Beattie, 1988). Viable parasites have been isolated from poultry (Jacobs and Melton, 1966). Its prevalence in commercially raised poultry is unknown but likely to be low. Most of poultry in the U.S. is frozen. Freezing kills *T. gondii* (Kotula et al., 1991).

*Table 3. Prevalence of Toxoplasma gondii antibodies\* in pigs in the United States*

| Year | Location | No. tested | % positive | Comment[‡] | Reference |
|------|----------|-----------|------------|---------|-----------|
| 1983-1984 | National sample | 11,229 | 42 | Sows | Dubey et al. (1991) |
|  |  | 613 | 23 | Market | Dubey et al. (1991) |
| 1990 | National sample | 3,479 | 20 | Sows | Patton et al. (1996) |
| 1991-1992 | Tennessee | 3,841 | 36 | Sows | Assadi-Rad et al., (1995) |
| 1992 | Illinois | 5,080 | 20.8 | Sows | Weigel et al. (1995) |
|  |  | 1,885 | 3.1 | Market | Weigel et al. (1995) |
| 1992-1993 | Illinois | 4,252 | 2.3 | Market | Patton et al. (Un.P) |
|  |  | 2,617 | 15.1 | Sows |  |
| 1994-1995 | North Carolina | 2,238 | 0.58 | Market | Davis et al. (1998) |
|  |  | 1,752 | 0.005 | Market (confinement) | Davis et al. (1998) |
| 1999 | New England states | 1,897 | 47.5 | Market | Gamble et al. (1999) |

\*Modified agglutination test titer @ 1:25
[‡]Market weight pigs (finishers, 5-8 months old)

Although *T. gondii* was isolated from 9.3% of 86 sheep diaphragms (Jacobs et al., 1960), 4% of 50 lamb chops (Remington, 1968) its prevalence in edible lamb is unknown. Overall, the consumption of lamb in the U.S. is low (Dubey, 1994) and adult sheep are not sold for human consumption. However, consumers should be aware of the danger of eating uncooked lamb because in one study viable *T. gondii* was isolated from legs of 8 of 8 congenitally-infected lambs (Dubey and Kirkbride, 1989). The prevalence of *T. gondii* in horses is very low (<5%).

Wild game is a potential source of *T. gondii* infection. Seroprevalence of *T. gondii* in white-tailed deer and black bears is very high (up to 80%) and viable *T. gondii* has been isolated from venison (Lindsay et al., 1991) and black bears (Dubey et al., 1995a).

The relative frequency of acquisition of toxoplasmosis from eating raw meat and that due to ingestion of food contaminated by oocysts from cat feces is very difficult to determine. Therefore, any statements on the subject are at best speculative.

There is little, if any, danger of *T. gondii* infection by drinking cow's milk and, in any case, cow's milk is generally pasteurized or even boiled, but infection has followed drinking unboiled goat's milk (Dubey and Beattie, 1988). Raw hens' eggs, although an important source of *Salmonella* infection, are extremely unlikely to transmit *T. gondii* infection. Transmission by sexual activity including kissing is probably rare and epidemiogically unimportant (Dubey and Beattie, 1988).

Transmission may occur through blood transfusions and organ transplants. Of these means, transmission by transplantation is most important. Toxoplasmosis may actually arise in 2 ways in people undergoing transplantation: *i.e.* 1) from implantation of an organ or bone marrow from an infected donor into a nonimmune immunocompromised recipient and, 2) from induction of disease in an immunocompromised, latently infected recipient. The tissue cysts in the transplanted tissue or in the latently infected are probably the source of the infection. In both cases, the cytotoxic and immunosuppressive therapy given to the recipient is the cause of the induction of the active infection and the disease (Frenkel, 1973; Dubey and Beattie, 1988).

## RESERVOIRS AND DETECTION OF *T. GONDII* IN THE ENVIRONMENT

As mentioned earlier, humans become infected by ingesting tissue cysts in undercooked or uncooked meat or by ingesting food and water contaminated with oocysts from infected cat feces. There are no tests at the present time to determine the source of infection in a given person. All the evidence is based on epidemiological surveys. For example, in certain areas of Brazil, approximately 60% of 6-8 year old children have antibodies to *T. gondii* linked to the ingestion of oocysts from the environment heavily contaminated with *T. gondii* oocysts (Bahia-Oliveira et al., 2003). Oocysts of *T. gondii* survive even in harsh environments for months. Infection in aquatic mammals indicate contamination and survival of oocysts in sea water (Cole et al., 2000). The largest outbreak of clinical toxoplasmosis in humans was epidemiologically-linked to drinking water from a municipal water reservoir in British Columbia, Canada (Bowie et al., 1997; Isaac-Renton et al., 1998). This water reservoir was thought to be contaminated with *T. gondii* oocysts excreted by cougars (*Felis concolor*) (Aramini et al., 1998, 1999).

Oocysts can be detected by examination of cat feces. Concentration methods (eg. flotation in high density sucrose solution) are often used because the number of *T. gondii* oocysts in cat feces may be too few to be detected by direct smear. For definitive identification, *T. gondii* oocysts should be sporulated and then bioassayed in mice to distinguish them from other related coccidians.

For epidemiological surveys, detection of *T. gondii* oocysts in cat feces is not very practical; at any given time only 1% of cats are found shedding oocysts because oocysts are shed for only a short period (1-2 weeks) in the life of the cat (Dubey and Beattie, 1988). Determining serological prevalence is a better measure of exposure of cats to *T. gondii* infection than detection of oocysts. It is a fair assumption that cats that are seropositive have already shed *T. gondii* oocysts. In an epidemiological survey on *T. gondii* infection on pig farms, *T. gondii* oocysts were detected in only 5 of 274 (1.8%) samples of cat feces, 2 of 491 (0.4%) samples of feed and 1 of 79 (1.3 %) samples of soil on some farms; 267 of 391 (68.3%) cats had antibodies to *T. gondii* (Dubey et al., 1995c). Serologic surveys indicate that about half of cats surveyed in the US had antibodies to *T. gondii* (Table 4).

*Table 4. Recent surveys of the seroprevalence of Toxoplasma gondii antibodies in domestic cats from the U.S.*

| Location | Type of cats | No. of cats | % seropositive | Serologic test* | Cut-off titer | Reference |
|---|---|---|---|---|---|---|
| Ohio | Urban | 1,000 | 39 | IFAT | 1:16 | Claus et al. (1977) |
| Ohio | Rural | 275 | 48 | MAT | 1:25 | Dubey et al. (2002) |
| Iowa | Rural | 74 | 42 | MAT | 1:32 | Smith et al. (1992) |
| Illinois | Rural | 391 | 68 | MAT | 1:25 | Dubey et al. (1995) |
| Rhode Island | Urban | 200 | 42 | MAT | 1:25 | DeFeo et al. (2002) |

* MAT = Modified agglutination test,
  IFAT = Indirect fluorescent antibody test.

Although *T. gondii* has been isolated from soil, there is no simple method for use on an epidemiological scale. Bioassay of soil samples in pigs and chickens may be more useful than direct determination of oocysts in soil; pigs can be infected by feeding as few as one oocyst (Dubey et al., 1996). In a study of feral chickens, *T. gondii* was isolated from 54% of 50 chickens by bioassay in mice (Ruiz and Frenkel, 1980a). Because feral chickens on small farms feed on the ground, finding *T. gondii* in chickens is a good indicator of infection in the environment. Although attempts to recover *T. gondii* oocysts from water samples in the British Columbia outbreak were unsuccessful, methods to detect oocysts were reported (Isaac-Renton et al., 1998). At present there are no commercial reagents available to detect *T. gondii* oocysts in the environment.

As stated earlier, among food animals, infection is more prevalent in sheep, goats, pigs, rabbits than in cattle and horses (Dubey and Beattie,

1988). *T. gondii* infection is also prevalent in game animals. Among wild game, *T. gondii* infection is most prevalent in black bears and in white-tailed deer. Approximately 80% of black bears are infected in the U. S. (Dubey and Odening, 2001), and about 60% of raccoons have antibodies to *T. gondii* (Dubey et al., 1995; Dubey and Odening, 2001). Because raccoons and bears scavenge for their food, infection in these animals is a good indicator of the prevalence of *T. gondii* in the environment.

The number of *T. gondii* in meat from food animals is very low. It is estimated that as few as 1 tissue cyst may be present in 100 grams of meat. Therefore, without using a concentration method, it is not practical to detect this low level of *T. gondii* infection. Therefore, digestion of meat samples in trypsin or pepsin is used to concentrate *T. gondii* in meat (Dubey, 1998). Digestion in trypsin and pepsin ruptures the *T. gondii* tissue cyst wall releasing hundreds of bradyzoites. The bradyzoites survive in the digests for several hours. Even in the digested samples, only a few *T. gondii* are present and their identification by direct microscopic examination is not practical. Therefore, the digested material is bioassayed in mice (Dubey, 1998). The mice inoculated with digested material have to be kept for 6-8 weeks before *T. gondii* infection can be detected reliably- this procedure is not practical for mass scale samples. The detection of *T. gondii* DNA in meat samples by PCR has been reported (Warnekulasuriya et al., 1998; Aspinall et al., 2002) but there are no data on specificity and sensitivity of this method to detect *T. gondii* in meat samples. A highly sensitive method using a Real-Time PCR and fluorogenic probe was found to detect *T. gondii* DNA from as few as 4 bradyzoites (Jauregue et al., 2001).

## PREVENTION AND CONTROL

To prevent infection of human beings by *T. gondii*, the hands of people handling meat should be washed thoroughly with soap and water before they go to other tasks (Dubey and Beattie, 1988; Lopez et al., 2000). All cutting boards, sink tops, knives, and other materials coming in contact with uncooked meat should be washed with soap and water also. Washing is effective because the stages of *T. gondii* in meat are killed by contact with soap and water (Dubey and Beattie, 1988).

*Toxoplasma gondii* organisms in meat can be killed by exposure to extreme cold or heat. Tissue cysts in meat are killed by heating the meat throughout to $67^{0}C$ (Dubey et al., 1990). *Toxoplasma gondii* in meat is killed by cooling to $-13^{0}C$ (Kotula et al., 1991). *Toxoplasma* in tissue cysts are also killed by exposure to 0.5 kilorads of gamma irradiation (Dubey and Thayer, 1994). Meat of any animal should be cooked to $67^{0}C$ before consumption, and tasting meat while cooking or while seasoning should be avoided. Pregnant women, especially, should avoid contact with cats, soil, and raw meat. Pet cats should be fed only dry, canned, or cooked food. The cat litter box should be emptied every day, preferably not by a pregnant woman. Gloves should be worn while gardening. Vegetables should be washed thoroughly before eating because they may have been contaminated with cat feces. Expectant mothers should be aware of the dangers of toxoplasmosis (Foulon et al., 2000). At present there is no vaccine to prevent toxoplasmosis in humans.

## REFERENCES

Aramini, J.J., C.Stephen, and J.P.Dubey. 1998. *Toxoplasma gondii* in Vancouver Island cougars (*Felis concolor vancouverensis*): serology and oocyst shedding. Journal of Parasitology **84**: 438-440.

_____, _____, _____, C.Engelstoft, H.Schwantje, and C.S.Ribble. 1999. Potential contamination of drinking water with *Toxoplasma gondii* oocysts. Epidemiology and Infection **122**: 305-315.

Aspinall, T.V., D.Marlee, J.E.Hyde, and P.F.G.Sims. 2002. Prevalence of *Toxoplasma gondii* in commercial meat products as monitored by polymerase chain reaction - food for thought? International Journal for Parasitology **32**: 1193-1199.

Assadi-Rad, A.M., J.C.New, and S.Patton. 1995. Risk factors associated with transmission of *Toxoplasma gondii* to sows kept in different management systems in Tennessee. Veterinary Parasitology **57**: 289-297.

Bahia-Oliveira, L.M.G., J.L. Jones, J. Azevedo-Silva, C.C.F. Alves, F. Orefice, and D.G. Addiss. 2003. Highly endemic, waterborne toxoplasmosis in North Rio de Janeiro State, Brazil. Emerging Infectious Diseases **9**: 55-62.

Bowie, W.R., A.S.King, D.H.Werker, J.L.Isaac-Renton, A.Bell, S.B.Eng, and S.A.Marion. 1997. Outbreak of toxoplasmosis associated with municipal drinking water. Lancet **350**: 173-177.

Claus, G.E., E.Christie, and J.P.Dubey. 1977. Prevalence of *Toxoplasma* antibody in feline sera. Journal of Parasitology **63**: 266.

Cole, R.A., D.S.Lindsay, D.K.Howe, C.L.Roderick, J.P.Dubey, N.J.Thomas, and L.A.Baeten. 2000. Biolgical and molecular characterizations of *Toxoplasma gondii* strains obtained from southern sea otters (*Enhydra lutris nereis*). Journal of Parasitology **86**: 526-530.

Cook, A.J.C., R.E.Gilbert, W.Buffolano, J.Zufferey, E.Petersen, P.A.Jenum, W.Foulon, A.E.Semprini, and D.T.Dunn. 2000. Sources of toxoplasma infection in pregnant women: European multicentre case-control study. British Medical Journal **321**: 142-147.

Davies, P.R., W.E.M.Morrow, J.Deen, H.R.Gamble, and S.Patton. 1998. Seroprevalence of *Toxoplasma gondii* and *Trichinella spiralis* in finishing swine raised in different production systems in North Carolina, USA. Preventive Veterinary Medicine **36**: 67-76.

DeFeo, M.L., J.P.Dubey, T.N.Mather, and R.C.Rhodes III. 2002. Epidemiology investigation of seroprevalence of antibodies to *Toxoplasma gondii* in cats and rodents. American Journal of Veterinary Research. **63**: 1714-1717.

Dubey, J.P. 1976. Re-shedding of *Toxoplasma* oocysts by chronically infected cats. Nature **262**: 213-214.

_____. 1994. Toxoplasmosis. Journal of the American Veterinary Medical Association **205**: 1593-1598.

_____. 1995. Duration of immunity to shedding of *Toxoplasma gondii* oocysts by cats. Journal of Parasitology **81**: 410-415.

_____. 1996. Infectivity and pathogenicity of *Toxoplasma gondii* oocysts for cats. Journal of Parasitology **82**: 957-960.

_____. 1998. Refinement of pepsin digestion method for isolation of *Toxoplasma gondii* from infected tissues. Veterinary Parasitology **74**: 75-77.

_____. 2001. Oocyst shedding by cats fed isolated bradyzoites and comparision of infectivity of bradyzoites of the VEG strain *Toxoplasma gondii* to cats and mice. Journal of Parasitology **87**: 215-219.

_____. 2002a. A review of toxoplasmosis in wild birds. Vèterinary Parasitology **106**: 121-153.

_____. 2002b. Tachyzoite-induced life cycle of *Toxoplasma gondii* in cats. Journal of Parasitology **88**: 713-717.

_____, and C.P.Beattie. 1988. Toxoplasmosis of animals and man. CRC Press, Boca Raton, Florida 220p.

_____, and J.L.Carpenter. 1993a. Neonatal toxoplasmosis in littermate cats. Journal of the American Veterinary Medical Association **203**: 1546-1549.

_____.and _____ 1993b. Histologically confirmed clinical toxoplasmosis in cats - 100 cases (1952-1990). Journal of the American Veterinary Medical Association **203**: 1556-1566.

Dubey JP., and J.K.Frenkel. 1972. Cyst-induced toxoplasmosis in cats. Journal of Protozoology **19**: 155-177.

_____, and _____. 1976. Feline toxoplasmosis from acutely infected mice and the development of *Toxoplasma* cysts. Journal of Protozoology **23**: 537-546.

_____, and I.Johnstone. 1982. Fatal neonatal toxoplasmosis in cats. Journal of the American Animal Hospital Association **18**: 461-467.

_____, and C.A.Kirkbride. 1989. Economic and public health considerations of congenital toxoplasmosis in lambs. Journal of the American Veterinary Medical Association **195**: 1715-1716.

_____, and K.Odening. 2001. Toxoplasmosis and related infections. Samuel,W.M., M.J.Pybus, and A.A.Kocan, eds. Parasitic Diseases of Wild Mammals. Iowa State University Press, Ames, 478-519.

_____, and D.W.Thayer. 1994. Killing of different strains of *Toxoplasma gondii* tissue cysts by irradiation under defined conditions. Journal of Parasitology **80**: 764-767.

_____, J.L.Carpenter, M.J.Topper, and A.Uggla. 1989. Fatal toxoplasmosis in dogs. Journal of the American Animal Hospital Association **25**: 659-664.

_____, E.Christie, and P.W.Pappas. 1977. Characterization of *Toxoplasma gondii* from the feces of naturally infected cats. Journal of Infectious Diseases **136**: 432-435.

_____, H.R. Gamble, D. Hill, C. Sreekumar, S. Romand, and P. Thulliez. 2003. High prevalence of viable *Toxoplasma gondii* infection in market weight pigs from a farm in Massachusetts. Journal of Parasitology **88**: 1234-1238.

_____, D.H.Graham, C.R.Blackston, T.Lehmann, S.M.Gennari, A.M.A.Ragozo, S.M.Nishi, S.K.Shen, O.C.H.Kwok, D.E.Hill, and P.Thulliez. 2002. Biological and genetic characterisation of *Toxoplasma gondii* isolates from chickens (*Gallus domesticus*) from São Paulo, Brazil: Unexpected findings. International Journal for Parasitology **32**: 99-105.

_____, J.G.Humphreys, and P.Thulliez. 1995a. Prevalence of viable *Toxoplasma gondii* tissue cysts and antibodies to *T. gondii* by various serologic tests in black bears (*Ursus americanus*) from Pennsylvania. Journal of Parasitology **81**: 109-112.

_____, A.W.Kotula, A.Sharar, C.D.Andrews, and D.S.Lindsay. 1990. Effect of high temperature on infectivity of *Toxoplasma gondii* tissue cysts in pork. Journal of Parasitology **76**: 201-204.

_____, J.C. Leighty, V.C. Beal, W.R. Anderson, C.D. Andrews, and P. Thulliez. 1991. National seroprevalence of *Toxoplasma gondii* in pigs. Journal of Parasitology **77**: 517-521.

_____, J.K.Lunney, S.K.Shen, O.C.H.Kwok, D.A.Ashford, and P.Thulliez.      1996. Infectivity of low numbers of *Toxoplasma gondii* oocysts to pigs. Journal of Parasitology **82:** 438-443.

_____, W.J.A.Saville, J.F.Stanek, and S.M.Reed. 2002. Prevalence of *Toxoplasma gondii* antibodies in domestic cats from rural Ohio. Journal of Parasitology **88:** 802-803.

_____, P.Thulliez, and E.C.Powell. 1995b. *Toxoplasma gondii* in Iowa sows: Comparison of antibody titers to isolation of *T. gondii* by bioassays in mice and cats. Journal of Parasitology **81:** 48-53.

_____, R.M.Weigel, A.M.Siegel, P.Thulliez, U.D.Kitron, M.A.Mitchell, A.Mannelli, N.E.Mateus-Pinilla, S.K.Shen, O.C.H.Kwok, and K.S.Todd. 1995c. Sources and reservoirs of *Toxoplasma gondii* infection on 47 swine farms in Illinois. Journal of Parasitology **81:** 723-729.

Foulon W., A.Naessens, and D.Ho-Yen. 2000. Prevention of congenital toxoplasmosis. Journal of Perinatal Medicine **28:** 337-345.

Frenkel, J.K. 1973. Toxoplasmosis: parasite life cycle, pathology and immunology. Hammond,D.M., and P.L.Long, eds. The Coccidia. *Eimeria, Isospora, Toxoplasma* and Related Genera. University Park Press, Baltimore, Maryland, 343-410.

_____, J.P.Dubey, and N.L.Miller. 1970. *Toxoplasma gondii* in cats: fecal stages identified as coccidian oocysts. Science **167:** 893-896.

Freyre, A., J.P.Dubey, D.D.Smith, and J.K.Frenkel. 1989. Oocyst-induced *Toxoplasma gondii* infections in cats. Journal of Parasitology **75:** 750-755.

_____, J.Falcon, J.Berdie, J.C.Cruz, V.de Oliviera, and I.Sampaio. 1981. Estudio inicial del huesped definitivo de la toxoplasmosis en Montevideo. Annali della Facolta di Medicina Veterinaria di Uruquay **18:** 77-88.

Gamble, H.R., R.C.Brady, and J.P.Dubey. 1999. Prevalence of *Toxoplasma gondii* infection in domestic pigs in the New England states. Veterinary Parasitology **82:** 129-136.

Garrido, A.J., J., I.Cour Boveda, A.M.Berzosa Aguilar, and J.Pareja Miralles. 1972. Estudios sobre la epidemiología de la toxoplasmosis. La infección del gato domestico en los alrededores de Madrid. Encuestas serólogica y coproparasitologica. Medicina Tropical **48:** 24-39.

Grover, C.M., P.Thulliez, J.S.Remington, and J.C.Boothroyd. 1990. Rapid prenatal-diagnosis of congenital *Toxoplasma* infection by using Polymerase Chain Reaction and amniotic fluid. Journal of Clinical Microbiology **28:** 2297-2301.

Isaac-Renton J., W.R.Bowie, A.King, G.S.Irwin, C.S.Ong, C.P.Fung, M.O.Shokeir, and J.P.Dubey. 1998. Detection of *Toxoplasma gondii* oocysts in drinking water. Applied and Enviromental Microbiology **64:** 2278-2280.

Ito, S., K.Tsunoda, H.Nishikawa, and T.Matsui. 1974. Small type of *Isospora bigemina*. Isolation from naturally infected cats and relations with *Toxoplasma* oocyst. National Institute of Animal Health Quarterly **14:** 137-144.

Jacobs, L., and M.L.Melton. 1966. Toxoplasmosis in chickens. Journal of Parasitology **52:** 1158-1162.

_____, J.S.Remington, and M.L.Melton. 1960. A survey of meat samples from swine, cattle, and sheep for the presence of encysted *Toxoplasma*. Journal of Parasitology **46:** 23-28.

Jakob-Hoff, R.M., and J.D.Dunsmore. 1983. Epidemiological aspects of toxoplasmosis in southern Western Australia. Australian Veterinary Journal **60:** 217-218.

Janitschke, K., and D.Kuhn. 1972. *Toxoplasma*-Oozysten im Kot naturlich infizierter Katzen. Berliner und Münchener Tierärztliche Wochenschrift **85:** 46-47.

Jauregui, LH., J.Higgins, D.Zarlenga, J.P.Dubey, and J.K.Lunney. 2001. Development of a real-time PCR assay for the detection of *Toxoplasma gondii* in pig and mouse tissues. Journal of Clinical Microbiology **39:** 2065-2071.

Kotula, A.W., J.P.Dubey, A.K.Sharar, C.D.Andrew, S.K.Shen, and D.S.Lindsay. 1991. Effect of freezing on infectivity of *Toxoplasma gondii* tissue cysts in pork. Journal of Food Protection **54:** 687-690.

Lindsay DS., B.L.Blagburn, J.P.Dubey, and W.H.Mason. 1991. Prevalence and isolation of *Toxoplasma gondii* from white-tailed deer in Alabama. Journal of Parasitology **77:** 62-64.

Lopez, A., V.J.Dietz, M.Wilson, T.R.Navin, and J.L.Jones. 2000. Preventing congenital toxoplasmosis. Morbidity and Mortality Weekly Report **49:** 59-75.

Nery-Guimaraes, F., and H.A.Lage. 1973. Producao irregular e inconstante de oocistos pela ministracao de cistos de "*Toxoplasma gondii*" Nicolle & Manceaux, 1909, em gatos. Memorias do Instituto Oswaldo Cruz **71:** 157-167.

Pampiglione, S., G.Poglayen, B.Arnone, and F.de Lalia. 1973. *Toxoplasma gondii* oocysts in the faeces of naturally infected cat. British Medical Journal **2:** 306

Patton, S., J.Zimmerman, T.Roberts, C.Faulkner, V.Diderrich, A.Assadi-Rad, P.Davies, and J.Kliebenstein. 1996. Seroprevalence of *Toxoplasma gondii* in hogs in the National Animal Health Monitoring System (NAHMS). Journal of Eukaryotic Microbiology **43:** 121S

Remington, J.S., R.McLeod, and G.Desmonts. 1995. Toxoplasmosis. Remington,J.S., and J.O.Klein, eds. Infectious diseases of the fetus and newborn infant. 4th edition: Saunders Company, Philadelphia, 140-267.

_____. 1968. Toxoplasmosis and congenital infection. Birth Defects **4:** 49-56.

Rifaat, M.A., M.S.Arafa, M.S.M.Sadek, N.T.Nasr, M.E.Azab, W.Mahmoud, and M.S.Khalil. 1976. *Toxoplasma* infection of stray cats in Egypt. Journal of Tropical Medicine and Hygiene **79:** 67-70.

Ruiz, A., and J.K.Frenkel. 1980a. Intermediate and transport hosts of *Toxoplasma gondii* in Costa Rica. American Journal of Tropical Medicine and Hygiene **29:** 1161-1166.

_____, and _____. 1980b. *Toxoplasma gondii* in Costa Rican cats. American Journal of Tropical Medicine and Hygiene **29:** 1150-1160.

Smith, K.E., J.J.Zimmerman, S.Patton, G.W.Beran, and J.T.Hill. 1992. The epidemiology of toxoplasmosis on Iowa swine farms with an emphasis on the roles of free-living mammals. Veterinary Parasitology **42:** 199-211.

Tenter, A.M., A.R.Heckeroth, and L.M.Weiss. 2000. *Toxoplasma gondii*: from animals to humans. International Journal for Parasitology **30:** 1217-1258.

Varga, I. 1982. Adatok a *Toxoplasma gondii* oocisztáinak budapesti macskák bélsarában és cisztáinak vágóhídi sertésekben észlelt gyakoriságához. Magyar Allatorvosok Lapja **37:** 733-736.

Wallace, G.D. 1973. The role of the cat in the natural history of *Toxoplasma gondii*. American Journal of Tropical Medicine and Hygiene **22:** 313-322.

Warnekulasuriya, M.R., J.D.Johnson, and R.E.Holliman. 1998. Detection of *Toxoplasma gondii* in cured meats. International Journal of Food Microbiology **45:** 211-215.

Weigel, R.M., J.P.Dubey, A.M.Siegel, D.Hoefling, D.Reynolds, L.Herr, U.D.Kitron, S.K.Shen, P.Thulliez, R.Fayer, and T.S.Todd. 1995. Prevalence of antibodies to *Toxoplasma gondii* in Illinois swine in 1992. Journal of the American Veterinary Medical Association **206:** 1747-1751.

# PROTEASES AS POTENTIAL TARGETS FOR BLOCKING *TOXOPLASMA GONDII* INVASION AND REPLICATION

V.B. Carruthers

*W. Harry Feinstone Department of Molecular Microbiology and Immunology, Johns Hopkins Bloomberg School of Public Health, 615 N. Wolfe Street, Baltimore, Maryland 21205, USA*

## ABSTRACT

Apicomplexan parasites including *Toxoplasma gondii* share a common mechanism for actively invading host cells in a unique process that critically relies on the orchestrated release of secretory proteins. The organellar targeting and activity of these secretory proteins is often proteolytically modulated as they traffic through the secretory pathway or on the parasite surface during invasion. Proteases also play key roles in parasite intracellular replication. This chapter reviews the recent developments in the characterization of *T. gondii* proteases and protein substrates and discusses potential opportunities for impairing invasion and replication by specifically blocking proteolytic processing events.

**Key words**: *Toxoplasma gondii*, toxoplasmosis, protease, processing, secretion, invasion, drug development

## INTRODUCTION

*Toxoplasma gondii* is a pervasive intracellular protozoan of humans and animals. Although the infection is chronic and inconsequential in most individuals, immunodeficient or congenitally infected individuals can suffer fatal encephalitis or pneumonia, blinding retinochoroiditis, or irreversible cognitive impairment. Current standard therapies using antifolates or macrolide antibiotics are usually effective for treating toxoplasmosis. However, these antibiotics have significant shortcomings including allergic reactions, toxic side effects, and treatment failures. Such limitations are especially problematic for long-term drug treatment, particularly prophylactic therapy for precluding reactivation of chronic infections in AIDS patients (Georgiev, 1994). Hence, there is a critical need for safer and more effective chemotherapeutic options and accordingly, much of the basic research on *T. gondii* is aimed at identifying and validating novel drug targets. This review focuses on the emerging evidence that proteases play key roles in *T. gondii* invasion and intracellular replication and are therefore attractive potential targets for anti-parasitic drug development.

## INVASION

*Toxoplasma gondii* is unable to replicate extracellularly. Extracellular zoites are susceptible to antibody neutralization (Mineo et al., 1993), complement fixation by the classical pathway (Schreiber and Feldman, 1980), and engulfment and destruction by host macrophages (Hauser and Remington, 1981). Consequently, parasites are rapidly eliminated if they are unable to invade a cell. In addition to being an essential process, cell invasion by *T. gondii* is highly unusual. Whereas most intracellular pathogens rely heavily on the host cell's participation in uptake, *T. gondii* and related apicomplexan parasites invade by actively penetrating into host cells (Morisaki et al., 1995). A striking example of this is *T. gondii*'s ability to invade human fibroblasts killed by formaldehyde fixation (V.B. Carruthers, unpublished observation).

*Toxoplasma gondii* invasion is energy dependent (Werk and Bommer, 1980), rapid (15-30 seconds) (Morisaki et al., 1995), and requires parasite motility (Dobrowolski and Sibley, 1996). *T. gondii* zoites display a specialized form of substrate-dependent movement known as gliding motility (Frixone et al., 1996; Hakansson et al., 1999). Unlike the crawling motility exhibited by amoeboid cells, gliding motility does not involve cell shape changes. Rather, the parasite slides along a substrate in circular or star-shaped patterns that can be visualized by the membrane and protein deposits left behind in the "slime trail" (Figure 1A). Gliding motility is driven by an actin-myosin linear motor underlying the parasite plasma membrane (Sibley et al., 1998). Since actin or myosin antagonists also block invasion (Ryning and Remington, 1978), gliding motility is thought to be an integral part of the cell entry process. Indeed, parasite gliding toward or on top of a cell often immediately precedes invasion (Morisaki et al., 1995) (Figure 1B). Gliding is essentially unidirectional, with the parasite's anterior or apical end leading. As a result, active invasion is also highly polarized; the parasite invariably penetrates with its apical end breaching the host membrane first (Figure 1C).

A hallmark morphogical feature of *T. gondii* invasion is the formation of a prominent constriction visible as the parasite squeezes through the host subpellicular cytoskeleton. The constriction corresponds to the position of a moving junction, a close apposition of the parasite and host plasma membranes (Figure 1D). The moving junction is formed during parasite apical attachment when parasite ligands engage host receptors.

The moving junction slides over the parasite from the anterior to posterior end as the parasite penetrates into the target cell. During penetration, the parasite creates a parasitophorous vacuole (PV) that is mainly composed of host-derived lipids from the invaginated host plasma membrane (Suss-Toby et al. 1996). However, integral membrane proteins from the host plasma membrane are excluded by the moving junction (Mordue et al., 1999). Consequently, the PV is non-fusigenic and does not acidify (Mordue et al., 1999), properties that are crucial to ensure the parasite's intracellular survival.

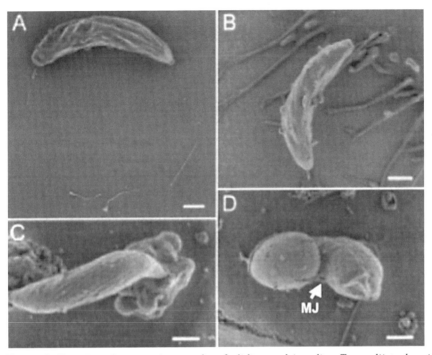

*Figure 1. Scanning electron micrographs of gliding and invading T. gondii tachyzoites. (A) Tachyzoite gliding in a circular pattern. Note the membrane deposits left behind in the "slime" trail. The discontinuous trail might be due to loss of material during processing. (B) Tachyzoite approaching a target cell. (C) Apically attached parasite. (D) Tachyzoite actively penetrating a human fibroblast cell. Note the prominent constriction representing the moving junction (MJ). Images were captured on a Leo Field Emission In-lens Scanning Electron (FEISEM) microscope at the Johns Hopkins University Core Microscopy Facility. Bar, 1 μm.*

To accomplish active invasion, the parasite has evolved several specialized cytoskeletal components and secretory organelles that directly participate in cell entry. Not surprisingly, this complex is strategically placed within the apical pole of the parasite, thus inspiring the phylogenic name, Apicomplexa. Situated at the extreme apical pole, the conoid is a thimble-shaped assemblage of microtubules (MT) connected to polar rings at either end. Radiating from the anterior polar rings, 22 singlet MTs extend approximately two-thirds the length of the parasite. The microtubular cytoskeleton is slightly twisted and this contributes to the helical motion displayed by gliding parasites and corkscrew action exhibited by invading parasites (Chiappino et al., 1984).

Micronemes (MIC) are small (~50 x 150 nm), cigar-shaped secretory organelles that densely populate the apical region, especially under the membrane complex (Figure 2). Microneme secretion is highly regulated. Although extracellular parasites release micronemal products at a low-level basal rate, rapid discharge is only associated with invasion (Carruthers and Sibley, 1997, 1999). Since discharge is regulated by parasite intracellular calcium (Carruthers et al., 1999; Carruthers and Sibley, 1999), initial contact with the host cell is thought to trigger a calcium-based signaling pathway that in turn activates release of micronemal products. Micronemes appear to fuse with the anterior vesicle, a 50 nm invagination of the membrane at the extreme apical tip of the parasite (Carruthers and Sibley, 1999). Among the contents of micronemes are a series of proteins containing recognizable adhesive domains that strikingly resemble sequences found in host proteins involved in cell-cell and cell-matrix interactions. These parasite adhesive proteins may have arisen from convergent evolution or by horizontal gene transfer from the host. Regardless, it appears that the parasite exploits these domains to bind corresponding receptors on the surface of target host cells. A subset of microneme proteins contain a single transmembrane (TM) segment and thus are candidates for connecting external host cell receptors (invasion) or substrate (motility) with the internal actin-myosin motor complex. Recent studies have revealed that microneme proteins assemble into quaternary complexes, each with one TM-anchored protein (Rabenau et al., 2001; Reiss et al., 2001; Meissner et al., 2002). This arrangement is presumably a mechanism to ensure that "soluble" MIC proteins are correctly targeted to the micronemes via sorting signals in the cytosolic domain of the TM MIC (Di Cristina et al., 2000). Multimerization of these proteins might also contribute towards tight binding of host receptors by promoting multivalent adhesion within the moving junction.

Rhoptries (ROP) are long (2-3 μm), club-shaped organelles with a narrow (~50 nm) ductile end situated within the conoid and a wider bulbous end that extends into the cytoplasm. *Toxoplasma gondii* tachyzoites typically possess 6-8 rhoptries that are clustered together, often to one side of the parasite. Rhoptries extrude their contents through the ductile end by a contractile process that results in a dramatic shortening and widening of the organelles. Based on this unusual secretory mechanism, rhoptries have been classified as "extrusomes" (Porchet-Hennere and Nicolas, 1983). Multiple rhoptries often appear to fuse during discharge, leaving an electron lucent (empty) heart-shaped sac. Rhoptries contain proteins and lipids, including an uncommon abundance of cholesterol and phosphatidylcholine (Foussard et al., 1991). Secretion of most rhoptry products is tightly regulated and only occurs during invasion. The rhoptry contents are injected through the host plasma membrane at the site of invasion where they form multiple vesicles, which rapidly fuse with the nascent parasitophorous vacuole membrane (Hakansson et al., 2001). Of the ten ROP proteins that have been described to date, only six have been molecularly cloned (Figure 2) and fewer still have been assigned specific functions. Among these proteins is a cysteine protease called toxopain-1 and a serine protease called SUB2. These enzymes will be further discussed below.

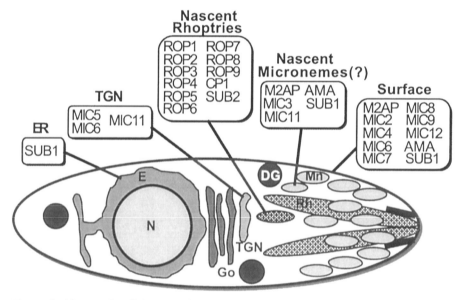

*Figure 2. Major sub-cellular sites for proteolytic processing of T. gondii secretory proteins. Sites were assigned based on data from metabolic labeling, immunostaining, and secretion experiments. It should be noted that no markers are available for the trans-Golgi network (TGN) of T. gondii. However, the TGN is a major site for protein maturation in other eukaryotic cells and is presumed to be the site for moderately fast ($T_{1/2}$ ~15 min) processing of secretory proteins in T. gondii. Also, almost nothing is known about the biogenesis of micronemes, but it is presumed that they develop through an immature stage (nascent micronemes), which might correspond to the slow ($T_{1/2}$=30-40 min) processing site. ER, endoplasmic reticulum; N, nucleus; Go, Golgi apparatus; DG, dense granules; Rh, rhoptry; Mn, microneme. CP1=Toxopain-1*

Dense granules (DG) are electron-dense, spherical vesicles that fuse with discrete lateral docking sites approximately 2 μm from the apical tip. Hence they are often referred to as apical secretory organelles despite being dispersed throughout the cytoplasm. *T. gondii* uses DGs as the default secretory pathway for proteins that do not have a specific forward targeting sequence. This is unusual, since in most other eukaryotes such proteins traffic directly to the cell surface without being stored in secretory granules. Although slightly upregulated after invasion (Carruthers and Sibley, 1997), DG secretion is essentially constitutive and GRA proteins are constantly released. DGs contain a collection of enigmatic proteins that bear little resemblance to other proteins of known function. Also, since none of the known GRA proteins are proteolytically processed, they will not be discussed in any further detail here.

Apical secretory compartments are sequentially discharged at distinct steps during invasion, allowing them to deploy their contents precisely where and when they are needed to fulfill their assigned functions (Carruthers and Sibley, 1997). In the first step, micronemes release adhesive proteins that bind host receptors for parasite attachment. Within a few seconds, ROP proteins are deployed to assist in the formation of the parasitophorous vacuole. Finally, a few minutes after invasion the DG contents cascade into the PV. There, in cooperation with membranous tubules liberated from the posterior end of the parasite, they are thought to modify the compartment for the uptake of nutrients from the host cytoplasm and organelles. This elaborate and sophisticated mechanism of cell entry allows the parasite to invade host cells on its own terms, thereby ensuring that it avoids the host's natural defense mechanisms such as destruction by acidification and degradation by lysosomal enzymes.

Although few studies have been done, the available data suggests that parasite proteases participate in, and are likely necessary for, invasion. Many parasite proteins involved in invasion are extensively modified by proteolytic processing during protein maturation or on the parasite surface during invasion (see below). Conseil et al. (1999) showed that two serine protease inhibitors, 3,4-DCI and AEBSF, impaired *T. gondii* tachyzoite invasion of human fibroblasts. Inhibition was dose-dependent and specific to treatment of parasites since pre-incubation with host cells had no effect. 3,4-DCI inhibited invasion >70% at 10 μM or higher. AEBSF was slightly less potent, blocking invasion >40% at 50 μM or higher. AEBSF is also effective in vivo since this compound substantially lengthened the survival time of mice infected with a large inoculum ($2.5 \times 10^3$ tachyzoites) of the highly virulent RH strain (Buitrago-Rey et al. 2002). While the proteases that are neutralized by 3,4-DCI and

AEBSF have not been identified, it is possible that the recently documented subtilisin-like protease SUB1 is a target for one or both of these compounds (Miller et al., 2001).     PRT2253, a cathepsin-B selective cysteine protease inhibitor, also significantly impaired tachyzoite invasion, possibly by targeting the recently identified enzyme toxopain-1 (Que et al., 2002).

## INTRACELLULAR REPLICATION

Once inside the host cell, the parasite divides asynchronously within a population but synchronously within each vacuole. This synchrony is the basis for why $2^N$ parasites (i.e., 2, 4, 8...) are almost invariably observed in each vacuole.  Parasites divide by endodyogeny, an unusual form of cell replication where two daughter cells develop within a mother cell. Assembly of a rudimentary conoid and its associated polar rings is the first sign of daughter cell development. MTs begin to polymerize from the anterior polar ring and are accompanied by synthesis of the innermembrane complex (IMC), a pair of closely apposed membranes formed from large flattened vesicles that are sutured together under the plasma membrane. The MT/IMC assembly grows toward the posterior end of the parasite and appears to bisect the dividing apicoplast (a vestigial, non-photosynthetic chloroplast), Golgi, nucleus, and endoplasmic reticulum (ER), which is an integral part of the nuclear membrane (Hager et al., 1999).  Finally, the two daughter cells acquire the plasma membrane from the mother cell as they are separated by a cleavage furrow.  Remnants of the mother cell are deposited at the base of the dividing parasites in the residual body.  Because new apical secretory organelles are synthesized de novo during each replicative cycle, secretory proteins involved in invasion are probably synthesized in a rapid burst of activity immediately preceding cell division.

A multitude of protease inhibitors have been tested for effects on parasite growth and replication.   3,4-DCI and AEBSF significantly inhibited tachyzoite replication at concentrations above 5 µM and 50 µM, respectively (Conseil et al., 1999). Cathepsin inhibitor III, TPCK (a chymotrypsin inhibitor), and subtilisin inhibitor III impaired parasite replication and caused marked swelling of the early secretory compartments including the ER and Golgi (Shaw et al., 2002).  Cathepsin inhibitor III and subtilisin inhibitor III also dramatically disabled rhoptry formation.  The chaotic disruption of the secretory pathway by these protease inhibitors is consistent with a key role for proteases in the maturation and activation of proteins destined for the apical secretory organelles.   Proteasome inhibitors including lactacystin, MG-132, proteosome inhibitor I, and gliotoxin also efficiently arrest parasite intracellular development and replication (Shaw et al., 2000; Paugam et al., 2002). In some cases these inhibitors induced the formation of membranous whorls derived from the ER, possibly a consequence of impaired breakdown of defective secretory proteins by the proteasome.

## PROTEOLYTIC PROCESSING OF SECRETORY PROTEINS

Proteases play crucial roles in a variety of cellular processes including protein trafficking, regulation, and degradation. This section will discuss the role of proteolytic processing in the function of secretory proteins, with particular emphasis on proteins associated with invasion.

Most secretory proteins are synthesized as preproteins with a hydrophobic N-terminal signal sequence that targets the protein to the ER, the gateway to the secretory system. The signal peptide is removed co-translationally by signal peptidase, an integral membrane protease that preferentially cleaves after small, non-polar amino acids. *Toxoplasma gondii* appears to closely follow this convention since virtually all of the known secretory proteins of the parasite contain conventional signal sequences. However, signal peptidase is highly conserved throughout eukaryotes and therefore is not considered to be a favorable target for drug design.

Some secretory proteins are synthesized as preproproteins. These proteins contain a propeptide (or proregion), usually at the N-terminus, that is proteolytically amputated as it transits through the secretory pathway. Propeptides can fulfill diverse roles including protein folding, targeting, and regulation. *Toxoplasma gondii* has an exceptional abundance of propeptide-containing proteins. For example, ROP1, a protein necessary for maintaining normal rhoptry ultrastructure (Kim and Boothroyd, 1993; Soldati et al., 1995), has a 66 amino acid propeptide that is removed in the nascent rhoptries during organellar biogenesis (Soldati et al., 1998). Although this propeptide is sufficient for rhoptry targeting when fused to a heterologous protein (Bradley and Boothroyd, 2001), it is not necessary for sorting to the rhoptries since ROP1 contains a secondary targeting sequence in its C-terminus (Striepen et al., 2001). Propeptide cleavage is not necessary for targeting since correct sorting to the rhoptries was observed for a recombinant ROP1 bearing a mutation in the cleavage site P1 residue (Bradley et al., 2002). (By convention, cleavage site residues are designated P4-P3-P2-P1 / P1'-P2'-P3'-P4, where / represents the scissile amide bond). Propeptides are also present on virtually all of the other ROP proteins identified to date, and they presumably function in rhoptry targeting in a manner similar to ROP1.

Propeptides are also a common feature of MIC proteins. Pulse/chase metabolic labeling experiments suggest that propeptide cleavage occurs in one of three main sites in the secretory system (Figure 3). These sites

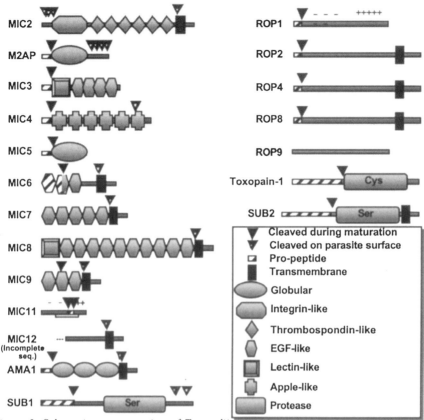

*Figure 3. Schematic representation of T. gondii secretory proteins with indicated sites of proteolytic cleavage. Only proteolytically processed proteins for which deduced amino acid sequences are available are shown.*

can be distinguished by the timing of processing and their sensitivity to the fungal metabolite brefeldin A (BFA), which collapses the medial- and cis-Golgi compartment upon the ER. The first site is the ER where the initial processing of SUB1 is resistant to BFA and occurs rapidly with a half-life time ($T_{1/2}$) of approximately 5 min. The second site where, for example, propeptide cleavage of MIC5 takes place, probably corresponds to the trans-Golgi network (TGN), the main sorting compartment in the secretory pathway. Processing at this site is sensitive to brefeldin A and is moderately fast ($T_{1/2}$ ~15 min). Cleavage within the third site is also sensitive to BFA, but is considerably slower ($T_{1/2}$=30-40 min) and therefore might correspond to the nascent micronemes. Propeptide processing of M2AP, for example, occurs here. In most cases, functions have not been ascribed for MIC propeptides. However, a recent study demonstrated that propeptide cleavage unmasked receptor binding activity of the N-terminal lectin-like domain of MIC3 (Cerede et al., 2002). This implies the MIC3 propeptide functions to inactivate MIC3 until it reaches the micronemes prior to mobilization to the parasite surface where it is thought to participate in invasion. MIC6 has an unusually long propeptide that includes one and one half of its three epidermal growth factor-like (EGF-like) domains. The MIC6 propeptide is removed in the TGN but is apparently not necessary for proper targeting since a propeptide deletion mutant was correctly sorted to the micronemes (Reiss et al., 2001). In contrast, propeptide deletion mutants of MIC5 and M2AP, a protein tightly associated with the MIC2 adhesin, are withheld in the ER and Golgi (S.D. Brydges, J.M. Harper, and V.B. Carruthers, unpublished). Although the basis of this retention phenotype is not known, it is possible the respective propeptides are necessary for proper folding of these proteins or assembly into protein complexes. A similar but distinct possibility is that cleavage of these propeptides is necessary in order to pass a quality control checkpoint in the secretory system. Creating and testing non-cleavable mutants should help distinguish between these possibilities.

The parasite surface is another major site of proteolytic activity. Upon release from the micronemes at the apical surface, MIC protein complexes begin translocating (capping) backwards by engaging their cytosolic sequences with the actin-myosin motor. As they travel through the membrane, they are subjected to two distinct processing events. The first type involves proteolytic events that do not release the complex from the surface. Functioning in this capacity is MPP2, a protease that has not been cloned but is defined by its activity, cleavage site specificity, and its sensitivity to chymostatin and calpain inhibitors (ALLN and ALLM). MPP2 trims off an N-terminal extension from MIC2 (Carruthers et al., 2000) and C-terminal sequences from MIC4 (Brecht et al., 2001) and M2AP (Rabenau et al., 2001). Although the functional significance of these processing events remains unknown, the pervasiveness of MPP2 processing suggests it plays multiple modulatory roles in invasion. Interestingly, MPP2 processing of its substrates is markedly enhanced by treatment with cytochalasin D. This observation could be explained if MPP2 is confined to the apical surface, since inactivation of the linear motor by cytochalasin D treatment would trap substrates in the vicinity of MPP2, thereby enhancing proteolysis.

The second type of surface processing occurs at a membrane proximal site near the base of TM-anchored proteins, releasing the extracellular portion of MIC protein complexes as soluble products. This processing event is executed by a second protease called MPP1 (Carruthers et al., 2000) and serves at least two purposes. First, it precludes antibody neutralization of MIC adhesive activity by preventing their accumulation on the parasite surface from basal secretion prior to invasion. Second, it provides a means of disconnecting MIC adhesive protein complexes from the parasite surface during the final step of invasion when the PV fuses behind the parasite. In a surprising recent development, Opitz et al. (2002) reported evidence that MIC6 and MIC12 are cleaved within their TM anchor sequences by MPP1. Intramembranous cleavage is very rare, but a few examples have been documented in bacterial and mammalian cells in a phenomenon known as regulated intramembrane proteolysis, or Rip (Brown et al., 2000). In each of these examples, cleavage is performed by an unusual integral membrane protease with an active site situated within the lipid bilayer. If MPP1 is a Rip protease, it could explain why it is resistant to all protease inhibitors tested to date (Carruthers et al., 2000). The MIC6 cleavage site was defined by mass spectroscopy and this site is highly conserved in other TM MIC proteins such as MIC2. However, Brossier et al. (2003) recently showed that a dibasic amino acid motif positioned outside MIC2's TM is crucial for MPP1-mediated cleavage. This raises the intriguing possibility that the MPP1 recognition site on MIC2 is separate from the cleavage site. It should soon be possible to test this hypothesis since Soldati and colleagues have recently identified a candidate for MPP1 that is expressed on the parasite surface and is a member of the rhomboid family of integral membrane serine proteases (D. Soldati, personal communication).

A recent string of studies have defined many of the proteolytic cleavage sites on *T. gondii* secretory proteins by protein microsequencing or mass spectroscopy (Table 1). In some cases, there are clear similarities among cleavage site sequences. For example, the MIC6 propeptide cleavage site (VQLS / ET) closely resembles that of M2AP (AQLS / TF). Also, a striking likeness is seen in the N-terminal cleavage sites of MIC2 and 3 of the 4 C-terminal cleavage sites of M2AP, with cleavage occurring after a small, non-polar amino acid. These similarities are consistent with both proteins being processed by the same protease, MPP2, especially since these cleavages all occur on the parasite surface and are blocked by the calpain inhibitors, ALLN and ALLM. As more data is collected, additional similarities will emerge, revealing a clearer picture of the diversity and breadth of proteases involved in processing secretory proteins.

## SECRETORY PROTEASES

Although several secretory proteases have been described, they vary tremendously in the level to which they have been characterized. For example, MPP1 and MPP2 were defined solely on the basis of their inhibitor profiles and their ability to cleave MIC protein substrates

(Carruthers et al., 2000). A cytosolic metallopeptidase of unknown function was purified from tachyzoites and partially characterized

Table 1. *Cleavage site properties of proteolytically processed secretory proteins from T. gondii.*

| Protein | Site | Site# | Sequence | T_ | Location | Inhibitors | References |
|---|---|---|---|---|---|---|---|
| MIC2 | N-term | 1 | ...ANGV[50]/DT... | NA | Surface | ALLN, ALLM, Chymost. | Carruthers et al. (2000) |
| | | 2 | ...VDTS[53]/GV... | | | | |
| | | 3 | ...SAIG[66]/AA.. | | | | |
| | C-term | 1 | ??? | NA | Surface | None found so far | Carruthers et al. (2000) |
| M2AP | N-term | 1 | ...AQLS[46]/TF... | 30-40' | Nasc. Mics(?) | DFP | Rabenau et al. (2001) |
| | C-term | 1 | ...NSMQ[302]/EV... | NA | Surface | ALLN, ALLM (Sites 2-4 only) | Zhou and Carruthers, in preparation |
| | | 2 | ...SLPL[268]/SQ... | | | | |
| | | 3 | ...ATPA[253]/EE... | | | | |
| | | 4 | ...NSGS[249]/AT... | | | | |
| MIC3 | N-term | 1 | ...SSVQ[66]/SP... | 30-40' | Nasc. Mics(?) | ND | Garcia-Réguet et al. (2000) |
| MIC4 | N-term | 1 | ...ANVT[57]/SS... | NA | Surface | None found so far | Brecht et al. (2001) |
| | C-term | 1 | ??? | NA | Surface | ALLN, ALLM, Chymost. | Brecht et al. (2001) |
| MIC5 | N-term | 1 | ...EAGR[43]/RT... | ~15' | TGN(?) | DFP | Brydges et al. (2000) |
| MIC6 | N-term | 1 | ...VQLS[94]/ET... | ND | TGN | ND | Reiss et al. (2001) |
| | C-term | 1 | ...GAIA[292]/GG... | NA | Surface | ND | Opitz et al. (2002) |
| MIC11 | Internal | 1 | ...LQER[146]/KF... | ~15' | TGN(?) | ND | Zhou and Carruthers, unpublished |
| | | 2 | ...DFSF[126]/LD... | 30-40' | Nasc. Mics(?) | ND | |
| ROP1 | N-term | 1 | ...SFVE[83]/AP... | | Nasc. Rhops | ND | Bradley and Boothroyd (1999) |

(Berthonneau et al., 2000). Proteases of 80, 70, and 42 kDa were identified in culture supernatants based on gelatinolytic activity (Ahn et al., 2001). By virtue of reacting with a monoclonal antibody (mAb), the

42 kDa protease has been characterized in greater detail. Since it was inhibited by TPCK and EGTA and stimulated by calcium, the 42 kDa enzyme is likely a calcium dependent serine protease. Calcium regulation might be a mechanism of activating the protease upon secretion where it would encounter a relatively high concentration (~1 mM) of calcium in the extracellular milieu. The 42 kDa protease-specific mAb reacted with the rhoptries of *T. gondii* tachyzoites and cross-reacted with an orthologous protein in the closely related parasite *Neospora caninum*.

Although this data demonstrate that it is a conserved component of the rhoptries, the precise function of this protease remains unknown. It is anticipated that further characterization of the 42 kDa protease will be substantially aided by cloning and sequencing of its cognate gene.

SUB1 is a subtilisin-like serine protease that is abundantly secreted from the micronemes of *T. gondii* (Miller et al., 2001). As stated earlier, SUB1 is synthesized as a preproprotein that is converted in the ER from a 120 kDa precursor to a 90 kDa product by propeptide removal in a step that is likely autocatalytic. The 90 kDa species is probably the form that is stored in the micronemes. Upon secretion the 90 kDa enzyme is converted to major 80 and 70 kDa species and minor 40 and 30 kDa species. The 80 and 70 kDa SUB1 species could be responsible for the 80 and 70 kDa gelatinolytic activities reported by Ahn et al. (2001). In this case, SUB1 is also activated upon secretion since the 90 kDa gelatinolytic species was not detected in tachyzoite lysates (Ahn et al., 2001). Although it is not known whether SUB1 is calcium dependent, this is a strong possibility since many subtilisins require calcium for activity (Steiner, 1998). Because of challenges associated with expression of active recombinant enzyme, the catalytic properties and specificities of SUB1 have not been determined and the natural substrates not known. However, since many eukaryotic subtilisins are pro-protein convertases, it is tempting to speculate that SUB1 could function in the targeting or activation of MIC proteins by propeptide processing.

A second subtilase, SUB2, was recently identified and shown to be present in the rhoptries of *T. gondii* (Miller et al., in press). Like SUB1, SUB2 is autocatalytically activated and undergoes a series of maturation steps. Genetic disruption of SUB2 was unsuccessful, suggesting that it is an essential enzyme. Interestingly, SUB2 forms a stable complex with ROP1, which is a likely substrate since the ROP pro-peptide processing site closely resembles the autocatalytic cleavage site on SUB2. Thus SUB2 is likely a rhoptry protein maturase responsible for activation of ROP proteins within nascent rhoptries.

Toxopain-1 is a rhoptry-derived cysteine protease belonging to the cathepsin-B family (Que et al., 2002). Indirect evidence based on inhibitor studies suggests that toxopain-1 also participates in propeptide processing of rhoptry proteins, particularly ROP2. Toxopain-1 prefers to cleave after a positively charged residue in the P1 position of its substrates. In higher eukaryotes cathepsin-B proteases are typically found in lysosomes where they function to degrade material taken up by endocytosis or phagocytosis. The available evidence suggests that rhoptries are lysosome-like organelles, and this is consistent with the localization of toxopain-1. It is interesting to note that toxopain-1 is uniquely confined to a narrow strip in the center of each rhoptry and, unlike most rhoptry proteins, it is released as a soluble species from extracellular parasites. This raises the possibility that toxopain-1 is also involved in the proteolytic processing of secretory proteins on the parasite surface.

## FUTURE PROSPECTIVES

With the recent highly publicized successes in using protease inhibitors to treat HIV infections (Deeks and Volberding, 1997) and hypertension (Nawarskas et al., 2001), much attention is now being focused on proteases for their potential as targets for treating infectious diseases. The extensive proteolytic processing of *T. gondii* secretory proteins supports the contention that proteases will be good therapeutic targets for treating toxoplasmosis. However, the characterization of proteases and secretory proteins in *T. gondii* is in its infancy and much work is yet to be done. First, the entire repertoire of parasite proteases must be defined. Although completion of the *T. gondii* genome and bioinformatics analysis will greatly aid in this goal, additional contemporary approaches will also be necessary. Protease activity profiling using chemical inhibitors that covalently label target proteases of a particular class, particularly serine and cysteine proteases, should be extremely valuable in identifying novel targets (Greenbaum et al., 2002). An important feature of these chemical inhibitors is that they are exceedingly versatile and can be tailored for high specificity, a property that is important for defining the function of individual proteases. Also they can be modified for affinity purification and identification of their specific targets. In cases where active recombinant enzyme is available, another valuable approach will be to screen phage peptide display libraries to define the cleavage specificity of each protease. This data can then be used to tentatively assign protein substrates based on their cleavage sites. The recent development of tools for conditional gene knockouts (Meissner et al., 2002) should be invaluable for determining the importance of each protease to parasite invasion or intracellular replication. This information can be used to prioritize the proteases for the rational design of inhibitors or large scale screening of small molecule inhibitor libraries.

In parallel, it will be necessary to determine specific functions of secretory proteins in invasion and intracellular survival and especially the role of proteolytic processing events in modulating their functions. The pace of analyzing proteolytic cleavage sites has been greatly accelerated by the recent wide-spread application of mass spectrometry to protein analysis. By defining cleavage sites and inhibitor profiles, investigators can determine which substrates are likely processed by the same protease. Knowledge of the cleavage site can also be used to synthesize specific reporter substrates for use in monitoring the purification of a target protease. Based on cleavage site information, non-cleavable mutants can also be expressed in transgenic parasites for testing detrimental effects on the secretory system, invasion, and parasite replication. Recently developed methods for regulated expression of toxic proteins might also be useful in this context (Meissner et al., 2001). It is anticipated that the genetic and biochemical tractability of *T. gondii* will greatly facilitate the identification, characterization, and validation of proteases as targets for the design of new generation therapeutics for treating toxoplasmosis.

## REFERENCES

Ahn, H. J., K. J. Song, E. S. Son, J. C. Shin and H. W. Nam. 2001. Protease activity and host cell binding of the 42-kDa rhoptry protein from *Toxoplasma gondii* after secretion. Biochemical Biophysical Research Communions 287: 630-635.

Berthonneau, J., M. H. Rodier, B. El Moudni and J. L. Jacquemin. 2000. *Toxoplasma gondii*: purification and characterization of an immunogenic metallopeptidase. Experimental Parasitology 95: 158-162.

Bradley, P. J. and J. C. Boothroyd. 2001. The pro region of *Toxoplasma* ROP1 is a rhoptry-targeting signal. International Journal for Parasitology 31: 1177-1186.

_____, C. L. Hsieh and J. C. Boothroyd. 2002. Unprocessed *Toxoplasma* ROP1 is effectively targeted and secreted into the nascent parasitophorous vacuole. Molecular and Biochemical Parasitology 125: 189-193.

Brecht, S., V. B. Carruthers, D. J. Ferguson, O. K. Giddings, G. Wang, U. Jaekle, J. M. Harper, L. D. Sibley and D. Soldati. 2001. The *Toxoplasma* micronemal protein MIC4 is an adhesin composed of six conserved apple domains. Journal of Biological Chemistry 276: 4119-4127.

Brossier, F., T.J. Jewett, J.L. Lovett, and L.D. Sibley. 2003. C-terminal processing of the *Toxoplasma* protein MIC2 is essential for invasion into host cells. Journal of Biological Chemistry 278: 6229-6234

Brown, M. S., J. Ye, R. B. Rawson and J. L. Goldstein. 2000. Regulated intramembrane proteolysis: a control mechanism conserved from bacteria to humans. Cell 100: 391-398.

Buitrago-Rey, R., J. Olarte and J. E. Gomez-Marin. 2002. Evaluation of two inhibitors of invasion: LY311727 [3-(3-acetamide-1-benzyl-2-ethyl-indolyl-5-oxy)propane phosphonic acid] and AEBSF [4-(2-aminoethyl)-benzenesulphonyl fluoride] in acute murine toxoplasmosis. Journal of Antimicrobial Chemotherapy 49: 871-874.

Carruthers, V. B., S. N. J. Moreno and L. D. Sibley. 1999. Ethanol and acetaldehyde elevate intracellular calcium and stimulate microneme discharge in *Toxoplasma gondii*. Biochemical Journal 342: 379-386.

_____, G. D. Sherman and L. D. Sibley. 2000. The *Toxoplasma* adhesive protein MIC2 is proteolytically processed at multiple sites by two parasite-derived proteases. Journal of Biological Chemistry **275**: 14346-14353.

_____, and L. D. Sibley. 1997. Sequential protein secretion from three distinct organelles of *Toxoplasma gondii* accompanies invasion of human fibroblasts. European Journal of Cell Biology **73**: 114-123.

_____. 1999. Mobilization of intracellular calcium stimulates microneme discharge in *Toxoplasma gondii*. Molecular Microbiology **31**: 421-428.

Cerede, O., J. F. Dubremetz, D. Bout and M. Lebrun. 2002. The *Toxoplasma gondii* protein MIC3 requires pro-peptide cleavage and dimerization to function as adhesin. EMBO Journal **21**: 2526-2536.

Chiappino, M. L., B. A. Nichols and G. R. O'Connor. 1984. Scanning electron microscopy of *Toxoplasma gondii*: parasite torsion and host-cell responses during invasion. Journal of Protozoology **31**: 288-292.

Conseil, V., M. Soete and J. F. Dubremetz. 1999. Serine protease inhibitors block invasion of host cells by *Toxoplasma gondii*. Antimicrobial Agents and Chemotherapy **43**: 1358-1361.

Deeks, S. G. and P. A. Volberding. 1997. HIV-1 protease inhibitors. AIDS Clin Rev: 145-185.

Di Cristina, M., R. Spaccapelo, D. Soldati, F. Bistoni and A. Crisanti. 2000. Two conserved amino acid motifs mediate protein targeting to the micronemes of the apicomplexan parasite *Toxoplasma gondii*. Molecular and Cellular Biology **20**: 7332-7341.

Dobrowolski, J. M. and L. D. Sibley. 1996. *Toxoplasma* invasion of mammalian cells is powered by the actin cytoskeleton of the parasite. Cell **84**: 933-939.

Foussard, F., M. A. Leriche and J. F. Dubremetz. 1991. Characterization of the lipid content of *Toxoplasma gondii* rhoptries. Parasitol **102**: 367-370.

Frixone, E., R. Mondragon and I. Meza. 1996. Kinematic analysis of *Toxoplasma gondii* motility. Cell Motility and the Cytoskeleton **34**: 152-163.

Georgiev, V. S. 1994. Management of toxoplasmosis. Drugs **48**: 179-188.

Greenbaum, D., A. Baruch, L. Hayrapetian, Z. Darula, A. Burlingame, K. F. Medzihradszky and M. Bogyo. 2002. Chemical approaches for functionally probing the proteome. Mol Cell Proteomics **1**: 60-68.

Hager, K. M., B. Striepen, L. G. Tilney and D. S. Roos. 1999. The nuclear envelope serves as an intermediary between the ER and golgi complex in the intracellular parasite *Toxoplasma gondii*. Journal of Cell Science **112**: 2631-2638.

Hakansson, S., A. J. Charron and L. D. Sibley. 2001. *Toxoplasma* evacuoles: a two-step process of secretion and fusion forms the parasitophorous vacuole. EMBO Journal **20**: 3132-3144.

_____, H. Morisaki, J. Heuser and L. D. Sibley. 1999. Time-lapse video microscopy of gliding motility in *Toxoplasma gondii* reveals a novel, biphasic mechanism of cell locomotion. Molecular Biology of the Cell **10**: 3539-3547.

Hauser, W. E., Jr. and J. S. Remington. 1981. Effect of monoclonal antibodies on phagocytosis and killing of *Toxoplasma gondii* by normal macrophages. Infection and Immunity **32**: 637-640.

Kim, K. and J. C. Boothroyd. 1993. Gene replacement in *Toxoplasma gondii* with chloramphenicol acetyltransferase as selectable marker. Science **262**: 911-914.

Meissner, M., S. Brecht, H. Bujard and D. Soldati. 2001. Modulation of myosin A expression by a newly established tetracycline repressor-based inducible system in *Toxoplasma gondii*. Nucleic Acids Research **29**: E115, 111-110.

_____, M. Reiss, N. Viebig, V. Carruthers, C. Toursel, S. Tomavo, J. Ajioka and D. Soldati. 2002. A family of transmembrane microneme proteins of *Toxoplasma gondii* contain EGF-like domains and function as escorters. Journal of Cell Science **115**: 563-574.

_____, D. Schluter and D. Soldati. 2002. Role of *Toxoplasma gondii* myosin A in powering parasite gliding and host cell invasion. Science **298**: 837-840.

Miller, S. A., E. M. Binder, M. J. Blackman, V. B. Carruthers and K. Kim. 2001. A conserved subtilisin-like protein TgSUB1 in microneme organelles of *Toxoplasma gondii*. Journal of Biological Chemistry **276**: 45341-45348.

_____, V. Thathy, J. W. Ajioka, M. J. Blackman and K. Kim. TgSUB2 is a *Toxoplasma gondii* rhoptry organelle processing proteinase (in press)

Mineo, J. R., R. McLeod, D. Mack, J. Smith, I. A. Khan, K. H. Ely and L. H. Kasper. 1993. Antibodies to *Toxoplasma gondii* major surface protein (SAG-1, P30) inhibit infection of host cells and are produced in murine intestine after peroral infection. Journal of Immunology **150**: 3951-3964.

Mordue, D., S. Hakansson, I. R. Niesman and L. D. Sibley. 1999. *Toxoplasma gondii* resides in a vacuole that resists fusion with host cell endocytic and exocytic vesicular trafficking pathways. Experimental Parasitology **92**: 87-99.

_____, N. Desai, M. Dustin and L. D. Sibley. 1999. Invasion by *Toxoplasma gondii* establishes a moving junction that selectively excludes host cell plasma membrane proteins on the basis of their membrane anchoring. Journal of Experimental Medicine **190**: 1783-1792.

Morisaki, J. H., J. E. Heuser and L. D. Sibley. 1995. Invasion of *Toxoplasma gondii* occurs by active penetration of the host cell. Journal of Cell Science **108**: 2457-2464.

Nawarskas, J., V. Rajan and W. H. Frishman. 2001. Vasopeptidase inhibitors, neutral endopeptidase inhibitors, and dual inhibitors of angiotensin-converting enzyme and neutral endopeptidase. Heart Dis **3**: 378-385.0

Opitz, C., M. Di Cristina, M. Reiss, T. Ruppert, A. Crisanti, and D. Soldati., 2002. Intramembrane cleavage of microneme proteins at the surface of the apicomplexan parasite *Toxoplasma gondii*. Embo Journal **21**: 1577-1585.

Paugam, A., C. Creuzet, J. Dupouy-Camet and P. Roisin. 2002. In vitro effects of gliotoxin, a natural proteasome inhibitor, on the infectivity and proteolytic activity of *Toxoplasma gondii*. Parasitology Research **88**: 785-787.

Porchet-Hennere, E. and G. Nicolas. 1983. Are rhoptries really extrusomes. Journal of Ultrastructure Research **84**: 194-203.

Que, X., H. Ngo, J. Lawton, M. Gray, Q. Liu, J. Engel, L. Brinen, P. Ghosh, K. A. Joiner and S. L. Reed. 2002. The cathepsin B of *Toxoplasma gondii*, toxopain-1, is critical for parasite invasion and rhoptry protein processing. Journal of Biological Chemistry **277**: 25791-25797.

Rabenau, K. E., A. Sohrabi, A. Tripathy, C. Reitter, J. W. Ajioka, F. M. Tomley and V. B. Carruthers. 2001. TgM2AP participates in *Toxoplasma gondii* invasion of host cells and is tightly associated with the adhesive protein TgMIC2. Molecular Microbiology **41**: 1-12.

Reiss, M., N. Viebig, S. Brecht, M. N. Fourmaux, M. Soete, M. Di Cristina, J. F. Dubremetz and D. Soldati. 2001. Identification and characterization of an escorter for two secretory adhesins in *Toxoplasma gondii*. Journal of Cell Biology **152**: 563-578.

Ryning, F. W. and J. S. Remington. 1978. Effect of cytochalasin D on *Toxoplasma gondii* cell entry. Infection and Immunity **20**: 739-743.

Schreiber, R. D. and H. A. Feldman. 1980. Identification of the activator system for antibody to *Toxoplasma* as the classical complement pathway. Journal of Infectious Diseases **141**: 366-369.

Shaw, M. K., C. Y. He, D. S. Roos and L. G. Tilney. 2000. Proteasome inhibitors block intracellular growth and replication of *Toxoplasma gondii*. Parasitology **121** ( Pt 1): 35-47.

_____, D. S. Roos and L. G. Tilney. 2002. Cysteine and serine protease inhibitors block intracellular development and disrupt the secretory pathway of *Toxoplasma gondii*. Microbes Infection **4**: 119-132.

Sibley, L. D., S. Hakansson and V. B. Carruthers. 1998. Gliding motility: an efficient mechanism for cell penetration. Current Biology **8**: R12-R14.

Soldati, D., K. Kim, J. Kampmeier, J.-F. Dubremetz and J. C. Boothroyd. 1995. Complementation of a *Toxoplasma gondii* ROP1 knock-out mutant using phleomycin. Molecular and Biochemical Parasitology **74**: 87-97.

_____, A. Lassen, J. F. Dubremetz and J. C. Boothroyd. 1998. Processing of *Toxoplasma* ROP1 protein in nascent rhoptries. Molecular and Biochemical Parasitology **96**: 37-48.

Steiner, D. F. 1998. The proprotein convertases. Current Opinion in Chemical Biology **2**: 31-39.

Striepen, B., D. Soldati, N. Garcia-Reguet, J. F. Dubremetz and D. S. Roos. 2001. Targeting of soluble proteins to the rhoptries and micronemes in *Toxoplasma gondii*. Molecular and Biochemical Parasitology **113**: 45-53.

Suss-Toby, E., J. Zimmerberg and G. E. Ward. 1996. *Toxoplasma* invasion: the parasitophorous vacuole is formed from host cell plasma membrane and pinches off via a fission pore. Proceedings of the National Academy of Sciences USA **93**: 8413-8418.

Werk, R. and W. Bommer. 1980. *Toxoplasma gondii*: membrane properties of active energy-dependent invasion of host cells. Tropenmedical Parasitology **31**: 417-420.

# TARGETING THE *TOXOPLASMA GONDII* APICOPLAST FOR CHEMOTHERAPY

Sunny C. Yung and Naomi Lang-Unnasch

*Division of Geographic Medicine, University of Alabama at Birmingham, Birmingham, Alabama. USA*

**ABSRACT:** The apicoplast represents a potential drug target to combat *Toxoplasma gondii* infection, as it is essential to the parasite and absent from host cells. Functions of the apicoplast include fatty acids synthesis, protein synthesis, DNA replication, electron transport, and heme biosynthesis. Each pathway and its potential for new chemotherapeutic leads will be discussed.

**Key words:** Apicoplast, *Toxoplasma gondii*, chemotherapy

## INTRODUCTION

*Toxoplasma gondii* causes severe neurological deficits in immunosuppressed patients (such as those with AIDS) and lymphadenopathy in healthy adults. It can cross the placenta (generally in women with no or low antibody levels) and cause congenital infections characterized by intracerebral calcifications, chorioretinitis, hydrocephaly or microcephaly, and convulsions (Kasper, 2002). This parasite is acquired primarily by ingestion of undercooked or raw meat. It may also be acquired from cat feces. A wide variety of animals carry *T. gondii* (Dubey et al., 1998).

The apicoplast is an apically located organelle in *T. gondii*. It contains its own 35 kb circular genome which consists of an inverted tandem repeat of large and small subunit rRNA genes, a complete set of tRNAs, most ribosomal proteins, clpC, tufA, and rpoB, C1, and C2 genes. In addition, it has six unidentified open reading frames (Wilson et al., 1996). All phylogenetic analyses have supported a plant-plastid like origin for the apicoplast rather than prokaryotic origin. However, there is no consensus on which plant it originated from. Molecular phylogenetic analysis based on the elongation translation factor Tu (tufA) suggest this 35 kb element is more related to green algal plastid than cyanobacteria or any free living prokaryote (Kohler et al., 1997). In contrast, gene arrangement and molecular phylogeny of various genes support other algal origin for the apicoplast

(Egea et al., 1995; Fast et al., 2001; Lang-Unnasch et al., 1998). In either case, the apicoplast clearly represents an organelle with algal origin and the potential for new chemotherapy leads since mammal cells lack such a plastid-like organelle.

*Toxoplasma gondii* likely acquired its apicoplast via secondary endosymbiosis whereby its ancestral heterotrophic eukaryote engulfed a photosynthetic eukaryote (Waller et al., 1998). The nucleus of the photosynthetic eukaryote has since disappeared but its plastid still remains within the heterotrophic host. Such a plastid would have more membranes than the plastids of primary endosymbiosis whereby an eukaryote engulfed a cynobacteria (McFadden, 1999). Indeed, the apicoplast of *T. gondii* is surrounded by four membranes whereas the chloroplasts of higher plants are surrounded by two membranes. Both the ultrastructural and phylogenetic data support the hypothesis that the apicoplast was acquired by secondary endosymbiosis (Roos et al., 1999).

Several lines of evidence suggest that the apicoplast is essential for parasite survival. Normally, when a tachyzoite invades a host cell, it forms a vacuole within the host cell called the parasitophorous vacuole. Within this vacuole, the parasite replicates synchronously and clonally by binary fission. As the parasite multiples, the vacuole enlarges. As many as 128 parasites can exist in a single vacuole. Eventually, the size of the vacuole becomes too great and the host cell is lysed. Each released parasite is just like the original parasite. It can reinfect another host cell and start the infective cycle again (Black and Boothroyd, 2000).

A segregation mutant of *T. gondii* was isolated that allowed only one tachyzoite in the parasitophorous vacuole to retain the apicoplast. In this segregation mutant, the apicoplast did not divide during binary fission (He et al., 2001). Therefore, only one parasite within the parasitophorous vacuole had an apicoplast. This parasite having the apicoplast could reinfect host cells indefinitely, but parasites that lacked apicoplasts could reinfect a host cell just one more time. Within the newly formed parasitophorous vacuole, parasites without an apicoplast did not multiply beyond eight parasites, and they eventually died inside the host cell. The apicoplast might therefore be essential for tachyzoites replication or release from the parasitophorous vacuole. This type of death profile has been termed "delayed death".

Another form of evidence for the importance of the apicoplast comes from the study of thiostrepton, an antibiotic that inhibits translation and ribosomal GTPase. Thiostrepton binds to the apicoplast 1srRNA of *Plasmodium falciparum*, another apicomplexan parasite with an apicoplast, and inhibits apicoplast-encoded RNA transcripts (Clough et al., 1997). This antibiotic inhibits the growth of *P. falciparum* at concentration similar to those required to impair bacteria (Rogers et al., 1997). Furthermore, antibiotics such as clindamycin, and chloramphenicol inhibit parasite replication by a "delayed death" phenomenon similar to what was observed in parasites that lacked apicoplasts (Fichera et al., 1995; Soldati et al., 1993), and a mutant resistant to clindamycin was shown to contain a point mutation in the apicoplast large subunit rRNA (Camps et al., 2002).

Although essential, the function of the apicoplast has been elusive since the apicoplast genome encodes mainly transcriptional and translational genes (Wilson et al., 1996). The function of plant and algae plastid include metabolic processes such as photosynthesis, lipid metabolism, protein synthesis, heme biosynthesis, amino acid synthesis, and electron transport chain (Alberts et al., 1994). If the apicoplast performs any of these functions, such genes are nuclear encoded, and their gene products are targeted back to the apicoplast. Gene transfers of endosymbiotic organelle genes to the nucleus are common for mitochondria and plastids. In fact, a vast majority of metabolic enzymes are nuclear encoded and post translationally imported into plastids (Alberts et al., 1994). Processes that most likely occur in the apicoplast will be discussed first.

## LIPID SYNTHESIS AND METABOLISM

Fatty acid synthesis is perhaps the best characterized metabolic pathway that occurs in the apicoplast. There are two types of fatty acid biosynthesis pathways. Type I is found in the cytosol of animals and fungi (Smith, 1994). Type II is widespread among bacteria but in eukaryotes is restricted to the plastids of plants and algae (Harwood, 1996). The biosynthesis of fatty acids consists of several two-carbon elongation cycles. During each cycle, an acetyl primer is condensed from malonyl-CoA into a fatty acid precursor, reduced, dehydrated, and reduced again to form a saturated fatty acid chain longer by two carbons (Smith, 1994). At least seven enzymes are involved in this pathway: acyl carrier protein, acyl transferase, ketoacyl synthase, ketoacyl reductase, dehydrase, enoyl reductase, and thioesterase. In mammalian cells, these cytoplasmic activities are encoded in a single multifunctional protein (Smith, 1994). In contrast, plant fatty acid biosynthesis occurs in the plastids, and the enzymes are encoded as discrete mono-functional polypeptides.

The apicoplast genome does not encode any fatty acid biosynthetic enzymes, but genes for this pathway, including acyl carrier protein (ACP), beta-hydroxyacyl-ACP dehydratase (FabZ), acetyl-CoA carboxylase (ACC), and enoyl acyl carrier protein (FabI) have been detected on the nuclear genome of *T. gondii*. They are nuclear encoded proteins destined for the apicoplast. The ACP gene product was detected in the apicoplast by anti-

ACP antibodies using immunofluorescence microscopy and immunoelectronmicroscopy (Waller et al., 1998). Other enzymes such as FabZ, ACC, and FabI, contain leaders sequences that are sufficient to target reporter proteins to the apicoplast (Jelenska et al., 2001; Waller et al., 2000). Molecular phylogenetic analysis of ACP, FabH and FabI suggest they are more related to plastid than bacterial enzymes (McLeod et al., 2001; Waller et al., 1998). Due to their differences with mammalian fatty acid biosynthetic enzymes, the apicoplast fatty acid enzymes represent good potential chemotherapeutic targets. Furthermore, inhibiting several different enzymes in the same metabolic pathway may provide synergistic effects.

Inhibitors that specifically block type II eubacterial fatty acid synthesis (Hayashi et al., 1983; Nishida et al., 1986) have shown efficacy against *Plasmodium falciparum* and *Toxoplasma gondii*. Triclosan, a potent and specific inhibitor of enoyl acyl carrier protein reductase (FabI) (McMurray et al., 1998; Perozzo et al., 2002), inhibited the growth of *P. falciparum* and *T. gondii* at IC 50 of 150-2000 ng/ml and 62 ng/ml, respectively (McLeod et al., 2001). These concentrations are comparable to antibacterial concentrations (Heath et al., 1998).

Thiolactomycin, an inhibitor of Beta-ketoacyl-ACP synthase (FabH), inhibited the growth of *P. falciparum* in cell culture at concentrations that was comparable with those that inhibit plants (Waller et al., 1998). The IC 50 of thiolactomycin against *P. falciparum in vitro* was about 50 µM. At this concentration, thiolactomycin did not affect type I fatty acid biosynthesis in *Saccharomyces cerevisiae, Candida albicans* and rat liver, suggesting inhibition is specific for the type II pathway (Hayashi et al., 1983).

Another class of type II fatty acid inhibitors, aryloxyphenoxy-propionates are herbicides that target plastid acetyl-CoA carboxylase (ACC) of grasses (Alban et al., 1994). Acetyl CoA carboxylase in plants is a chloroplast-localized, biotin-containing enzyme that catalyses the carboxylation of acetyl-CoA to malonyl-CoA. It represents the first committed step in fatty acid biosynthesis (Rawsthorne, 2002). Two ACC homologues were found in *T. gondii*, one of which possesses an apicoplast targeting leader (Jelenska et al., 2001). A biotinated protein, presumed to be ACC, was detected in the apicoplast by immunofluorescence microscopy. The ACC antagonist clodinafop inhibited the growth of *T. gondii* 70% over 2 days at 10 µM. Moreover, it eliminated parasites within 4 days at 100 µM (Zuther et al., 1999). Clodinafop was only mildly toxic to the host cells at 400 µM.

A mevalonate-independent pathway for isoprenoid synthesis was identified in *P. falciparum* (Jomaa et al., 1999). This 1-deoxy-D-xylulose 5-phosphate (DXP) pathway, by which sterols and ubiquinones are synthesized, was only seen previously in eubacteria, algae, and plants (Eisenreich et al., 1998). Instead of a pathway in which three molecules of acetyl coenzyme A are converted to 3-hydroxy-3-methylglutaryl coenzyme A (HMG-CoA), the DXP pathway is characterize by condensation of glyceraldehyde 3-phosphate and pyruvate to form DXP (Lichtenthaler, 1999). Both the DXP synthase and the DXP reductoisomerase genes were identified and cloned in *P. falciparum*. This pathway likely exists in the apicoplast since the leader peptide of *P. falciparum* DOXP reductoisomerase could target a reporter to the apicoplast (Jomaa et al., 1999). Both DXP reductoisomerase [TGG_29344, TGG_2562] and DXP synthase [TGG_15694, TGG_16566, TGG_4904] are present in the *T. gondii* genome database (http://toxodb.org), suggesting that this pathway may be present in all Apicomplexa.

Fosmidomycin, an inhibitor of DXP reductoisomerase in plants and bacteria (Lichtenthaler, 2000), inhibited the growth of *P. falciparum*. It had an IC 50 of about 350 nM depending on the *P. falciparum* strain. Mice infected with the rodent malaria parasite *P. vinckei* were cured after oral treatment with fosmidomycin (Jomaa et al., 1999). All ten catalytic enzymes associated with the DOXP pathway are potential chemotherapeutic targets (Lichtenthaler et al., 2000).

## PROTEIN SYNTHESIS

Genes on the apicoplast genome of *T. gondii* mainly encode for components of organelle protein synthesis. The complete apicoplast genome of *T. gondii* does not contain any genes which indicate an obvious function besides transcription and translation. Its genome encodes a comprehensive set of tRNAs and the translational elongation factor tufA. The apicoplast genome is actively transcribed (Gardner et al., 1991; Wilson, 1993). Transcripts of ssrRNA (Egea et al., 1995) and the ribosomal protein gene rps4 (Kohler et al., 1997) have been detected in *T. gondii*. Ribosome-like particles have been detected within the apicoplast of *T. gondii* in electron micrographs with size corresponding closely to bacterial 70S ribosomes (McFadden et al., 1996). Finally, hybridization studies showed erythrocytic stages of *P. falciparum* possess a subset of polysomes carrying apicoplast-specified rRNAs and mRNA (Roy et al., 1999).

A wealth of antibiotics inhibits the prokaryotic transcription and translation machinery. Antibiotics such as erythromycin, chloramphenicol, clindamycin, rifampin, and tetracycline show antiparasitic effects at clinically appropriate concentrations (McFadden et al., 1999; Pfefferkorn et al., 1992). The apicoplast large subunit rRNA has been identified as the target for clindamycin (Camps et al., 2002). Clindamycin resistant mutants possess a guanine to uridine point mutation at position 1857 of the apicoplast large rRNA. This position corresponds to position 2061 in the rRNA of *Escherichia coli* that is predicted to bind clindamycin (Douthwaite, 1992). In addition, this guanine is predicted to be a critical contributor to the transpeptidation reaction and is conserved in all kingdoms (Nissen et al., 2000).

Another antibiotic, thiostrepton, inhibits translation and ribosomal GTPase activity by binding to the 1srRNA (McConkey et al., 1997). The region of 1srRNA that thiostrepton binds is highly conserved among bacteria and organelles, but is absent from eukaryotic 1srRNA. In *P. falciparum*, treatment with thiostrepton inhibited apicoplast-encoded RNA transcripts within 6 hours. Furthermore, thiostrepton eliminated infection with the erythrocytic forms of *Plasmodium berghei* in mice (Sullivan et al., 2000). The clearance of these infected red blood cells followed the delayed death kinetics seen with drugs that interact with the apicoplast. However, the 1srRNA of *T. gondii* possesses a mutation at a critical nucleotide position that prevents thiostrepton binding (Clough et al., 1997). Similarly, a single point mutation in *P. falciparum* could confer thiostrepton resistance (Rogers et al., 1997). A related drug, micococcin was found to be 100 time more potent than thiostrepton with an IC50 of 35 nM (Rogers et al., 1998).

## DNA REPLICATION

As discussed above, the apicoplast contains a circular 35-kb genome which must be replicated. Trovafloxacin, a fluoroquinolone, was found to be active against *Toxoplasma gondii* (Khan et al., 1996). Later, it was shown that ciprofloxacin, a fluoroquinolone, induces the cleavage of this 35 kb DNA (Weissig et al., 1997). Ciprofloxacin inhibits bacterial type II DNA topoisomerases and progressively reduces the copy number of the apicoplast genome as detected by the extranuclear DAPI (4',6-diamidiono-2-phenyl-indole) staining and by the strength of hybridization signals to an apicoplast specific probe (Fichera et al., 1997). The rate of actual parasite replication was not affected by drug exposure while inside the first parasitophorous vacuole, but slowed dramatically in the second vacuole (Fichera et al., 1997). As discussed above, this "delayed death" phenotype has been observed in parasites lacking apicoplasts and with other antibiotics that target the apicoplast. Furthermore, a structural-activity relationship model was developed based on the anti-*Toxoplasma* activity of 24 different quinolones (Gozalbes et al., 2000). Fluorine at position Carbon-6 was essential for activity, and the presence of a methyl group at Carbon-5 or an azabicyclohexane at Carbon-7 had an enhancing effect. Such studies suggest better drug designs are possible for antibiotics that affect the apicoplast.

## ELECTRON TRANSPORT CHAIN

Nuclear encoded but apicoplast localized proteins involved in electron transfer were discovered in *T. gondii* and *P. falciparum*. The ferredoxin-NADP+ reductase from *T. gondii* (FNR) and a [2Fe-2S] ferredoxin from *P. falciparum* (Fd) were cloned and functionally expressed (Vollmer et al., 2001). Both enzymes contain N-terminal leaders that look like apicoplast targeting sequences. The role of this redox system may be to provide reduced ferredoxin for the desaturation of fatty acids. In nonphotosynthetic plastids such as those that are found in roots, electrons flow from NADPH to ferredoxin (Onda et al., 2000). Such reduced ferredoxin can then be used by nitrite reductase, sulfite reductase, glutamate synthase or lipid desaturase (Neuhaus et al., 2000; Shanklin et al., 1998). Phylogenetic analysis of *T. gondii* FNR and *P. falciparum* Fd suggests they are more related to nonphotosynthetic than photosynthetic isoforms (Vollmer et al., 2001).

## HEME BIOSYNTHESIS

A delta-aminolevulinic acid dehydratase (ALAD) homolog in *P. falciparum* has been reported (Sato et al., 2002; Van Dooren et al., 2002). ALAD catalyses the second reaction in the heme biosynthetic pathway, and in plants is plastid localized (Smith, 1988). The *P. falciparum* homolog possesses a putative apicoplast targeting leader, is actively transcribed, and rescues an ALAD-null mutant of *Escherichia coli*. (Sato et al., 2002; Van Dooren et al., 2002). These results suggest that the malarial ALAD is targeted to the apicoplast and is functional. Phylogenetic analysis shows this enzyme to be most similar to chloroplast ALADs (Sato et al., 2002). Succinylacetone, a structural analogue of delta-aminolevulinic acid (ALA) is an inhibitor of ALAD. It has been shown to inhibit the growth and heme biosynthesis of *P. falciparum* (Surolia et al., 1992). The genes for delta-aminolevulinic acid dehydratase and synthase are present in the *T. gondii* database (www.toxodb.org), although their functional significance will require further studies. Inhibitors to this pathway should be tested against *T. gondii* as they might inhibit its growth.

## CONCLUSIONS

In summary, some drugs that were previously known to be effective against apicomplexan parasites have now been shown to act in the apicoplast. Antibiotics such as ciprofloxacin and clindamycin disrupt prokaryotic DNA replication and protein synthesis in the apicoplast. These drugs have already been approved for human consumption. As new antibiotics are discovered, their anti-parasitic effects should be tested, as they may inhibit the apicoplast.

While more functions of the apicoplast remain to be discovered, some metabolic pathways in the apicoplast have already been identified. For example, fatty acid biosynthesis likely occurs in the apicoplast, and this pathway has provided new chemotherapeutic targets. Inhibitors of this pathway such as triclosan and clodinafop have shown activity against *T. gondii*. As protein structures of enzymes in this and other pathways become available, better inhibitors could be designed specifically for anti-parasitic activities. Moreover, inhibition of different enzymes in the same metabolic pathway will likely provide synergistic effects against *T. gondii* as seen in folate biosynthesis (Derouin et al., 1989). Other potential pathways such as electron transport chain and heme biosynthesis may provide additional drug targets.

It is possible that the apicoplast likely performs other functions in *T. gondii* besides those discussed in this article and that some of these functions might be specific for *T. gondii*. Depending on its environment and life cycle, members of the apicomplexan phylum may retain or lose specific functions of the ancestral apicoplast. Complete sequencing of the *T. gondii* genom should help to elucidate additional functions of the *T. gondii* apicoplast. These metabolic pathways should provide additional and perhaps specific chemotherapeutic targets to combat Toxoplasmosis.

## REFERENCES

Alban, C., P. Baldet, and R. Douce. 1994. Localization and characterization of two structurally different forms of acetyl-CoA carboxylase in young pea leaves, of which one is sensitive to aryloxyphenoxypropionate herbicides. Plant Molecular Biology **24**: 35-49.

Alberts, B., D. Bray, J. Lewis, M. Raff, K. Roberts, and J.D. Watson. 1994. Molecular Biology of the Cell, 3rd edn. Garland Publishing.

Black, M.W., and J.C. Boothroyd. 2000. Lytic cycle of *Toxoplasma gondii*. Microbiology and Molecular Biology Reviews **64**: 607-623.

Camps, M., G. Arrizabalaga, and J. Boothroyd. 2002. An rRNA mutation identifies the apicoplast as the target for clindamycin in *Toxoplasma gondii*. Molecular Microbiology **43**: 1309-1318.

Clough, B., M. Strath, P. Preiser, P. Denny, and I.R. Wilson. 1997. Thiostrepton binds to malarial plastid rRNA. FEBS Letters **406**: 123-125.

Derouin, F., and C. Chastang. 1989. In vitro effects of folate inhibitors on *Toxoplasma gondii*. Antimicrobial Agents and Chemotherapy **33**: 1753-1759.

Douthwaite, S. 1992. Interaction of the antibiotics clindamycin and lincomycin with *Escherichia coli* 23S ribosomeal RNA. Nucleic Acids Research **20**: 4717-4720.

Dubey, J.P., D.S. Lindsay, and C.A. Speer. 1998. Structures of *Toxoplasma gondii* tachyzoites, bradyzoites, and sporozoites and biology and development of tissue cysts. Clinical Microbiology Reviews **11**: 267-299.

Egea, N., and N. Lang-Unnasch. 1995. Phylogeny of the large extrachromosomal DNA of organisms in the phylum apicomplexa. Journal of Eukaryotic Microbiology **42**: 679-684.

Eisenreich, W., M. Schwarz, A. Cartayrade, D. Arigoni, M.H. Zenk, and A. Bacher. 1998. The deoxyxylulose phosphate pathway of terpenoid biosynthesis in plants and microorganisms. Chemistry and Biology **5**: 221-233.

Fast, N.M., J.C. Kissinger, D.S. Roos, and P.J. Keeling. 2001. Nuclear-encoded, plastid-targeted genes suggest a single common origin for apicomplexan and dinoflagellate plastids. Molecular Biology and Evolution **18**: 418-426.

Fichera, M.E., M.K. Bhopale, and D.S. Roos. 1995. In vitro assays elucidate peculiar kinetics of clindamycin action against *Toxoplasm gondii*. Antimicrobial Agents and Chemotherapy **39**: 1530-1537.

_____, and D.S. Roos. 1997. A plastid organelle as a drug target in apicomplexan parasites. Nature **390**: 407-409.

Gardner, M., J. Feagin, and D. Moore. 1991. Organization and expression of small subunit ribosomal RNA genes encoded by a 35 kbase circular DNA in *Plasmodium falciparum*. Molecular and Biochemical Parasitology **48**: 77-88.

Gozalbes, R., M. Brun-Pascaud, R. Garcia-Domenech, J. Galvez, P.M. Girard, J.P. Doucet, and F. Derouin. 2000. Anti-*Toxoplasma* activities of 24 quinolones and fluoroquinolones in vitro: prediction of activity by molecular topology and virtual computational techniques. Antimicrobial Agents and Chemotherapy **44**: 2771-2776.

Harwood, J.L. 1996. Recent advances in the biosynthesis of plant fatty acids. Biochimica Biophysica Acta **1301**: 7-56.

Hayashi, T., O. Yamamoto, H. Sasaki, A. Kawaguchi, and H. Okazaki. 1983. Mechanism of action of the antibiotic thiolactomycin inhibition of fatty acid synthesis of *Escherichia coli*. Biochemica Biophysica Research Communication **115**: 1108-1113.

He, C.Y., M.K. Shaw, C.H. Pletcher, B. Streipen, L.G. Tilney, and D.S. Roos. 2001. A plastid segregation defect in the protozoan parasite *Toxoplasma gondii*. EMBO Journal **20**: 330-339.

Heath, R.J., Y.T. Yu, M.A. Shapiro, E. Olson, and C.O. Rock. 1998. Broadspectrum antimicrobial biocides target the Fab I component of fatty acid synthesis. Journal of Biological Chemistry **273**: 30316-30320.

Jelenska, J., M. Crawford, O. Harb, E. Zuther, R. Haselkorn, D. Roos, and P. Gornicki. 2001. Subcellular localization of acetyl-CoA carboxylase in the apicomplexan parasite *Toxoplasma gondii*. Proceedings of the National Academy of Science USA **98**: 2723-2728.

Jomaa, H., J. Wiesner, S. Sanderbrand, B. Altincicek, C. Weidemeyer, M. Hintz, T. I., M. Eberl, J. Zeidler, H.K. Lichtenthaler, D. Soldati, and E. Beck. 1999. Inhibitors of the nonmevalonate pathway of isoprenoid biosynthesis as antimalarial drugs. Science **285**: 1573-1576.

Kasper, L.H. 2002. *Toxoplasma* Infection. McGraw-Hill Companies, www.harrisonsonline.com.

Khan, A., T. Slifer, F. Araujo, and J. Remington. 1996. Trovafloxacin is active against *Toxoplasma gondii*. Antimicrobial Agents and Chemotherapy **40**: 1855-1859.

Kohler, S., C.F. Delwiche, P.W. Denny, L.G. Tilney, P. Webster, R.J.M. Wilson, J.D. Palmer, and D.S. Roos. 1997. A plastid of probable green algal origin in apicomplexan parasites. Science **275**: 1485-1489.

Lang-Unnasch, N., M.E. Reith, J. Munholland, and J.R. Barta. 1998. Plastids are widespread and ancient in parasites of the phylum Apicomplexa. International Journal of Parasitology **28:** 1743-1754.

Lichtenthaler, H.K. 1999. The 1-deoxy-d-xylulose-5-phosphate pathway of isoprenoid biosynthesis in plants. Annual Review of Plant Physiology and Plant Molecular Biology **50:** 47-65.

_____. 2000. Non-mevalonate isoprenoid biosynthesis: enzymes, genes and inhibitors. Biochemical Society Transactions **28:** 785-789.

_____, J. Zeidler, J. Schwender, and C. Muller. 2000. The non-mevalonate isoprenoid biosynthesis of plants as a test system for new herbicides and drugs against pathogenic bacteria and the malaria parasite. Z Naturforsch **55:** 305-313.

McConkey, G.A., M.J. Rogers, and T.F. McCutchan. 1997. Inhibition of *Plasmodium falciparum* protein synthesis: targeting the plastid -like organelle with thiostrepton. Journal of Biological Chemistry **272:** 2046-2049.

McFadden, G.I. 1999. Plastids and protein targeting. Journal of Eukaryotic Microbiology **46:** 339-346.

_____, M.E. Reith, J. Munholland, and N. Lang-Unnasch. 1996. Plastid in human parasites. Nature **381:** 482.

_____, and D.S. Roos. 1999. Apicomplexan plastid as drug targets. Trends in Microbiology **7:** 328-333.

McLeod, R., S.P. Muench, J.B. Rafferty, D.E. Kyle, E.J. Mui, M.J. Kirisits, D.G. Mack, C.W. Roberts, B.U. Samuel, R.E. Lyons, M. Dorris, W.K. Milhous, and D.W. Rice. 2001. Triclosan inhibits the growth of *Plasmodium falciparum* and *Toxoplasma gondii* by inhibition of Apicomplexan Fab I. International Journal of Parasitology **31:** 109-113.

McMurray, L.M., M. Oethinger, and S.B. Levy. 1998. Triclosan targets lipid synthesis. Nature **394:** 531-532..

Neuhaus, E.H., and M.J. Emes. 2000. Nonphotosynthetic metabolism in plastids. Annual Review of Plant Physiology and Plant Molecular Biology **51:** 111-140.

Nishida, I., A. Kawaguchi, and M. Yamada. 1986. Effect of thiolactomycin on the individual enzymes of the fatty acid synthase system in Escherichia coli. Journal of Biochemistry (Tokyo) **99:** 1447-1454.

Nissen, P., J. Hansen, N. Ban, P.B. Moore, and T.A. Steitz. 2000. The structural basis of ribosome activity in peptide bond synthesis. Science **289:** 920-930.

Onda, Y., T. Matsumura, Y. Kimata-Ariga, H. Sakakibara, T. Sugiyama, and T. Hase. 2000. Differential Interaction of Maize Root Ferredoxin: NADP+ Oxidoreductase with Photosynthetic and Non-Photosynthetic Ferredoxin Isoproteins. Plant Physiology **123:** 1037-1046.

Perozzo, R., M. Kuo, A.S. Sidhu, J.T. Valiyaveettil, R. Bittman, W.R. Jacobs, JR., D.A. Fidock, and J.C. Sacchettini. 2002. Structural elucidation of the specificity of the antibacterial agent triclosan for malarial enoyl acyl carrier protein reductase. Journal of Biological Chemistry **277:** 13106-13114.

Pfefferkorn, E.R., R.F. Nothnagel, and S.E. Borotz. 1992. Parasiticidal effect of clindamycin on *Toxoplasma gondii* grown in cultured cells and selection of a drug-resistant mutant. Antimicrobial Agents and Chemotherapy **36:** 1091-1096.

Rawsthorne, S. 2002. Carbon flux and fatty acid synthesis in plants. Progress in Lipid Research **41:** 182-196.

Rogers, M.J., Y.V. Bukhman, T.F. McCutchan, and D.E. Draper. 1997. Interaction of thiostrepton with an RNA fragment derived from the plastid-encoded ribosomal RNA of the malaria parasite. RNA **3:** 815-820.

_____, E. Cundliffe, and T.F. McCutchan. 1998. The antibiotic micrococcin is a potent inhibitor of growth and protein synthesis in the malaria parasite. Antimicrobial Agents and Chemotherapy **42:** 715-716.

Roos, D.S., M.J. Crawford, R.G.K. Donald, J.C. Kissinger, L.J. Klimczak, and B. Striepen. 1999. Origin, targeting, and function of the apicomplexan plastid. Current Opinion in Microbiology **2:** 426-432.

Roy, A., R. Cox, D. Williamson, and R. Wilson. 1999. Protein synthesis in the plastid of *Plasmodium falciparum*. Protist **150:** 183-188.

Sato, S., and R.J.M. Wilson. 2002. The genome of *Plasmodium falciparum* encodes an active delta-aminolevulinic acid dehydratase. Current Genetics **40:** 391-398.

Shanklin, J., and E.B. Cahoon. 1998. Desaturation and related modifications of fatty acids. Annual Review of Plant Physiology and Plant Molecular Biology **49:** 611-641.

Smith, A.G. 1988. Subcellular localization of two porphyrin-synthesis enzymes in *Pisum sativum* (pea) and *Arum* (cuckoo-pint) species. Biochemistry Journal **249:** 423-428.

Smith, S. 1994. The animal fatty acid synthase: one gene, one polypeptide, seven enzymes. FASEB Journal **8:** 1248-1259.

Soldati, D., and J.C. Boothroyd. 1993. Transient transfection and expression in the obligate intracellular parasite *Toxoplasma gondii*. Science **260:** 349-352.

Sullivan, M., J. Li, S. Kumar, M.J. Rogers, and T.F. McCutchan. 2000. Effects of interruption of apicoplast function on malaria infection, development, and transmission. Molecular and Biochemical Parasitology **109:** 17-23.

Surolia, N., and G. Padmanaban. 1992. de novo biosynthesis of heme offers a new chemotherapeutic target in the human malarial parasite. Biochemical Biophysica Research Communication **187:** 744-750.

Van Dooren, G.G., V. Su, M.C. Diombrain, and G.I. McFadden. 2002. Processing of an apicoplast leader sequence in Plasmodium falciparum, and the identification of a putative leader cleavage enzyme. Journal of Biological Chemistry **25:** 25.

Vollmer, M., N. Thomsen, S. Wiek, and F. Seeber. 2001. Apicomplexan parasites possess distinct nuclear-encoded, but apicoplast-localized, plant-type ferredoxin-NADP+ reductase and ferredoxin. Journal of Biological Chemistry **276:** 5483-5490.

Waller, R.F., P.J. Keeling, R.G.K. Donald, B. Striepen, E. Handman, N. Lang-Unnasch, A.F. Cowman, G.S. Besra, D.S. Roos, and G.I. McFadden. 1998. Nuclear-encoded proteins target to the plastid in *Toxoplasma gondii* and *Plasmodium falciparum*. Proceedings of the National Academy of Science USA **95:** 12352-12357.

_____, M.B. Reed, A.F. Cowman, and G.I. McFadden. 2000. Protein trafficking to the plastid of *Plasmodium falciparum* is via the secretory pathway. EMBO Journal **19:** 1794-1802.

Weissig, V., T.S. Vetro-Widenhouse, and T.C. Rowe. 1997. Topoisomerase II inhibitors induce cleavage of nuclear and 35-kb plastid DNAs in the malarial parasite *Plasmodium falciparum*. DNA and Cell Biology **16:** 1483-1492.

Wilson, I. 1993. Plastids: better red than dead? Nature **366:** 638.

Wilson, R.J.M., P.W. Denny, P.R. Preiser, K. Roberts, A. Roy, A. Whyte, M. Strath, D.J. Moore, and D.H. Williamson. 1996. Complete gene map of the plastid-like DNA of the malaria parasite *Plasmodium falciparum*. Journal of Molecular Biology **261:** 155-172.

Zuther, E., J.J. Johnson, R. Haselkorn, R. McLeod, and P. Gornicki. 1999. Growth of *Toxoplasma gondii* is inhibited by aryloxyphenoxypropionate herbicides targeting acetyl-CoA carboxylase. Proceedings of the National Academy of Science USA **96:** 13387-13392.

# FACTORS DETERMINING RESISTANCE AND SUSCEPTIBILITY TO INFECTION WITH *TOXOPLASMA GONDII*

Yasuhiro Suzuki

*Center for Molecular Medicine and Infectious Diseases, Department of Biomedical Sciences and Pathobiology, Virginia-Maryland Regional College of Veterinary Medicine, Virginia Polytechnic Institute and State University, 1410 Prices Fork Road, Blacksburg, Virginia 24061. USA*

## ABSTRACT

*Toxoplasma gondii* is an intracellular protozoan parasite that invades a variety of host cells in various organs including the central nervous system. IFN-γ-dependent, cell-mediated immunity is crucial for controlling the parasites during the acute stage of infection and for preventing development of toxoplasmic encephalitis (TE) during the later stage of infection. Multiple populations of both T and non-T cells are important sources of IFN-γ in the resistance. IL-12, IL-18, Bcl-3, NF-κB(2), and CD40-CD40L ligand interaction upregulate the IFN-γ production. Down-regulation of IFN-γ-mediated immune responses is also important for host resistance to prevent development of immunopathology caused by overly stimulated responses. IL-10. TGF-β and lipoxin A4 are involved in such down-regulation. IFN-γ-mediated immune responses control tachyzoites in both phagocytic and non-phagocytic cells through at least five different mechanisms, most likely depending on the types of cells responding to IFN-γ. Such effector functions involve production of nitric oxide by inducible nitric oxide synthase, tryptophan degradation by indolamine 2,3-dioxygenase, unidentified mechanism(s) mediated by 47- to 48-kDa proteins (e.g. IGTP) encoded by an IFN-γ responsive gene family, limiting the availability of intracellular iron to the parasite, and production of reactive oxygen intermediates. Host genes affect resistance/susceptibility to infection with this parasite. Genes that regulate resistance against acute infection differ from those that regulate resistance to development of TE during the late stage of infection. In mouse, at least five genes are involved in determining survival during the acute stage whereas the $L^d$ gene within the D region of the major histocompatibility (H-2) complex confers resistance to development of TE.

In AIDS patients, HLA-DQ3 appears to be a genetic marker of susceptibility to development of TE, and HLA-DQ1 appears to be a resistance marker. Strains (genetic variation) of *T. gondii* are another factor that affects the susceptibility of the host to both acute and late stages of infection. The genotypes of the parasite are important for determining the susceptibility. Some combination of alleles at two or more loci appears to be pivotal to determine virulence of the parasite during acute stage of infection. It is possible that the genes of the host and genetic variation of *T. gondii* affect immune response of the host to the parasite, thereby contribute to determining resistance to the infection.

**Key words:** *Toxoplasma gondii*, Parasite infection, Encephalitis, Cell-mediated immunity, IFN-γ, Antibody, Genetic regulation, Susceptibility, Virulence


**INTRODUCTION**

   *Toxoplasma gondii* is a ubiquitous, obligate intracellular protozoan parasite in humans and animals. Chronic (latent) infection with this parasite is likely one of the most common infections of humans. During the acute stage of the infection, tachyzoites quickly proliferate within a variety of nucleated cells and spread throughout host tissues. The acute infection/disease is caused by infection with this form of the parasite. Following the acute stage, the parasite forms cysts (latent stage) in various organs, especially the brain, heart and skeletal muscle, establishing chronic infection. Infection in immunocompetent individuals usually is unnoticed or is a benign self-limiting illness (e.g. lymphadenitis), and results in chronic infection. Immunosuppression in the chronically infected individuals may result in reactivation of a latent infection; which is initiated by disruption of cysts, then followed by proliferation of tachyzoites. Such reactivation of *T. gondii* infection usually presents as toxoplasmic encephalitis (TE). TE is an important opportunistic infectious disease in the central nervous system in AIDS patients as well as immunocompromised non-AIDS patients, such as those with organ-transplants. Since immunocompetent individuals do not usually suffer apparent untoward effects, including development of encephalitis, it is clear that the immune response is critical for controlling the parasite and preventing the diseases. Interferon-gamma (IFN-γ)-dependent, cell-mediated immunity, the genetic background of the host and the strain of *T. gondii* have been shown to be critical for determining the resistance/susceptibility to infection with *T. gondii*. Each of these factors is addressed below. Cells and molecules involved in the IFN-γ-mediated resistance are summarized in Table 1.

## A. IMPORTANCE OF IFN-γ-DEPENDENT, CELL-MEDIATED IMMUNITY

IFN-γ is the central cytokine in resistance against acute acquired infection with *T. gondii* and recrudescence of chronic infection (TE). NK cells and T cells are important producers of this cytokine following infection. Non-T cells that do not appear to be NK cells also produce IFN-γ in the brain during the chronic stage of infection, and they are required for prevention of TE under collaboration with T cells (Kang and Suzuki, 2001). Multiple mechanisms have been identified to control the parasite within the effector cells activated by IFN-γ, and the effective mechanisms appear to differ depending on the types of the effector cells. The effector mechanisms are summarized below.

### Nitric oxide (NO) produced by inducible NO synthase (iNOS)

Macrophages become quickly activated to kill intracellular tachyzoites following infection with *T. gondii*. This activation is mediated by IFN-γ since neutralization of the activity of this cytokine by treatment with anti-IFN-γ mAb blocks the activation (Suzuki et al., 1988). In the absence of activity of endogenous IFN-γ, mice die in one week after intraperitoneal infection and their mortality is associated with numerous tachyzoites in their peritoneal cavities (Suzuki et al., 1988). Thus, IFN-γ-mediated activation of macrophages is critical for resistance against acute infection with this parasite.

Murine peritoneal macrophages become activated after treatment with a combination of IFN-γ and TNF-α in vitro, and such activated cells inhibit intracellular replication of tachyzoites through generation of NO by iNOS. However, the situation appears to differ in vivo. Mice lacking tumor necrosis factor (TNF) receptor type 1 (R1) and type 2 (R2) (Yap et al., 1998) and those lacking iNOS (Scharton- Kersten et al., 1997) are able to control parasite growth in the peritoneal cavity following intraperitoneal infection, indicating that TNF-α and iNOS are not essential for controlling acute infection in mice. Consistent with these findings are experiments in mice that lack IFN-γ regulatory factor 1 (IRF-1) which is essential for iNOS induction by IFN-γ (Khan et al., 1996).

Table Table. 1. Immunological factors important for determining the resistance and susceptibility to infection with *T. gondii*.

| Functions | Cells or molecules |
|---|---|
| Production of IFN-g | ab T cells |
| | gd T cells |
| | NK cells |
| | Unidentified non-T (non-NK) cells |
| Up-regulation of IFN-g production | IL-12 |
| | IL-18 |
| | Bcl-3 |
| | NF-kB(2) |
| | CD40L |
| Down-regulation of IFN-g production | IL-10 |
| | TGF-b |
| | Lipoxin A4 |
| Controlling *T. gondii* following activation by IFN-g | |
| | Macrophages: NO* (mouse) |
| | Macrophages, monocytes: ROI** (humans) |
| | Enterocytes: iron (rat) |
| | Fibroblasts: IDO*** (humans) |
| | Retinal pigment epithelial cells: IDO (humans) |
| | Microglia: NO, NO-independedt (mouse) |
| | Microglia: NO-independent (humans) |
| | Astrocutes: IGTP (mouse) |
| | Astrocytes: IDO, NO (humans) |
| | Brain microvascular endothelial cells: IDO (humans) |

* Nitric oxide,
** Reactive oxygen intermediates
*** Indolamine 2,3-dioxygenase

Although the IRF-1-deficient animals are more susceptible to infection with *T. gondii* than control animals, they survive the acute stage of the infection through iNOS-independent mechanism(s). These results indicate that the protective mechanism(s) that require IFN-γ but do not require TNF-α or iNOS is sufficient for mice to control parasite growth during the acute stage of the infection. These results do not rule out a possibility that TNF-α and iNOS play a partial role in resistance to *T. gondii* in this stage of infection. The iNOS-independent mechanisms described in the sections below (such as tryptophan degradation by indolamine 2,3-dioxygenase, IGTP, etc) may be involved in this resistance.

In contrast to the acute stage of infection, mice deficient in TNF R1/R2 (Yap et al., 1998) or iNOS (Scharton- Kersten et al., 1997) succumbed to necrotizing TE during the late stage of infection. These results are consistent with those of earlier studies; treatments of chronically infected wild-type mice with anti-TNF-α mAb or aminoguanididine (an iNOS inhibitor) resulted in development of TE (Gazzinelli et al., 1993; Hayashi et al., 1996). Thus, TNF-α and iNOS are critical for prevention of proliferation of tachyzoites in the brain although there is a possibility in these studies using the deficient mice that the absence of TNF-α andiNOS-mediated resistance during the acute stage of infection may have resulted in increased cyst burden in organs and may have partially contributed increased mortality during the late stage of infection. As mentioned earlier, IFN-γ plays the central role in resistance of the brain against this parasite (Gazzinelli et al., 1992; Suzuki et al., 1990). Since neutralization of IFN-γ or TNF-a results in decreased iNOS expression and development of severe TE (Gazzinelle et al., 1993), activation of iNOS mediated by IFN-γ and TNF-α appears to play a key role in prevention of TE.

Microglia are likely important effector cells involved in iNOS-mediated protective mechanism in the brain of mice. Murine microglia become activated in vitro to inhibit intracellular proliferation of tachyzoites following treatment with IFN-γ plus lypopolysaccharide (LPS) (Chao et al., 1993). NO was shown to mediate the inhibitory effect of activated murine microglia on intracellular replication of tachyzoites since treatment of these cells with $N^G$-monomethyl-L-arginine (which blocks the generation of NO) ablates their inhibitory activity (Chao et al., 1993). Recently, Freund et al (Freund et al., 2001) reported an involvement of both NO-dependent and –independent mechanisms in the resistance of murine microglia activated by a combination of IFN-γ and TNF-α. Human microglia also become activated in vitro to inhibit intracellular proliferation of tachyzoites following treatment with IFN-γ plus LPS (Chao et al., 1994). TNF-α and interleukin

(IL)-6 are involved in activation of human microglia. However, in contrast to murine microglia, NO is not involved in the inhibitory effect of human microglia against tachyzoites (Chao et al., 1994). In vivo, following *T. gondii* infection, microglia become activated to produce TNF-α, and IFN-γ mediates the activation (Deckert-Schuter et al., 1999). IFN-γ-mediated activation of microglia in collaboration with autocrine TNF-α is likely one of the resistance mechanisms of the brain against *T. gondii* although the role of NO in resistance differs between humans and mice.

Astrocytes appear to be another effector type of IFN-γ-responding effector cells involved in prevention of *T. gondii* growth in the brain. Peterson et al (Peterson et al., 1995) reported that human astrocytes become activated following treatment with IFN-γ plus IL-1β to inhibit intracellular proliferation of tachyzoites, and that inhibitory effect is mediated by NO. Pelloux et al (Pelloux et al., 1996) reported that TNF-α induced a significant reduction in intracellular multiplication of the parasite in human astrocytoma-derived cells whereas IL-1a induced an increase in the parasite multiplication. Importance of IDO, but not of NO, was demonstrated in human astrocytes and gliablastome cells activated by a combination of IFN-γ and TNF-α (Daubener et al., 1996) (see "Tryptophan degradation by IDO" section below).

In vivo, iNOS-independent mechanism(s), in addition to the iNOS-mediated mechanism, plays an important role in resistance against TE. Yap et al (Yap et al., 1999) demonstrated an involvement of iNOS-independent mechanisms in prevention of mortality in *T. gondii*-infected mice using bone marrow chimera. In order to address the mechanisms for restricting the growth of *T. gondii* within cells of nonhematopoietic origin, they developed chimeric mice expressing IFN-γ receptors, TNF R1/R2, or iNOS either on hematopoietic or nonhematopoietic cells. Resistance to acute and persistent infection was displayed only by mice in which IFN-γ receptors and TNF R1/R2 were expressed in both hematopoietic and nonhematopoietic cells. In contrast, expression of iNOS by only hematopoietic cells was sufficient for host resistance. These results suggest that in concert with bone marrow-derived effector cells, nonhematopoietic cells can directly mediate IFN-γ- and TNF-a-dependent resistance to the parasite. This resistance does not require expression of iNOS in nonhematopoietic cells. Requirement of these multiple mechanisms for resistance to *T. gondii* appears to be due to its infection of not only mononuclear phagocyte lineage but also a wide variety of host cells.

Resistance to development of TE is under genetic control in both humans and mice (see "Host genes involved in regulating resistance" section below). Our recent studies suggested a crucial role of IFN-γ-dependent, but iNOS-independent, mechanism in the genetic resistance of BALB/c mice to this disease. BALB/c-background IFN-γ-deficient mice infected and treated with sulfadiazine developed severe TE after discontinuation of sulfadiazine treatment although these animals expressed large amounts of mRNA for TNF-α and iNOS and in their brains. The amounts of the mRNA expressed in these animals were equivalent to those expressed in the brains of infected control mice which prevented development of TE (Suzuki et al., 2000). Thus, expression of TNF-α and iNOS is insufficient for prevention of TE in the absence of IFN-γ.

### Reactive oxygen intermediates (ROI)

ROI have been implicated in the toxoplasmacidal activity of normal human monocytes and IFN-γ-activated human macrophages in vitro (Murray et al., 1995; Wilson et al., 1979). Involvement of ROI in their antimicrobial activity was suggested because; 1) impairing the ability of the cells to generate oxygen intermediates (by glucose deprivation or treatment with superoxide dismutase, catalase, or mannitol) inhibited toxoplasmacidal activity by greater than 80% (Murray et al., 1995) and 2) killing of *T. gondii* by monocytes obtained from chronic granulomatous disease patients was impaired (Murray et al., 1995). However, the physiological significance of the ROI pathway still remains unclear, especially in mice. It was reported that the parasites are resistant to the oxygen metabolites produced in murine macrophages (Cahng et al., 1989). Recently, p47phox-deficient mice, which lack an inducible oxidative burst, were described to control both the acute and chronic stages of *T. gondii* infection (Scharton- Kersten et al., 1997).

### Tryptophan degradation by indolamine 2,3-dioxygenase (IDO)

IDO is an enzyme which catalyzes the initial rate-limiting step of tryptophan catabolism to N-formylkynurenine and kynurenine. The depletion of intracellular tryptophan pools by IDO is an important mechanism by which IFN-γ controls the intracellular replication of *T. gondii* tachyzoites in various types of human cells. Therefore, IFN-γ-mediated induction of IDO appears to be important for resistance against the parasite in various organs in humans during the acute stage of infection. During the chronic stage of infection, IDO induced by IFN-γ likely plays an important role in controlling *T. gondii* in the brain. TNF-α and IFN-γ were shown to be synergistic in activation of IDO in human glioblastoma cell lines and native astrocytes (Däubener et al., 1996). This IDO activity resulted in a strong toxoplasmostatic effect of the activated glioblastoma cells (Däubener et al., 1996). Däubener et al (2001) recently reported that stimulation of human brain microvascular endothelial cells (HBMEC) with IFN-γ resulted in the induction of toxoplasmostasis. The capacity of HBMEC to restrict *Toxoplasma* growth after IFN-γ stimulation was enhanced in the presence of TNF-α, and such capability correlated with their IDO activities. Furthermore, the addition of excess amounts of tryphophan to the HBMEC cultures resulted in a complete abrogation of the IFN-γ-TNF-α-mediated toxoplasmostasis, indicating that their protective activity is mediated by IDO.

In contrast to human models, the role of IDO in resistance to *T. gondii* is unclear in mice. IFN-γ reportedly fails to induce IDO and toxoplasmastatic activity in mouse fibroblasts (Turco et al., 1986). IDO is not important for controlling *T. gondii* in murine astrocytes (Halonen et al., 2001). However, two groups recently reported IFN-γ-dependent expression of IDO in the brains and lungs of mice during acute stage of infection (Fujigaki et al., 2002; Silva et al., 2002). More studies are needed to assess a possible involvement of IDO in resistance to *T. gondii* in mice.

### Limiting the availability of intracellular iron to the parasite

Dimier and Bout (1998) reported that activation of rat enterocytes with rat recombinant IFN-γ resulted in an inhibition of intracellular replication of *T. gondii*. Neither nitrogen and oxygen derivatives nor tryphophan starvation are involved in the inhibitory effect. Experiments using $Fe_2^+$ salt, as well as carrier and chelator, suggested that IFN-γ-treated enterocytes inhibit *T. gondii* replication by limiting the availability of intracellular iron to the parasite. It is unknown whether this iron-dependent mechanism is also involved in resistance of other types of cells activated by IFN-γ.

### IGTP

*IGTP* is one of IFN-γ responsive genes which has recently been identified. It is representative of a family of at least six genes encoding 47- to 48-kDa proteins that contain a GTP-binding sequence and which are expressed at high levels in immune and nonimmune cells after exposure to IFN-γ. Several of these proteins, including IGTP, localize to the endoplasmic reticulum of cells, suggesting that they may be involved in the processing or trafficking of immunologically relevant proteins, such as antigens or cytokines. Taylor et al (2000) recently generated IGTP-deficient mice and found that despite normal immune cell development and normal clearance of *Listeria monocytogenes* and cytomegalivirus, the mice displays a profound loss of host resistance to acute infection with *T. gondii*. Thus, IGTP is an essential mediator of specialized antimicrobial activities of IFN-γ.

In vitro studies demonstrated an importance of IGTP for prevention of *T. gondii* replication in murine astrocytes. Following pre-treatment with IFN-γ or a combination of this cytokine with either TNF-α, IL-1, or IL-6, murine astrocytes are able to inhibit proliferation of tachyzoites in vitro. The inhibitory effect of activated astrocytes is not mediated by IDO, NO, reactive oxygen intermediates or iron deprivation. Halonen et al (2001) recently reported that astrocytes obtained from IGTP-deficient mice did not cause a significant inhibition of *T. gondii* growth following treatment with IFN-γ whereas wild-type astrocytes inhibited the growth. Therefore, IGTP plays a central role in the IFN-γ-induced inhibition of the parasite in murine astrocytes.

In relation to IGTP, the roles of other members of the gene family in resistance against *T. gondii* were studied in knockout mice that lacked expression of the genes LRG-47 and IRG-47 (Callazo et al., 2001). LRG-47–deficient mice succumbed uniformly and rapidly during the acute stage of the infection; in contrast, IRG-47-deficient mice displayed only partially decreased resistance that was not manifested until the chronic phase. Thus, LRG-47 and IRG-47 have vital, but distinct roles in immune defense against *T. gondii*.

## B. IMPORTANCE OF UP-REGULATION AS WELL AS DOWN-REGULATION OF IFN-γ-MEDIATED IMMUNITY

IL-12, IL-18, Bcl-3, NF-κb(2) and CD40-CD40L ligand interaction are important for up-regulation and maintaining IFN-γ-mediated immune responses during the course of *T. gondii* infection. However, down-regulation of the immune responses is also crucial for preventing development of IFN-γ-mediated immunopathology during the course of infection. Genetic susceptibility of mice to acute peroral infection was found to be associated with development of severe necrosis of the villi and mucosal cells in the small intestine (Liesendfeld et al., 1996), and the necrosis is caused by IFN-γ, TNF-α and iNOS-mediated immune responses (Khan et al., 1997; Liesendfeld et al., 1996; Liesendfeld et al., 1999). IL-10 plays a crucial role in prevention of development of IFN-γ-mediated intestinal pathology and mortality (Suzuki et al., 2000). TGF-β also appears to be involved in the down-regulation (Buzoni-Gatel et al., 2001). Lipoxin A4 may play an important down-regulatory role during the chronic stage of infection (Aliberti et al., 2002).

## C. INVOLVEMENT OF HUMORAL IMMUNITY IN RESISTANCE

Antibodies are involved in resistance against *T. gondii* although cell-mediated immunity plays the major role as mentioned above. Frenkel and Taylor (Frenkel et al., 1982) examined the effect of depletion of B cells by treatment with anti-m antibody on toxoplasmosis in mice infected with a virulent strain and treated with sulfadiazine. They observed mortality associated with pneumonia, myocarditis and/or encephalitis in infected anti-m-treated mice after discontinuation of sulfadiazine treatment. Administration of antisera to *T. gondii* reduces mortality in these animals. These results suggest that antibody production by B cells may be important for controlling the latent persistent infection. However, these studies do not provide conclusive information because of the potential side effects of anti-m antibody treatment on the immune system.

Recently, we examined the role of B cells in resistance to *T. gondii* by using B cell-deficient (μMT) mice generated by disruption of one of the membrane exon of μ-chain gene (Kang et al., 2000). All B cell-deficient mice died between 3 and 4 weeks after infection whereas no mortality was observed in the control mice until 8 weeks. At the stage during which μMT animals succumbed to the infection, large numbers of tachyzoites were detected only in their brains. Furthermore, treatment of infected μMT mice with anti-*T. gondii* IgG antibody reduced mortality and prolonged time to death. These results indicate that B cells play an important role through production of specific antibodies in prevention of TE in mice.

## D. HOST GENES INVOLVED IN REGULATING RESISTANCE

Resistance against *T. gondii* is under genetic control in both acute and chronic stages of infection. Of interest, genes involved in resistance differ between these two stages. Susceptibility of inbred strains of mice to acute infection does not correlate with that to chronic infection (Suzuki et al., 1993). A minimum of five genes are involved in determining survival of mice during acute stage (McLeod et al., 1989). One of these genes is linked to the major histocompatibility complex (H-2) (McLeod et al., 1989).

During the chronic stage of infection, development of TE in mice is regulated by the gene(s) within the D region of the major histocompatibility complex (H-2) (Brown et al., 1995; Suzuki et al., 1991; Suzuki et al., 1994). Mice with the *d* haplotype in the D region are resistant to development of TE and those with the *b* or *k* haplotypes are susceptible. Freund et al (Freund et al., 1992) found that polymorphisms in the *TNF-α* gene located in the D region of the H-2 complex correlate with resistance against development of TE and with levels of TNF-α mRNA in brains of infected mice. However, more recent studies using deletion mutant mice (Suzuki et al., 1994) and transgenic mice (Brown et al., 1995) demonstrated that the $L^d$ gene in the D region of the H-2 complex, but not the $TNF-\alpha$ gene, is important for resistance against development of TE. Resistance of mice to development of TE is observed in association with resistance to formation of *T. gondii* cysts in the brain (Brown et al., 1995; Suzuki et al., 1991; Suzuki et al., 1994). McLeod et al. (1989) reported that although the $L^d$ gene has the primary effect on cyst number in the brain, the *Bcg* locus on chromosome 1 may also effect it.

In humans, HLA-DQ3 was found to be significantly more frequent in white North American AIDS patients with TE than in the general white population or randomly selected control AIDS patients who had not developed TE (Suzuki et al., 1996). In contrast, the frequency of HLA-DQ1 was lower in TE patients than in healthy controls (Suzuki et al., 1996). Thus, HLA-DQ3 appears to be a genetic marker of susceptibility to development of TE in AIDS patients, and DQ1 appears to be a resistance marker. HLA-DQ3 also appears to be a genetic marker of susceptibility to cerebral toxoplasmosis in the fetus when infected congenitally. Significantly higher frequency of DQ3 was observed in infected infants with hydrocephalus than infected infants without hydrocephalus or normal controls (Mack et al., 1999). The role of the *HLA-DQ3* and *–DQ1* genes in regulation of the susceptibility/resistance of the brain to *T. gondii* infection is supported by the results from a transgenic mouse study (Mack et al., 1999). Expression of the *HLA-DQ1* transgene conferred greater protection against parasite burden and necrosis in brains in mice than did the *HLA-DQ3* transgene (Mack et al., 1999). Expression of the *HLA-B27* and *–Cw3* transgenes had no effects on the parasite burden (Brown et al., 1994). Since the $L^d$ gene in mice and the *HLA-DQ* genes in humans are a part of the MHC which regulate the immune responses, the regulation of the responses by these genes appears to be important to determine the resistance/susceptibility of the hosts to development of TE.

## E. GENETIC FACTORS OF *T. GONDII* IN DETERMINING DEVELOPMENT OF TE AND VIRULENCE

The strain of *T. gondii* has been shown to be an important determining factor for susceptibility to development of TE in murine models (Suzuki et al., 1989; Suzuki and Joh, 1994). Following infection of mice with either the ME49, Beverley or C56 strain of the parasite, the ME49 strain formed significantly greater numbers of *T. gondii* cysts in their brains than did the other strains, with no difference in the numbers of cysts between the Beverley and C56 strains (Suzuki and Joh, 1994). Following treatment with anti-IFN-γ mAb, mice infected with the ME49 strain developed significantly greater numbers of areas of acute focal inflammation in their brains than those infected with the other strains (Suzuki and Joh, 1994). Since the ME49 strain formed the largest numbers of cysts in the brains and induced the most severe encephalitis, the numbers of cysts in the brain appears to be an important factor in determining the susceptibility of the host to development of TE.

In addition, following treatment with anti-IFN-γ mAb, mice infected with the Beverley strain developed foci of acute focal inflammation in their brains whereas animals infected with the C56 strain did not develop such inflammatory changes (Suzuki and Joh., 1994). As mentioned above, mice infected with the Beverley or C56 strain had similar numbers of cysts in their brains before they were treated with anti-IFN-γ mAb. Therefore, a factor(s) which is related to the strain of *T. gondii* but not related to numbers of cysts in the brain is also important in determining the susceptibility of the host to development of TE. Strains of *T. gondii* have been classified into three genotypes, type I, II and III, based on polymorphisms of their genes (Sibley et al., 1992). It is noteworthy that both the ME49 and Beverley strains, which induced severe inflammatory changes in brains of mice following immunosuppressive treatment, belong to the same genotype, type II, whereas the C56 strain, which did not induce such inflammatory changes, belongs to type III (Sibley et al., 1992). The genotypes of the parasite may be an important factor for determining the susceptibility to development of TE. In relation to this, type II strains have been reported to be predominant in *T. gondii* strains isolated from AIDS and non-AIDS immunocompromised patients and those with congenital infections (Ajzanberg et al., 2002; Honore et al., 2000; Howe et al., 1997) whereas isolates from outbreaks of acute toxoplasmosis, which showed a tendency to cause severe ocular disease, were found to be type I (Lehmann et al., 2000; Sibley et al., 1992). Thus, parasite genotypes may influence development of clinical illness in humans.

In our attempt to determine whether certain loci might be responsible for the difference in development of TE between type II and III strains, we analyzed 16 F1 progeny from a genetic cross between a type II and a type III strain (ME49 and CEP strain, respectively) (Grigg et al., 2001). These progeny had previously been genotyped at 71 loci spanning all 11 known linkage groups. Mice were infected with $1 \times 10^3$ tachyzoites of each of the F1 strains and the two parental lines. The inoculum used was below the $LD_{50}$ for the parental strains, and therefore all mice were expected to survive the acute stage and establish a chronic infection as do their parental strains. However, two (S23 and CL11) of the 16 F1 strains were surprisingly found to be virulent during the acute stage of infection and killed most of mice. Since *T. gondii* is haploid and neither parent showed such virulence, it appears that some combination of alleles at two or more loci are pivotal to determine acute virulence of the parasite (Grigg et al., 2001). More recently, Su et al (Su et al., 2002) reported the virulence of recombinant progeny between a type I and a type III strain. Analysis of independent recombinant progeny identified several quantitative trait loci that contributed to acute virulence, including a major locus located on chromosome *VII* and a minor locus on chromosome *IV* (Su et al., 2002).

## ACKNOWLEDGEMENTS

This work was supported in part by Public Health Service grant AI47730. The author thanks Jennifer Claflin for her assistance with preparation of this manuscript.

## REFERENCES

Ajzanberg, D., N. Cogne, L. Paris, M.H. Bessieres, P. Thulliez, D. Filisetti, H. Pelloux, P. Marty, and M.L. Darde. 2002. Genotype of 86 *Toxoplasma gondii* isolates associated with human congenital toxoplasmosis, and correlation with clinical finding. Journal of Infectious Diseases **186**: 684-689.

Aliberti, J., S. Hieny, C. Reis, E. Sousa, C. Serhan, and A. Sher. 2002. Lipoxin-mediated inhibition of IL-12 production by dendritic cells: a mechanism for the regulation of microbial immunity. Nature Immunology **3**: 76-82.

Brown, C.R., C.S. David, S.J. Khare, and R. McLeod. 1994. Effects of human class I transgenes on *Toxoplasma gondii* cyst formation. Journal of Immunology **152**: 4537-4451.

_____, C.A. Hunter, R.G. Estes, E. Beckmann, J. Forman, C. David, J.S. Remington, and J. R. McLeod. 1995. Definitive identification of a gene that confers resistance against *Toxoplasma* cyst burden and encephalitis. Immunology **85**: 419-428.

Buzoni-Gatel, D., H. Debbabi, F.J. Mennechet, V. Martin, C. Lepage, J.D. Schwartzman, L.H. Kasper. 2001. Murine ileitis after intracellular parasite infection is controlled by TGF-β producing intraepithelial lymphocytes. Gastroenterology **120**: 914-924.

Cahng, H.R., and J.C. Pechere. 1989. Macrophage oxidative metabolism and intracellular *Toxoplasma gondii*. Microbial Pathology **7**: 37-44.

Callazo, C.M., G.S. Yap, G.D. Sempowski, K.C. Lushy, L. Tessarollo, G.F. Woude, A. Sher, G.A. Taylor. 2001. Inactivation of LRG-47 and IRG-47 reveals a family of interferon gamma-inducible genes with essential, pathogen-specific roles in resistance to infection. Journal of Experimental Medicine **194**: 181-188.

Chao, C.C., W.R. Anderson, S. Hu, A. Martella, G. Gekker, and P. K. Peterson. 1993. Activated microglia inhibit *Toxoplasma gondii* via a nitric oxide-mediated mechanism. Clinical Immunology and Immunopathology **67**: 178-183.

_____, G. Gekker, S. Hu, and P.H. Peterson. 1994. Human microglia cell defense against *Toxoplasma gondii*. Journal of Immunology **152**: 1246-1252.

Däubener, W., C. Remscheid, S. Nockemann, Korinna Pilz, S. Seghrouchni, C. Mackenzie, and U. Hadding. 1996. Anti-parasitic effector mechanisms in human brain tumor cells: role of interferon-γ and tumor necrosis factor-α. European Journal of Immunology **26**: 487-492.

_____, B. Spors, C. Hucke, R. Adam, M. Stins, K. S. Kim, H. Schroten. 2001. Restriction of *Toxoplasma gondii* growth in human brain microvascular endothelial cells by activation of indoleamine 2,3-dioxygenase. Infection and Immunity **69**: 6527-6531.

Deckert-Schlüter, M., H. Bluethman, N. Kaefer, A. Rang, and D. Schlüter D. 1999. Interferon-γ receptor-mediated but not tumor necrosis factor receptor type-1- or type 2-mediated signaling is crucial for the activation of cerebral blood vessel endothelial cells and microglia in murine *Toxoplasma* encephalitis. American Journal of Pathology **154**: 1549-1561.

Dimier, I.H., and D.T. Bout. 1998. Interferon-gamma-activated primary enterocytes inhibit *Toxoplasma gondii* replication: a role for intracellular iron. Immunology **94**: 488-495.

Frenkel, J.K., and D.W. Taylor DW. 1982. Toxoplasmosis in immunoglobulin M-suppressed mice. Journal of Immunology **38**: 360-367.

Freund, Y.R., G. Sgarlato, C.O. Jacob, Y. Suzuki, and J.S. Remington. 1992. Polymorphisms in the tumor necrosis factor α (TNF-α) gene correlate with murine resistance to development of toxoplasmic encephalitis and with levels of TNF-α mRNA in infected brain tissue. Journal of Experimental Medicine **175**: 683-688.

_____, N.T. Zaveri, and H.S. Javitz. 2001. In vitro investigation of host resistance to *Toxoplasma gondii* infection in microglia of BALB/c and CBA/Ca mice. Infection and Immunity **69**: 765-772.

Fujigaki, S., K. Saito, M. Takemura N. Maekawa, Y. Yamada, H. Wada, M. Seishima. 2002. L-tryptophan-L-kynurenine pathway metabolism accelerated by *Toxoplasma gondii* infection is abolished in gamma interferon-gene-deficient mice: cross-regulation between inducible nitric oxide synthase and indoleamine-2,3-dioxygenase. Infection and Immunity **70**: 779-786.

Gazzinelli, R.T., Y. Xu, S. Hieny, A. Cheever, and A. Sher. 1992. Simultaneous depletion of CD4[+] and CD8[+] T lymphocytes is required to reactivate chronic infection with *Toxoplasma gondii*. Journal of Immunology **149**: 175-180.

_____, I. Eltoum, T.A. Wynn, and A. Sher. 1993. Acute cerebral toxoplasmosis is induced by in vivo neutralization of TNF-α and correlates with the down-regulated expression of inducible nitric oxide synthase and other markers of macrophage activation. Journal of Immunology **151**: 3672-3681.

Grigg, M.E., S. Bonnefoy, A.B. Hehl, Y. Suzuki, and J.C. Boothroyd. 2001. Success and virulence in *Toxoplasma* as the result of sexual recombination between two distinct ancestries. Science **294**: 161-165.

Halonen, S.K., G.A. Taylor, and L.M. Weiss. 2001. Gamma interferon-induced inhibition of *Toxoplasma gondii* in astrocytes is mediated by IGTP. Infection and Immunity **69**: 5573-5576.

Hayashi, S., C.C. Chan, R.T. Gazzinelli, and F.G. Roberge 1996. Contribution of nitric oxide to the host parasite equilibrium in toxoplasmosis. Journal of Immunology **156**: 1476-1481.

Honore, S., A. Couvelard, Y.J. Garin, C. Bedel, D. Henin, M.L. Darde, and F. Derouin. 2000. [Genotypnig of *Toxoplasma gondii* strains from immuncompromised patients]. Pathologie Biologie (Paris) **448**: 541-547.

Howe, D.K., S. Honore, F. Derouin, and L.D. Sibley. 1997. Determination of genotypes of *Toxoplasma gondii* strains isolated from patients with toxoplasmosis. Journal of Clinical Microbiology **35**: 1411-1414.

Kang, H., J.S. Remington, and Y. Suzuki. 2000. Decreased resistance of B cell-deficient mice to infection with *Toxoplasma gondii* despite unimpaired expression of IFN-γ, TNF-α, and inducible nitric oxide synthase. Journal of Immunology **164**: 2629-2634.

_____, and Y. Suzuki. 2001. Requirement of non-T cells that produce gamma interferon for prevention of reactivation of *Toxoplasma gondii* infection in the brain. Infection and Immunity **69**: 2920-2927.

Khan, I.A, T. Matsuura, S. Fonseka, and L.H. Kasper. 1996. Production of nitric oxide (NO) is not essential for protection against acute *Toxoplasma gondii* infection in IRF-1-/- mice. Journal of Immunology **156**: 636-643.

_____, J.D. Schwartzman, T. Matsuura, and L. H. Kasper. 1997. A dichotomous role for nitric oxide during acute *Toxoplasma gondii* infection in mice. Proceedings of the National Academy of Sciences USA **94**: 13955-13960.

Lehmann, T., C.R. Blackston, S.F. Parmley, J.S. Remington, and J.P. Dubey. 2000. Strain typing of *Toxoplasma gondii*: comparison of antigen coding and housekeeping genes. Journal of Parasitology **86**: 960-971.

Liesendfeld, O., J. Kosek, J.S. Remington, and Y. Suzuki. 1996. Association of CD4$^+$ T cell-mediated, interferon-gamma-mediated necrosis of the small intestine with genetic susceptibility of mice to peroral infection with *Toxoplasma gondii*. Journal of Experimental Medicine **184**: 597-607

_____, H. Kang, D. Park, T.A. Nguyen, C.V. Parkhe, H. Wantanab, T. Abo, A. Sher, J.S. Remington, Y. Suzuki. 1999. TNF-$\alpha$, nitric oxide and IFN-$\gamma$ are all critical for development of necrosis in the small intestine and early mortality in genetically susceptible mice infected perorally with *Toxoplasma gondii*. Parasite Immunology **21**: 365-376.

Mack, D.G., J.J. Johnson, F. Roberts, C.W. Robert, R. G. Estes, C. David, F.C. Gumet, R. McLeod. 1999. HLA-class II genes modify outcome of *Toxoplasma gondii* infection. International Journal for Parasitology **29**: 1351-1358.

McLeod, R., P. Eisenhauer, D. Mack, G. Filice, and G. Spitalny. 1989. Immune responses associated with early survival after peroral infection with *Toxoplasma gondii*. Journal of Immunology **142**: 3247-3255.

_____, Skamene, C.R. Brown, P.B. Eisenhauer, and D.G. Mack. 1989. Genetic regulation of early survival and cyst number after peroral *Toxoplasma gondii* infection of AxB/BxA recombinant inbred and B10 congenic mice. Journal of Immunology **143**: 3031-3034.

Murray, H.W., B.Y. Rubin, S.M. Carriero, A. H. Harris, and E.A. Jaffee. 1995. Human mononuclear phagocyte antiprotozoal mechanisms: oxygen-dependent vs. oxygen-independent activity against intracellular *Toxoplasma gondii*. Journal of Immunology **134**: 1982-1988.

Pelloux, H., G. Pernod, B. Polack, E. Coursange, J. Ricard, J.M. Verna, and P. Ambroise-Thomas. 1996. Influence of cytokines on *Toxoplasma gondii* growth in human astrocytoma-derived cells. Parasitology Research **82**: 598-603.

Peterson, P.K,, G. Gekker, S. Hu, and C.C. Chao. 1995. Human astrocytes inhibit intracellular multiplication of *Toxoplasma* by a nitric oxide-mediated mechanism. Journal of Infectious Diseases **171**: 516-518.

Scharton-Kersten, T.M., G. Yap, J. Magram, and A, Sher. 1997 Inducible nitric oxide is essential for host control of persistent but not acute infection with the intracellular pathogen *Toxoplasma gondii*. Journal Experimental Medicine **185**: 1261-1273.

Sibley, L.D., and J.C. Boothroyd. 1992. Virulent strains of *Toxoplasma gondii* compromise a single clonal lineage. Nature **359**: 82-85.

Su, C., D.K. Howe, J.P. Dubey, J.W. Ajioka, and L.D. Sibley. 2002. Identification of quantitative trait loci controlling acute virulence in *Toxoplasma gondii*. Proceedings of the National Academy of Sciences USA. **99**: 10753-10758..

Suzuki, Y., M.A. Orellana, R.D. Schreiber, J.S. Remington. 1988 Interferon-$\gamma$: The major mediator of resistance against *Toxoplasma gondii*. Science **240**: 516-518.

_____, F.K. Conley, and J.S. Remington. 1989. Differences in virulence and development of encephalitis during chronic infection vary with the strain of *Toxoplasma gondii*. Journal of Infectious Diseases **159**: 790-794.

_____, F.K. Conley, and J.S. Remington. 1990. Importance of endogenous IFN-$\gamma$ for prevention of toxoplasmic encephalitis in mice. Journal of Immunology **143**: 2045-2050.

_____, K. Joh, M.A. Orellana, F.K. Conley, and J.S. Remington. 1991. A gene(s) within H-2D region determines the development of toxoplasmic encephalitis in mice. Immunology **74**: 732-739.

_____, M.A. Orellana, S.Y. Wong, F.K. Conley, and J.S. Remington. 1993. Susceptibility to chronic infection with Toxoplasma gondii does not correlate with susceptibility to acute infection in mice. Infection and Immunity **61**: 2284-2288.

_____, and K. Joh 1994. Effect of the strain of *Toxoplasma gondii* on the development of toxoplasmic encephalitis in mice treated with antibody to interferon-gamma. Parasitology Research **80**: 125-130.

_____, _____, O-C. Kwon, Q. Yang, F.K. Conley, and J.S. Remington. 1994. MHC class I gene(s) in the D/L region but not the TNF-α gene determines development of toxoplasmic encephalitis in mice. Journal of Immunology **153**: 4651-4654.

_____, S.Y. Wong, F.C. Grumet, J. Fessel, J.G. Montoya, A.R. Zolopa, A. Portimore, F. Schmacher-Perdreau, M. Schrappe, S. Köppen, B. Ruf, B.W. Brown, and J.S. Remington. 1996. Evidence for genetic regulation of susceptibility to toxoplasmic encephalitis in AIDS patients. Journal of Infectious Diseases **173**: 265-268.

_____, H. Kang, S. Parmley, S. Lim, and D. Park. 2000. Induction of tumor necrosis factor-α and inducible nitric oxide synthase fails to prevent toxoplasmic encephalitis in the absence of interferon-γ in genetically resistant BALB/c mice. Microbes and Infection **2**: 455-462.

_____, A. Sher, G. Yap, D. Park, L.E. Neyer, O. Liesenfeld, H. Kang, and E. Gufwoli. 2000. IL-10 is required for prevention of necrosis in the small intestine and mortality in both genetically resistant BALB/c and susceptible C57BL/6 mice following peroral infection with *Toxoplasma gondii*. Journal of Immunology **164**: 5375-5382.

Taylor, G.A., C.M. Collazo, G.S.Yap, K. Nguyen, T.A. Gregono, L.S. Taylor, B. Eagleson, L. Secrest, EA Southon, S.W. Reid, L. Tessarollo, M. Bray, D.W. McVicar, K.L. Komschlies, H.A. Young, CA Biron, A. Sher, and G. F. Vande Woude. 2000. Pathogen-specific loss of host resistance in mice lacking the IFN-γ-inducible gene IGTP. Proceedings of the National Academy of Sciences USA **97**: 751-755.

Turco, J., and H.H. Winkler. 1986. Gamma-interferon-induced inhibition of the growth of *Rickettsia prowazekii* in fibroblasts cannot be explained by the degradation of tryptophan or other amino acids. Infection and Immunity **53**: 38-46.

Wilson, C.B., and J.S. Remington. 1979. Activity of human blood leukocytes against *Toxoplasma gondii*. Journal of Infectious Diseases **140**: 890-895.

Yap, G.S., T. Scharton-Kersten, H. Charest, and A. Sher. 1998. Decreased resistance of TNF receptor p55- and p75-deficient mice to chronic toxoplasmosis despite normal activation of inducible nitric oxide synthase in vivo. Journal of Immunology **160**: 1340-1345.

Yap, G.S., and A. Sher. 1999. Effector cells of both nonhematopoietic and hematopoietic origin are required for interferon (IFN)-γ- and tumor necrosis factor (TNF)-α-dependent host resistance to the intracellular pathogen, *Toxoplasma gondii*. Journal of Exponential Medicine **189**: 1083-1091.

# IMMUNE RESPONSE TO *TOXOPLASMA GONDII* IN THE CENTRAL NERVOUS SYSTEM

Sandra K. Halonen

*Department of Microbiology, Montana State University, Bozeman, MT 59717, USA.*
*Visiting Assistant Professor, Department of Pathology, Albert Einstein College of Medicine, Bronx, New York 10461, USA*

## ABSTRACT

*Toxoplasma gondii* is a major opportunistic infection of the central nervous system (CNS) in AIDS patients and increasingly is a problem in other immunocompromised patients. In these patients clinical disease results from a reactivation of the latent tissue cyst stage in the brain. Tissue cysts contain bradyzoites, a slowly replicating form of the parasite. Tissue cysts are located primarily in muscle and brain and persist throughout the lifetime of the host. Cyst rupture and bradyzoite differentiation in to the rapidly replicating tachyzoite stage are thought to occur intermittently in the brain in the immunocompetant host, but replication of the parasite is limited due to the hosts' immune response. In the immunocompromised host however, when tissue cysts rupture, parasites replicate freely in the brain resulting in necrotic foci leading to encephalitis and other neurologic complications. In this review, an overview of the immune response, the role and function of the cells involved in the cell mediated immune response and the humoral response in the brain, cytokine regulation of the immune response and the impaired action in AIDS and other immunocompromised patients will be discussed.

**Key words:** *Toxoplasma gondii*, tissue cyst, bradyzoite, interferon-γ (IFNγ)

## INTRODUCTION

*Toxoplasma gondii* is an obligate intracellular protozoan parasite of worldwide distribution affecting almost all warm-blooded animal species. Humans are typically infected from ingesting the tissue cyst stage present in meat or via accidental ingestion of oocysts present in cat feces. There are two phases of infection, the acute phase, which is usually asymptomatic and self-limiting and the latent phase, which lasts for the lifetime of the host. The acute infection is characterized by the tachyzoite stage, a rapidly replicating form of the parasite that infects all nucleated cells and spreads the infection throughout the body. The latent infection is characterized by the bradyzoite, a slowly replicating form of the parasite that forms tissue cysts located primarily in muscle and brain that persist throughout the lifetime of the host.

The latent form of infection is usually asymptomatic due to an effective host immune response. *T. gondii* is a major opportunistic infection of the central nervous system (CNS) in AIDS patients and increasingly is a problem in other immunocompromised (IC) patients such as transplant recipients and patients receiving chemotherapy. In patients with AIDS and other immunocompromised individuals, infection results from a reactivation of the tissue cyst stage in the brain (Luft and Remington, 1992). Tissue cysts persist within astrocytes and neurons of the brain and elicit little to no inflammatory reaction (Ferguson et al., 1987). Tissue cyst rupture and bradyzoite differentiation to the tachyzoite stage are thought to occur intermittently in the brain in the immunocompetent host, but replication of the tachyzoites is limited due to the hosts' immune response (Ferguson et al., 1989; Frenkel, 1988). In the immunocompromised host however, when tissue cysts rupture, tachyzoites replicate freely in the brain. Unrestrained replication of tachyzoites causes lysis of host cells and results in necrotic foci in the brain leading to encephalitis and other neurologic complications.

In AIDS patients, toxoplasmic encephalitis (TE) results from a progressive impairment of immune function (Luft and Remington, 1992). In human immunodeficiency virus (HIV) infected patients reactivated toxoplasmosis occurs when $CD4^+$ lymphocytes $< 100/mm^3$ occur. Protection against *T. gondii* in the brain is primarily via a cell-mediated immunity (CMI), although evidence indicates humoral immunity may also be involved. Reactivation of the infection occurs only in approximately 30% of those individuals who are seropositive for *T. gondii*. It has been suggested that the neurological impairment of HIV leads to cytokine dysregulation in the CNS and that this in turn alters the balance of the immune response within the CNS and reactivation occurs as a consequence. In this review, an overview of the immune response, the role and function of the cells involved in the CMI response and the humoral response in the brain, cytokine regulation of the immune response and the impaired action in AIDS and other immunocompromised patients will be discussed.

## INITIATION OF THE IMMUNE RESPONSE TO *TOXOPLASMA GONDII* AND INNATE MECHANISMS

Infection with *T. gondii* usually begins via ingestion of the oocyst or the tissue cysts that release sporozoites or bradyzoites respectively, in the small intestine. Parasites are disseminated from the gut to other organs via the bloodstream. Studies in mice and rats after oral infection with oocysts, indicate the parasite first infects mesenteric lymph nodes, then the spleen, and subsequently the brain, heart and other organs via the bloodstream (Zenner et al., 1998).

In the innate phase of the immune response, neutrophils, monocytes and dendritic cells are elicited to the site of infection and all can be infected by *T. gondii* but dendritic cells and monocytes are more permissive to parasite growth than neutrophils (Channon et al., 2000). Neutrophils are the first cells to be recruited to site of infection and it has been suggested that they play a dual critical role both by retarding intracellular parasite growth and lysing extracellular parasites (Channon et al., 2000). The lysed and damaged parasites may provide a pool of extracellular antigen that is presented to the dendritic cells or monocytes, which arrive at the site of infection a few hours after neutrophils, and present antigen in the context of MHC I or II. The importance of neutrophils in initiating the specific immune response to *T. gondii* was recently demonstrated by a study which found that neutrophil depletion at the time of infection lead to an increase in parasite levels and an impaired ability to produce IFNγ, TNFα and IL12 (Bliss et al., 2001). Likewise, mice in which neutrophils had impaired migration had increased parasite loads and decreased levels of IFNγ (Del Rio et al., 2001).

Both macrophages and dendritic cells can function as antigen presenting cells (APC) but evidence now indicates dendritic cells serve as the APC early in infection (Reis e Sousa et al., 1997). Dendritic cells can produce IL-12 in the absence of IFNγ stimulation and studies show dendritic cells produce high levels of IL-12 upon interaction with *T. gondii*. Dendritic cells probably then transport antigen from infected tissue sites to the T cell areas of draining lymph nodes where naïve T cells are located (Reis e Sousa et al., 1997; Johnson and Sayles, 1997; Austyn, 1996). Dendritic cell production of IL-12 promotes differentiation of CD4[+] T cells to Th1 effector cells producing IFNγ and thus ensuring the appropriate adaptive immune response. Human dendritic cells upon stimulation with viable parasites have also been found to produce IL-12 and to stimulate naïve T cell stimulation (Subauste and Wessendarp, 2000). IL-12 also activates natural killer cells (NK), which are abundant in the spleen, to produce the cytokine, interferon-gamma (IFNγ) (Denkers and Gazzinelli, 1998). NK cells are activated in the spleen and migrate to the sites of infection via the blood stream. NK cell production of IFNγ primes macrophages to produce IL12, which further amplifies IFNγ production by NK cells locally (Alexander et al., 1997). IFNγ and IL-12 act synergistically to promote T cell differentiation into a Th1 type response (Denkers and Gazzinelli, 1998).

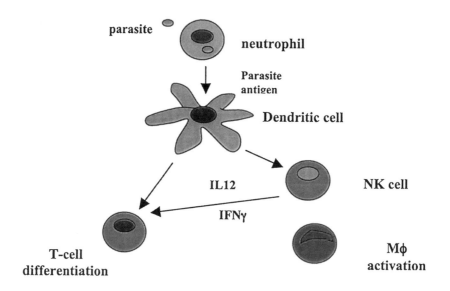

***Figure 1*** – *Role of innate immune mechanisms in the initiation of cell mediated immunity in T. gondii. Neutrophils and dendritic cells are the first cells recruited to the site of infection and both can be infected with T. gondii. Neutrophils are less permissive to growth and their primary role may be to stimulate dendritic cells, which secrete IL12 and traffic to lymph nodes where they will stimulate NK cells to secrete IFNγ. IFNγ stimulates macrophage activation and induces IL12 production that in the presence of IFNγ stimulates T cell differentiation into Th1 cells.*

During the early stage of infection with *T. gondii*, components of the innate immune system are the key cells for initiating the acquired immune response. Collectively, neutrophils, NK cells, dendritic cells and macrophages respond by the initial processing of antigen and releasing the cytokines, IL-12 and IFNγ. These innate mechanisms appear to serve two main functions. First, activation of the innate immune cells serves to limit tachyzoite replication prior to recruitment of T cell mediated immunity. Secondly, the proinflammatory cytokines produced by dendritic cell activation, induce an appropriate T-cell immunity by driving the differentiation of the Th precursor (Thp) cells into Th1 response. The role of these cells in initiating the acquired immune response is depicted in Figure 1.

## CELL MEDIATED IMMUNE RESPONSE AND THE ROLE OF IFNγ

It is well established that immunity to *T. gondii* is maintained by a T cell response. Athymic nude mice that lack functional T cells are extremely susceptible to both virulent and avirulent parasite strains (Gazzinelli et al., 1993; Lindberg and Frenkel, 1977). Additionally, numerous studies have shown adoptive transfer of immune T cells, and specifically CD8[+] T cells, to naïve mice protects animals against challenge with *T. gondii* (Denkers and Gazzinelli, 1998). A strong Th1 immune response to *T. gondii* has been demonstrated in both resistant and susceptible mouse strains and in humans (Denkers and Gazzinelli, 1998).

IFNγ is the main cytokine controlling both the acute and chronic stages of infection. Mice given IFNγ neutralizing antibodies or mice deficient in IFNγ receptor, are unable to control acute *Toxoplasma* infection (Suzuki et al., 1988; Suzuki and Remington, 1990; Scharton-Kerston et al., 1996). Likewise, treatment of mice chronically infected with *T. gondii* with IFNγ neutralizing antibody, results in reactivation of toxoplasmosis (Suzuki et al., 1989). IFNγ has also been shown to be the main cytokine controlling infection in the brain. Elevated levels of IFNγ, and the cytokines, IL-1 and TNFα, are present in the brains of *T. gondii* infected mice (Hunter et al., 1992; Hunter et al., 1994; Deckert-Schluter et al., 1995). Administration of anti-IFNγ results in the increased severity of TE in brains of infected mice (Suzuki et al., 1989; Gazzinelli et al., 1993a). Numerous studies indicate the major sources of IFNγ are CD4[+] and CD8[+] T cells that are recruited into the brain during an infection (Gazzinelli et al., 1993a; Ely et al., 1999).

Thus IFNγ and the immune effector cells they stimulate comprise the immune response to *T. gondii* in the brain. The effector cell populations in the brain include CD4[+] and CD8[+] T cells, microglia, astrocytes, dendritic cells, B cells and an IFNγ producing non-T cell population, which has just recently been described. A brief summary of each of these effector cell populations and their role in the intracerebral immune response to *T. gondii* is included below.

## EFFECTOR CELLS IN THE BRAIN

*T cells* – The role of T cells in immunity to *T. gondii* infection has been well studied (Denkers and Gazzinelli, 1998). Athymic nude mice, which lack T cells, succumb to acute infection (Suzuki et al., 1991). Mortality in athymic mice was associated with proliferation of large numbers of tachyzoites in various organs, including the brain. Both CD4[+] and CD8[+] T cells are recruited to the CNS during infection of mice with TE (Hunter et al., 1994). CD8[+] T cells are thought to be the primary effector cell against *T. gondii* in the brain with CD4[+] T cells playing a synergistic role (Denkers and Gazzinelli, 1998; Gazzinelli et al., 1991; Gazzinelli et al., 1992a; Suzuki and Remington, 1990a).

Studies have shown mice depleted of CD8[+] T cells succumb to lethal infection and host resistance can be adoptively transferred by primed CD8[+] T cells or by CD8[+] T cell clones specific for the *T. gondii* major surface antigen, SAG-1 (Parker et al., 1991; Khan et al., 1988).    Additionally, in *T. gondii* vaccinated mice, CD8[+] T cells were found to be necessary to induce an immune response and to be required at the time of challenge (Khan et al., 1990a).    CD8[+] T cells have also been shown to regulate the number of *T. gondii* tissue cysts that form in the brain (Brown and McLeod, 1990).

CD8[+] T cells secrete IFNγ in response to parasite antigen and are probably the primary producers of IFNγ during an acute infection (Suzuki and Remington, 1990b). The predominant protective effect of CD8[+] T cells is probably via their production of IFNγ (Gazzinelli et al., 1991; Gazzinelli et al., 1992b; Suzuki and Remington, 1990b). IFNγ stimulates microglia, the dominant immune effector cell in the brain and probably other effector cell populations (see sections below). IFNγ is also necessary for the induction and maintenance of CD8[+] T cells in infection with *T. gondii* (Ely et al., 1999).

CD8[+] T cells exhibit cytotoxic lymphocyte activity (CTL's). Both human and murine CD8[+] T cells can kill *T. gondii* infected target cells in vitro in a MHC-restricted manner (Hakim et al., 1991; Khan et al., 1990b; Subauste et al., 1991; Montoya et al., 1996). It is unclear however as to whether the primary function of CD8[+] T cells in vivo, is via IFNγ secretion or CTL activity.  Cytotoxic CD8[+] T cells induce apoptosis in target cells either by perforin and granzyme B or via the engagement of Fas (Kagi et al., 1994; Lowin et al., 1994).   A study using perforin deficient mice addressed the role of perforin-dependent cytotoxicity in protective immunity to *T. gondii* (Denkers et al., 1997). The perforin deficient mice were able to survive acute infection with *T. gondii* indicating that perforin-mediated CTL activity is not required for resistance to acute infection.  However, perforin-deficient mice showed a higher mortality during chronic infection and contained three to five times the number of tissue cysts than did brains of wild-type mice. This suggests CTL activity may play a role in the latent stages of infection. It has been suggested that CTL activity may play a role in prevention of tissue cyst reactivation or alternatively in limiting the number of parasites initially encysting within the brain. There is also some controversy as to whether cytotoxic $CD8^+$ T cells, if they do occur, can kill intracellular *T. gondii*. $CD8^+$ T cells have been shown to kill intracellular pathogens such as mycobacteria (Stenger et al., 1997; Stenger et al., 1999), however, one study found that $CD8^+$ CTL-mediated lysis of infected host cells did not result in killing of intracellular *T. gondii* (Yamashi et al., 1998). This study raises the possibility that CTL mediated killing of *T. gondii* infected cells may release live parasites that may have a chance to spread and therefore may not be beneficial to the host. A more recent study however, found $CD8^+$ T cells could inhibit proliferation of the parasites at least in part through a granule-dependent cytotoxic pathway (Nakano et al., 2001). The result of cytotoxic $CD8^+$ T cells on intracellular parasites and their possible role in the tissue cyst stage in the brain needs to be clarified.

$CD4^+$ T cells are thought to play a synergistic role in the immune response to *T. gondii* in the brain. $CD4^+$ T cells are a major source of IFNγ production during *T. gondii* infection and they have been suggested to be essential for priming of $CD8^+$ T cell effector immunity (Gazzinelli et al., 1991; Gazzinelli et al., 1992a; Denkers et al., 1993). However, in a recent study using mice lacking $CD4^+$ T cells ($CD4^{-/-}$), $CD4^+$ T were not found to be necessary for induction of $CD8^+$ T cell immunity, but were necessary for maintenance of $CD8^+$ T cell effector immunity against *T. gondii* (Casciotti et al., 2002). The lack of a need for $CD4^+$ T for the early $CD8^+$ T cell response against *T. gondii* may be explained by the presence of IFNγ - producing NK cells. The authors suggested that the function of IFNγ-producing $CD4^+$ T cells in the chronic state might be to maintain the antigen-presenting cells in an activated state so that they can constantly re-stimulate memory $CD8^+$ T cells. A recent study found CD4-deficient mice developed more brain cysts and had a shorter survival time than did wild-type controls (Johnson and Sayles, 2002). *Toxoplasma gondii* specific serum antibody levels were lower than wild-type and survival of $CD4^+$ deficient mice could be prolonged by immune serum. Evidence from this study indicates $CD4^+$ T cells, in addition to their role in maintenance of $CD8^+$ memory, may also contribute to protection against chronic infection via their role in humoral immunity.

*NK cells* –NK cells are also a source of IFNγ and are important in controlling *T. gondii* during the early stage of infection (Gazzinelli et al., 1993b). IL-12 is necessary to initiate IFNγ production by NK cells and IL-12 in conjunction with TNFα, IL-1β and IL-15, potentiates NK cell IFNγ production (Hunter et al., 1995a; Hunter et al., 1995b). The source of this IL-12 appears to be dendritic cells. NK cells play a critical role in induction of the protective T cell response. IL-12 is important in the induction of a Th-1 type T cell development. IFNγ is necessary for the optimum production of IL-12 and NK cells likely provide the initial IFNγ required for optimal IL-12 production during the early stage of the infection (Suzuki, 2002). NK cells do not appear to be necessary for the chronic stage of the infection (Kang and Suzuki, 2001).

***IFN-γ- producing non-T cells.*** A recent study found a requirement of non-T cells that produce IFNγ for prevention of reactivation of *T. gondii* infection in the brain (Kang and Suzuki, 2001). In this study adoptive transfer of immune T cells prevented development of TE in athymic or SCID, which lack T cells, but not in IFNγ - knockout (IFN-γ$^{-/-}$) mice. Large amounts of IFNγ in the brains were detected in the brains of infected nude and SCID mice but not in the IFN-γ$^{-/-}$ mice indicating that a non-T cell in the brain that produces IFNγ is required to prevent reactivation of TE. NK cells were not found to be the source of IFNγ.  The source of non-T cell IFNγ is not known. Neurons, astrocytes, and microglia have all been shown to produce IFNγ in vitro and any of these cells could be the source of IFNγ (De Simone et al., 1998; Neumann et al., 1997). Dendritic cells are another possible source of IFNγ as they have also been reported to produce IFNγ in vitro (Ohteki et al., 1999) and a dendritic-type cell has also been reported in the brains of *T. gondii*–infected mice.

***Dendritic Cells*** - Cellular immunity controls *T. gondii* infection in the central nervous system, but it is unclear which APC directs the intracerebral T cell response.  Presumptive local APC are astrocytes and microglia both of which require IFNγ (Sedgwick et al., 1991). Recent studies in mice have demonstrated that the brain also contains dendritic cells (DC) and that the brain will generate DC upon challenge with *T. gondii* (Fischer and Bielinsky, 1999; Fischer et al., 2000). Cells bearing the DC markers, CD11c and 33D1, are expressed during the chronic phase of the infection. These brain DC are mature as indicated by the expression of cell surface markers, MHC class II, CD40, CD54, CD80 and CD86, and by the ability to trigger antigen-specific T cell responses in vitro. The development of the DC's is apparently parasite dependent in that IL-12 secreting DC were found in response to parasite growth and DC produced high levels of IL-12 in response to parasite lysate. Soluble parasite products also induce DC maturation and GM-CSF, a key stimulus of DC differentiation, is produced by *T. gondii* infected astrocytes (Fischer and Bielinsky, 1999). Thus the parasite, may both induce DC differentiation, via GM-CSF secretion from infected astrocytes, as well as be involved in the maturation of and continual activation of DC. Brain DC represents a novel type of APC, because both astrocytes and microglia, the other known APC's require IFNγ, whereas brain DC do not. This feature of brain DC may resolve the question of how intracerebral T cell activation occurs before IFNγ is produced. These results suggest that brain DC are an important APC in the initiation of the immune response in the brain and that astrocytes and dendritic cells may be essential to prime the immune response in the brain. Dendritic cells may also be part of the mechanism that contributes to the chronicity of the intracerebral immune response in TE.

**Microglia** – Microglia are the resident macrophage population in the CNS and are probably the major effector cell in the prevention of *T. gondii* tachyzoite proliferation in the brain. Microglia are activated to inhibit growth of *T. gondii* in vitro with IFNγ plus lipopolysaccharide (Chao et al., 1993a; Chao et al., 1993b; Chao et al., 1994). TNFα and IL6 are also involved in the inhibition of human microglia (Chao et al., 1994). Nitric oxide (NO) was found to mediate the inhibitory effect in murine microglia but not human microglia (Chao et al., 1994). IFNγ has also been shown to be the major activator of microglia in vivo and CD8$^+$ T cells have been found to be necessary to regulate microglial cytokine production (Decker-Schluter et al., 1999; Schluter et al., 2001). IFNγ stimulates microglia to produce TNFα. The role of TNFα has been suggested to play a regulatory role in microglia stimulation via up-regulation of MHC antigens and ICAM in vitro and IFNγ mediated microglia, in combination with autocrine TNFα secretion, probably comprises one of the major resistance mechanisms against *T. gondii* in the brain (Deckert-Schluter et al., 1999). In addition mice with TE, activated microglia produce the cytokines, IL-1β, IL-12 and IL-15, and expression of MHC I and II, LFA-1 and ICAM-1 is upregulated (Schluter et al., 2001). IFNγ activated microglia thus have direct anti-toxoplasmacidal effects via secretion of IL1, TNFα, and iNOS induction or regulation of anti-parasite immune response via production of IL-10, I-12 and IL-15 possibly interacting with CD4$^+$ and CD8$^+$ T cells.

**Astrocytes** - Astrocytes can also inhibit growth of *T. gondii* in the brain. In human astrocytes IFNγ plus IL-1β, stimulated astrocytes to inhibit the growth of *T. gondii* via a NO mediated mechanism (Peterson et al., 1995). TNFα and IFNγ has been shown to inhibit growth of *T. gondii* via induction of indolamine 2,3-diozygenase (IDO) in human glioblastoma cell lines, human astrocytoma derived cells and native astrocytes (Daubener et al., 1993).

Murine astocytes have also been shown to inhibit the growth of *T. gondii* in vitro (Halonen et al., 1998). IFNγ alone induced a significant inhibition in growth of *T. gondii* in murine astrocytes. IFNγ in combination with TNFα, IL1 or IL6 acted synergistically to enhance this inhibition. Neither TNFα, IL1, nor IL6 alone, had any effect on growth of *T. gondii* in murine astrocytes. The inhibitory effect of IFNγ activated astrocytes was not mediated via NO or IDO (Halonen and Weiss, 2000) as was the case for human astrocytes or astrocytoma cells. The mechanism of IFNγ induced inhibition in murine astrocytes was also independent of reactive oxygen intermediates and iron deprivation, other known anti-microbial mechanisms. The mechanism of inhibition, however, appears to involve an IFNγ response gene, IGTP (Halonen et al., 2001). In murine astrocytes deficient in IGTP (IGTP$^{-/-}$) the IFNγ induced inhibition was reversed. IGTP has been shown to be necessary for protection of acute toxoplasmosis in mice (Taylor et al., 2000). The mechanism of IGTP mediated inhibition is not understood.

Infected astrocytes also become activated to produce IL-1 and IL-6 (Fischer et al., 1997). These proinflammatory cytokines together with the cytokines produced by activated microglia probably play an important role in inducing the infiltration of immune cells into the brain. An *in vivo* study found prominent astrocyte activation only early in TE phase of infection and low expression of MHC I, TNFα, MHC II and ICAM.

Astrocytes are an important immune effector cell in the brain. IFNγ stimulated astrocytes are likely to play a pivotal role in controlling the parasite in the brain. Astrocytes are the most abundant glia cell in the brain. Unstimulated astrocytes can support prolific replication of the tachyzoite stage as well as serve as host cells for the cysts (Halonen et al., 1998). IFNγ mediated inhibition of parasite growth in astrocytes probably limits tachyzoite replication in the brain and may also affect re-encystment or maintenance of the parasite in cysts in the brain. Infected astrocytes produce IL-1 and IL-6 that together with cytokines produced by microglia, induce the infiltration of immune cells in the brain. Infected astrocytes produce factors that enable dendritic cells to differentiate into APC and thus help initiate the intracerebral immune response. Infected astrocytes have recently been suggested to play a role in the down regulation of the immune response in the brain via secretion of prostaglandins and induction of microglia IL-10 secretion minimizing neuropathology.

**Endothelial Cells** – *T. gondii* is also able to infect human brain microvascular endothelial cells and IFNγ has been shown to induce inhibition of *T. gondii* in a dose-dependent (Daubener et al., 2001). This antiparasitic effect was enhanced by TNFα but TNFα alone had no effect on parasite growth. The effector mechanism in these cells was due to the induction of IDO. Since one of the first steps in the development of cerebral toxoplasmosis is penetration of blood-brain barrier, IFNγ induced inhibition in these cells may also prove important in limiting the replication of *T. gondii* in the brain.

**B cells** - Early studies examined the effects of B cell depletion on toxoplasmosis in mice by treatment with anti-μ antibody (Frenkel and Taylor, 1982) and suggested that antibody was not important in resistance against the acute infection but contributed to the control of chronic infection. Results of these experiments were inconclusive due to the large amounts of anti-μ antibody raising the possibility that the observed effects could be due to non-specific effects and the fact that anti-μ antibody does not completely deplete B cell populations in vivo. To more directly address the role of B cells in resistance to *T. gondii*, B cell deficient (μMT) mice generated by disruption of the μ-chain gene have been studied (Kang et al., 2000). B cell deficient mice have no detectable B cells or circulating antibody but retain normal development of T cells. In these studies, all μMT mice survived the acute phase of infection but died between 3 and 4 weeks after infection. Significantly greater numbers of tissue cysts and areas of inflammation associated with tachyzoites, and areas of necrosis were observed in μMT mice than in controls. IFNγ, IL-10, and iNOS did not differ between the μMT and control mice. A more recent study showed anti-*Toxoplasma* antisera could prolong survival of chronic infection in CD4-deficient mice (Johnson and Sayles, 2002). Both of these recent studies (Johnson and Sayles, 2002; Kang et al., 2000) support the conclusion of Frenkel and Taylor (1982) that B cells play a role in controlling the chronic toxoplasmosis. The precise role of antibody in the control of the chronic infection is not clear at this time, but it has been suggested antibodies may limit the ability of parasites to infect host cells or alternatively contribute to antibody coated parasites which may be more readily destroyed by host effector cells. Thus, B cells may also be involved in prevention of reactivated toxoplasmosis via antibody production.

## IMMUNE AND NON-IMMUNE EFFECTOR CELLS IN THE INTRACEREBRAL IMMUNE RESPONSE

Protection against both the acute and chronic stages of *T. gondii* requires cells from both the hemopoeitic and nonhemopoietic lineages (Yap et al., 1999). Hemopoeitic cells include cells of the immune system such as macrophages/microglia, T cells and B cells while non-hemopoeitic cells include fibroblasts, endothelial cells, epithelial cells and in the brain, astrocytes and neurons. Both cell populations are stimulated by IFNγ to exert anti-*Toxoplasmic* activity. The mechanism of IFNγ activity in the hemopoeitic cells is NO mediated while in the non-hemopoeitic cells it is a non-NO mediated mechanism. The role of non-hempoeitic cells in the immune response to *T. gondii* may be especially important since *T. gondii* infects cells of epithelial, endothelial, mesodermal and neuronal origin. The intracerebral immune response against *T gondii* likewise appears to be an interaction of immune effector cells such as T cells, B cells, and microglia and non-immune effector cells such as astrocytes, endothelial cells and possible other cell types. IFNγ stimulates microglia, in the brain to exert anti-*Toxoplasma* activity via NO mediated mechanisms while IFNγ stimulates the non-hemopoietic cells to control *T. gondii* via non-NO mediated mechanisms. The primary function of microglia may primarily be anti-microbial killing, while endothelial cells may limit entry of the parasite into the brain and astrocytes limit replication within the brain. CD8$^+$ T cells are the primary source of IFNγ and are probably necessary to continually activate microglia, astrocytes and endothelial cells. Further studies into the precise role of each of these cell populations and their interactions may yield insights into control of intracerebral immune response against *T. gondii*. A summary of these cell populations and their respective roles in the intracerebral immune response is presented in Figure 2.

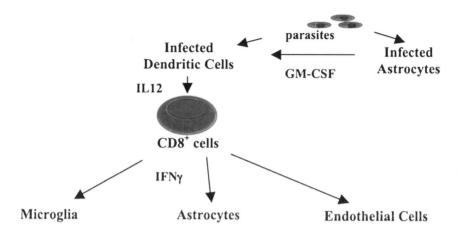

**Figure 2 – Intracerebral Immune Response.** *Parasites will infect both astrocytes and dendritic cells in the brain. The parasite in conjunction with GM-CSF produced by infected astrocytes stimulates dendritic cells to produce IL-12 and differentiate into APC's that activate CD8⁺ cells. Activated CD8⁺ produce IFNγ that stimulates anti-Toxoplasma activity in microglia, astrocytes, endothelial cells and possibly other effector cell populations in the brain. Maintenance of CD8⁺ cell stimulation may require IFNγ production by CD4⁺ cells.*

## Cytokine Regulation of the Immune Response in the Brain

Cytokines have been shown to play an important role in protection against *T. gondii* with IFNγ being the key cytokine mediating protection against both the acute and chronic phases of the infection. IL-12 produced by dendritic cells early in infection stimulates IFNγ production by NK cells. IFNγ and IL-12 drives the T cell precursor to differentiate into a Th1 phenotype. The Th1 cells produce the cytokines, IFNγ, TNFα, and IL-2. These Th1 cytokines then orchestrate the effector cells of the immune system. IFNγ, with TNFα and TGFβ, activate macrophages and microglia in the brain. IFNγ activates anti-*Toxoplasma* activity in endothelial cells and astrocytes in the brain and a variety of other non-hemopoietic effector cells, such as fibroblasts and epithelial cells, throughout the body. During the acute phase of the infection the cytokines IL12 and IFNγ and the proinflammatory cytokines TNFα, IL6 and IL1 are produced. The anti-inflammatory cytokine IL-10 is also produced during the acute phase of infection. During the chronic stage of the infection the Th1 cytokines IL1, IL6, TNFα and IFNγ decrease while IL10 increases in the brain (Gazzinelli et al., 1993a). IL4 is also briefly found during the chronic stage of toxoplasmosis.

IL-12 and IL-10 are the key regulatory cytokines mediating the immune response to *T. gondii*. IL-12 is important in the initiation of Th1 type immune response that protects against *T. gondii*, while IL-10 exerts an immuno-modulatory role down regulating Th1 cytokines and preventing immunopathological effects of the immune response. For example, during the acute phase of infection, mice depleted of IL12 quickly succumb to infection with *T. gondii* (Hunter et al., 1995; Khan et al., 1994; Gazzinelli et al. ,1994) while lack of IL-10 results in development of lethal inflammatory pathological changes (Gazzinelli et al., 1996; Neyer et al., 1997). In the absence of IL-10, mice acutely infected with *T. gondii* show no evidence of enhanced parasite proliferation but have elevated levels of IL-12, IFN-γ and TNFα. The actions of IL-10 include inhibition of IL-12, IL-1β and TNFα synthesis by macrophages, down regulation of IFNγ production by CD4$^+$ T cells and NK cells and inhibition of effector functions of macrophages activated with IFNγ such as NO secretion (Fiorentio et al., 1991a; Fiorentio et al., 1991b; Gazzinelli et al., 1992). The action of IL-10 appears to be to down regulate monokine and IFNγ responses to acute intracellular infection, thereby preventing immunopathological effects.

IL-12 and IL-10 also regulate the chronic phase of infection with *T. gondii*. In the chronic phase of infection, prevention of TE depends upon CD8$^+$ T cells and IFNγ (Gazzinelli et al., 1993a). IL-12 deficient mice, which are allowed to develop a chronic infection by the transient administration of IL-12, survive the acute phase of infection but upon IL-12 withdrawal exhibit increased levels brain cysts and succumb to toxoplasmic encephalitis (Yap et al., 2000). Reactivation is associated with a loss of T-dependent IFNγ production indicating IL-12 is required for the maintenance of IFNγ-dependent resistance in the chronic phase of *T. gondii*. Likewise, IL-10 has been shown to down regulate the intracerebral immune response, decreasing the levels IFNγ and IL-12 in the brain. Neutralization of IL-10 increases the number of immune cells and proinflammatory cytokines but also decreases the intracerebral parasitic load and (Deckert-Schluter et al., 1997). IL-10 thus exerts an immuno-modulatory role, down regulating cytokines in brain that are necessary to prevent immunopathological effects of the immune response. IL-10 however may concomitantly facilitate the persistence of *T. gondii* in the brain and it has been suggested that induction of IL-10 may be a strategy by which parasites evades IFNγ dependent cell mediated immune destruction (Deckert-Schluter et al., 1997).

Immunopathological effects include neurotoxicity. The pro-inflammatory byproducts of microglia activation while necessary to inhibit parasite replication in the brain are themselves detrimental to neurons. NO in particular is known to have a potent neurotoxic activity. Yet surprisingly there is little neurodegeneration associated with toxoplasmic encephalitis. A recent study suggests *T. gondii* infected astrocytes may participate in down regulation of the intracerebral immune response and neuroprotective effects in the brain. *T. gondii* infected astrocytes were found to secrete prostaglandin $E_2$ ($PGE_2$) and induced IL-10 secretion by microglia which collectively caused an inhibition of NO secretion by microglia (Rozenfeld, C., personal communication). These results suggest the presence of a *T. gondii* triggered mechanism that contributes both to host and parasite survival and results in the asymptomatic persistence of *T. gondii* in the brain.

IL-4 plays a major role in development of the CMI response through its effect in determining differentiation of Thp cells into the Th2 phenotype. IL-4 also potentiates the effects of IL-10. An early study found IL4 has a dual role in that IL-4 deficient mice exhibited a reduction in mortality during acute *T. gondii* infection as compared to wild-type animals but exhibited an increased susceptibility during the chronic stage (Roberts et al., 1996). During the acute phase the effect of IL-4 may be to due to the down regulation of IFNγ production by $CD4^+$ T cells thus antagonizing the Thp-cell differentiation towards a Th1 phenotype and the induction of an effective Th1 immune response. The detrimental effects of IL-4 in the chronic infection may be due to IL-4 inhibitory effects on pro-inflammatory cytokines and which may allow cysts to persist in the brain. However, another study using IL-4 deficient mice, found IL-4 was protective against development of TE (Suzuki et al., 1996). In this study IL-4 was found to prevent formation of *T. gondii* tissue cysts and proliferation of tachyzoites in the brain (Suzuki et al., 1996). Results from this study suggest that IL-4 may enhance IFNγ production by differentiated T cells in the late stage of the infection. IL-4 may have some effect on IFNγ production but this effect in the intracerebral immune response need to be clarified. The main effect of IL-4 in the chronic stage of infection is probably via its potentiation of the effects of IL-10 in down regulating the intracerebral immune response (Suzuki et al., 1996).

IL-6 is a multifunctional cytokine that regulates various aspects of the immune response. IL-6 mRNA is expressed in the brains of mice infected with *T. gondii* and is detected in the cerebrospinal fluid of infected mice. IL-6 deficient mice have significantly greater numbers of *T. gondii* tissue cysts and areas of inflammation associated with tachyzoites in their brains than did wild-type controls (Zhang and Denkers, 1999). IL-6 may be protective against development of TE by preventing tissue cyst development and tachyzoite proliferation by inducing IFNγ and lymphocyte proliferation. In relation to the protective role of IL-6 against TE, IL-6 has been shown to inhibit tachyzoite replication in human fetal microglia (Chao et al., 1994).

IL-5 has been suggested to play a role in protection against the chronic infection (Nickdel et al., 2001). IL5 deficient mice display increased tissue cyst and tachyzoite burdens and accelerated mortality during chronic infection. IL-5 is a Type 2 cytokine that is induced in the spleen, mesenteric lymph nodes and brain during murine infection. IL-5 enhances B cell proliferation and differentiation. Spleen cells from IL-5 deficient mice produced less IL-12 when stimulated with parasite antigen and therefore it has been suggested that IL-5 may a role in the production of IL-12. Recently IL5 has been found to have a detrimental role during the acute infection associated with an increase in small intestine pathology, eosinophilia and a decrease in IL12 and IFNγ levels (Khan and Casciotti, 1999). Thus, IL5, like IL4 may have a dual role depending upon the phase of infection.

CD8$^+$ T cells have been shown to be the main effector cell against *T. gondii*. The cytokine, IL-15, has been found to be necessary to stimulate this effector cell population. IL-15 has been found to augment the CD8$^+$ T cell response in mice against *T. gondii* as well as prolong the duration of CD8$^+$ T cell immunity against *T. gondii* (Khan et al., 2002; Khan et al., 1996). Inhibition of IL-15 activity blocks the development of memory CD8$^+$ T in both the acute and chronic infections indicating that the role of IL-15 is to maintain specific memory CD8$^+$ T cells (Kuniyoshi et al. 1999). IL-15 is closely related to the cytokine IL-2. IL-15, however, is produced by multiple cell types, including both immune and nonimmune cells such as dentritic cells, macrophages and placental cells, as opposed to IL-2, which is produced primarily by CD4$^+$ T cells (Kuniyoshi et al., 1999; Doherty et al., 1996; Sarcicron and Gherardi, 2000). Interestingly, IL-10, which is inhibitory for most monokines, stimulates the production of IL-15 in macrophages (Sarcicron and Gherardi, 2000).

## IMMUNE RESPONSE TO *T. GONDII* IN THE BRAIN IN AIDS AND IMMUNOCOMPROMISED INDIVIDUALS

γ is essential to prevent reactivation of toxoplasmosis in the brain. Humans infected with *T. gondii* normally develop resistance, resulting in an asymptomatic chronic infection. In AIDS patients, the decrease of IFNγ due to a decline in CD4$^+$ lymphocytes could break the balance necessary to control the parasite in the brain. IFNγ probably induces the quiescence of parasites in the brain by reducing replication of tachyzoites in astrocytes, endothelial cells, microglia and other cells in the brain. Studies in HIV infected patients indicate that IL2 and IFNγ responses are greatly impaired before the development of AIDS. Thus a global decrease in CD4$^+$

positive T cells and the balance in the cellular response could result in reactivated toxoplasmosis.

In AIDS patients there is also evidence of an unbalance between the Th1 and Th2 responses. IL-10 and IL-4 down regulates the cellular immune response. An increase in IL-10 and IL-4 was found to be associated with the decrease of TNF$\alpha$ and less significantly with IFN$\gamma$ and an increase in the IL10/IL12 ratio (Gazzinelli et al., 1995; Maggi et al., 1994). Thus, an alteration of the Th1/Th2 balance could further aggravate the decline the Th1 response due to the depletion of CD4$^+$ T cells. This proposed mechanism could have importance in regards to the possibilities of a human vaccine for *T. gondii* (Luder and Gross, 1998). For example, the route of vaccine administration and use of adjuvents have been found to be important in eliciting different types of immune responses such as a Th1 vs. Th2 type immunity (Luder and Gross, 1998).

## REFERENCES

Alexander, J., T.M. Scharton-Kerston, G. Yap, C.W. Roberts, F.Y. Liew and A. Sher. 1997. Mechanisms of innate resistance to *Toxoplasma gondii* infection. Philosophical Transactions of the Royal Society London Biology **352**: 1355-1359.

Austyn, J.M. 1996. New insights into the mobilization and phagocytic activity of dendritic cells. Journal of Experimental Medicine **183**: 1287-1292.

Bliss, S.K., L. C. Gavrilescu, A. Alcaraz and E.Y. Denkers. 2001. Neutrophil depletion during *Toxoplasma gondii* infection leads to impaired immunity and lethal systemic pathology. Infection and Immunity **69**: 4898-4905.

Brown, C.W. and R. McLeod. 1990. Class I MHC genes and CD8 T cells determine cyst number in *Toxoplasma gondii* infection. Journal of Immunology **145**: 3438-3441.

Casciotti, L., K.H. Ely, M.E. Williams, and I.A. Khan. 2002. CD8$^+$ T- cell immunity against *Toxoplasma gondii* can be induced but not maintained in mice lacking conventional CD4$^+$ T cells. Infection and Immunity **70**: 434-443.

Chao, C.C., W.R. Anderson, S. Hu, A. Martella, G. Gekker, and P.K. Peterson. 1993a. Activated microglia inhibit *Toxoplasma gondii* via a nitric oxide-mechanism. Clinical Immunology and Immunolpathology. **67**: 178-83.

_____, S. Hu, G.Gekker, W.J. Novick, Jr. J.S. Remington and P.K. Peterson. 1993b. Effect of cytokines on multiplication of *Toxoplasma gondii* in microglia cells. Journal of Immunology **150**: 3403-3410.

_____, G. Gekker, S. Hu, and P.H. Peterson. 1994. Human microglia cell defense against *Toxoplasma gondii*. Journal of Immunology **152**: 1246-52.

Channon, J.Y., R. M. Sequin and L.H. Kasper. 2000. Differential infectivity and division of *Toxoplasma gondii* in human peripheral blood leukocytes. Infection and Immunity **68**: 4822-4826.

Daubener, W., K. Pilz, S. Zennati, T. Bilzher, H.G. Fischer, and U. Hadding. 1993. Induction of toxoplasmosis in a human glioblastoma by interferon $\gamma$. Journal of Neuroimmunology **43**: 31-38.

_____, B. Spors, C. Hucke, R. Adam, M. Stins, K.S. Kim and H. Schroten. 2001. Restiction ation of indoleamine 2,3-dioxygenase. Infect. Immun. **69**: 6527-6531.

Deckert- Schluter, M., S. Albrecht, H. Hof, O.D. Wiestler, and D. Schluter. 1995. Dynamics of the intracerebral and splenic cytokine mRNA production in *Toxoplasma gondii*-resistant and –susceptible congenic strains of mice. Immunology **85**: 408-418.

_____, C. Buck and D. Weiner, et.al. 1997. Interleukin-10 downregulates the intracerebral immune response in chronic Toxoplasma encephalitis. Journal of Neuroimmunology **76**: 167-176.

_____, M., H. Bluethmann, N. Kaefer, A. Rang, and D. Schluter. 1999. Interferon-γ receptor-mediated but not tumor necrosis factor receptor type 1 or type 2-mediated signaling is crucial for the activation of cerebral blood vessel endothelial cells and microglia in murine Toxoplasma encephalitis. American Journal of Pathology **154**: 1549-1561.

Del Rio, L., S. Bennouna, J. Salinas, and E.Y. Denkers. 2001. CXCR2 deficiency confers impaired neutrophil recruitment and increased susceptibility during *Toxoplasma gondii* infection. Journal of Immunology **167**: 6503-6509.

Denkers, E.Y., G. Yap, T. Scharton-Kerston, et.al. 1997. Perforin-mediated cytolysis plays a limited role in host resistance to *Toxoplasma gondii*. Journal of Immunology **159**: 1903-1908.

_____, A. Sher, and R.T. Gazzinelli. 1993. Emergence of NK1.1 cells as effectors of IFN-gamma dependent immunity to *Toxoplasma gondii* in MHC class)-deficient mice. Journal of Experimental Medicine **178**: 1465-1472.

_____, and R.T. Gazzinelli. 1998. Regulation and function of T-cell mediated immunity during *Toxoplasma gondii* infection. Clinical Microbiology Reviews **11**: 569-588.

De Simone, R., G. Levi and F. Aloisi. 1998. Interferon gamma gene expression in rat central nervous system glia cells. Cytokine **10**: 418-422.

Doherty, T.M., R.A. Seer and A. Sher. 1996. Induction and regulation of IL-15 expression in murine macrophages. Journal of Immunology **156**: 735-741.

Ely, K.H., L.H. Kasper and I. A. Khan. 1999. Augmentation of the CD8 T cell response by IFN-γ in IL-12-deficient mice during *Toxoplasma gondii* infection. Journal of Immunology **62**: 5449-5454.

Ferguson, D.J.P. and W.M. Hutchison. 1987. The host-parasite relationship of *Toxoplasma gondii* in the brains of chronically infected mice. Virchows Archives A **411**: 39-43.

_____, _____, and E. Peterson. 1989. Tissue cyst rupture in mice chronically infected with *Toxoplasma gondii*. An immunocytochemical and ultrastructural study. Parasitology Research **75**: 599-603.

Fiorentino, D.F., A. Zlotnick, P. Vieria, T.R. Mossmann, M. Howard, K. Moore, and A. O'Garra. 1991a. IL-10 acts on the antigen-presenting cell to inhibit cytokine production by Th1 cells. Journal of Immunology **146**: 3444-3451.

_____, _____, T.R. Zlotnik, T.R. Mosmann, M. Howard, and A. O'Garra. 1991b. IL-10 inhibits cytokine production by activated macrophages. Journal of Immunology **147**: 3815-3822.

Fischer, H. G., B. Nitzgen, G. Reichmann and U. Hadding. 1997. Cytokine responses induced by *Toxoplasma gondii* in astrocytes and microglia cells. European Journal of Immunology **27**: 1539-1548.

_____, and A.K. Bielinsky. 1999. Antigen presentation function of brain derived dendritic cells depends on astrocyte help. International . Immunology. **11**: 1265-1273.

_____, U. Bonifas and G. Reichmann. 2000. Phenotype and functions of brain dendritic cells emerging during chronic infection of mice with *Toxoplasma gondii*. Journal of Immunology **164**: 4826- 4834.

Frenkel, J.K. 1988. Pathophysiology of Toxoplasmosis. Parasitology Today **4**: 273-278.

_____, and D.W. Taylor. 1982. Toxoplasmosis in immunoglobulin M-suppressed mice. Infection and Immunity **38**: 360-367.

Gazzinelli, R., F.T. Hakim, S. Hieny, G.M. Schearer, and A. Sher. 1991. Synergistic role of CD4[+] and CD8[+] T lymphocytes in IFN-gamma production and protective immunity induced by an attenuated *Toxoplasma gondii* vaccine. Journal of Immunology **146**: 286-292.

_____, Y. Xu, S. Hieny, A. Cheever, and A. Sher. 1992a. Simulataneous depletion of CD4[+] and CD8[+] T lymphocytes is required to reactivate chronic infection with *Toxoplasma gondii*. Journal of Immunology **149**: 175-180.

_____, I. P. Oswald, S. L. James, and A. Sher. 1992b. IL-10 inhibits parasite killing and nitrogen oxide production by IFN-γ-activated macrophages. Journal of Immunology **148**: 1792-1796.

_____, I. Eltoum, T.A. Wynn and A. Sher. 1993a. Acute cerebral toxoplasmosis is induced by in vivo neutralization of TNFα and correlates with the down-regulated expression of inducible nitric oxide synthase and other markers of macrophage activation. Journal of Immunology **151**: 3672-3681.

_____, S. Hieny, T.A. Wynn, S. Wolf, and A. Sher. 1993. Interleukin 12 is required for the T-lymphocyte-independent induction of interferon γ by an intracellular parasite and induces resistance in T-cell-deficient hosts. Proceedings of the National Academy of Sciences **90**: 6115-6119.

_____, M. Wysocka, S. Hayashi, E.Y. Denkers, S. Hieny, P. Casper, G. Trinchieri and A. Sher. 1994. Parasite-induced IL-12 stimulates early IFNγ synthesis and resistance during acute infection with *Toxoplasma gondii*. Journal of Immunology **153**: 2533-2543.

_____, S. Bala, R. Stevens, et. al. 1995. HIV infection suppresses type I lymphokine and IL-12 responses to *Toxoplasma gondii* but fails to inhibit the synthesis of other parasite-induced monokines. Journal of Immunology **155**: 1565-1574.

_____, M. Wysocka, S. Hieny, T. Scharton-Kerston, A. Cheever, P. Kuhn, W. Muller, G. Trinchier and A. Sher. 1996. In the absence of endogenous IL-10, mice acutely infected with *Toxoplasma gondii* succomb to a lethal immune response dependent upon CD4[+] T cells and accompanied by overproduction of IL-12, IFNγ and TNFα. Journal of Immunology **157**: 798-805

Hakim, F.T., R.T. Gazzinelli, E. Denkers, S. Hieny, G.M. Shearer and A. Sher. 1991. CD8[+] T cells from mice vaccinated against *Toxoplasma gondii* are cytotoxic for parasite-infected or antigen-pulsed host cells. Journal of Immunology **147**: 2310-2316.

Halonen, S. K., and L.M. Weiss. 2000. Investigation into the mechanism of gamma interferon-mediated inhibition of *Toxoplasma gondii* in murine astrocytes. Infection and Immunity **68**: 3426-3430.

_____, F.C. Chiu and L.M. Weiss. 1998. Effect of cytokines on growth of *Toxoplasma gondii* in murine astrocytes. Infection and Immunity **66**: 4989-4993.

_____, G.A. Taylor, and L.M. Weiss. 2001. Gamma interferon-induced inhibition of *Toxoplasma gondii* in astrocytes is mediated by IGTP. Infection and Immunity **69**: 5573-5576.

Hunter, C.A., C.W. Roberts, M. Murray, and J. Alexander. 1992. Detection of cytokine mRNA in the brains of mice with toxoplasmosis. Parasite Immunology **14**: 405-413.

_____, M.J. Litton, J.S. Remington, and J.S. Abrams. 1994. Immunochemical detection of toxoplasmic encephalitis in interleukin-6-deficient mice. Infection and Immunity **65**: 2339-2345.

_____, L. Bermudez, H. Beernink, W. Waegell and J.S. Remington. 1995a. Transforming growth factor-β inhibits interleukin-12-induced production of interferon-γ by natural killer cells: a role for transforming growth factor-β in the regulation of T cell-independent resistance to *Toxoplasma gondii*. European Journal of Immunology **25**: 994-1000.

_____, R. Chizzonite, and J.S. Remington. 1995b. IL-1β is required for IL-12 to induce production of IFN-γ by NK cells. A role for IL-1β in the T cell-independent mechanism of resistance against intracellular pathogens. Journal of Immunology **155**: 4347-4354.

_____, E. Candolfi, C. Sabauste, V. Van Cleave and J.S. Remington. 1995. Studies on the role of interleukin-12 in acute murine toxoplasmosis. Immunology **84**: 16-20.

Johnson, L.L. and P.C. Sayles. 1997. Interleukin-12, dendritic cells, and the initiation of host-protective mechanisms against *Toxoplasma gondii*. Journal of Experimental Medicine **11**: 1799-1802.

_____, and _____. 2002. Deficient humoral responses underlie susceptibility to *Toxoplasma gondii* in CD4-deficient mice. Infection and Immunity **70**: 185-191.

Kagi, D., F. Vignaux, K. Ledermann, V. Burki, V. Depraetere, S. Nagata, H. Hengartner and P. Golstein. 1994. Fas and perforin pathways as major mechanisms of T cell mediated cytotoxicity. Science **265**: 528-530.

Kang, H., J. S. Remington and Y. Suzuki. 2000. Decreased resistance of B cell-deficient mice to infection with *Toxoplasma gondii* despite unimpaired expression of IFN-γ, TNF-α, and inducible nitric oxide synthase. Journal of Immunology **164**: 2629-2634.

_____, and Y. Suzuki. 2001. Requirement of non-T cells that produce gamma interferon for prevention of reactivation of *Toxoplasma gondii* infection in the brain. Infection and Immunity **69**: 2920-2927.

Khan, I.A., K.A. Smith and L.H. Kasper. 1988. Induction of antigen-specific parasiticidal cytotoxic T cell splenocytes by a major membrane protein (p30) of *Toxoplasma gondii*. Journal of Immunology **141**: 3600-3605.

_____, _____, and _____. 1990a. Induction of antigen-specific human cytotoxic T cells by *Toxoplasma gondii*. Journal of Clinical Investigation **85**: 1879-1886.

Khan, I.A. and L.H. Kasper. 1996. IL-15 augments CD8⁺ T cell -mediated immunity against *Toxoplasma gondii* infection in mice. Journal of Immunology **157**: 2103-2108.

_____, T. Matsuma and L.H. Kasper. 1994. Interleukin-12 enhances murine survival against acute toxoplasmosis. Infection and Immunity **62**: 1639-1642.

_____, and L. Casciotti. 1999. IL-15 prolongs the duration of CD8⁺ T cell mediated immunity in mice infected with a vaccine strain of *Toxoplasma gondii*. Journal of Immunology **163**: 4503-09.

_____, M. Moretto, X. Wei, M. Williams, J.D. Schwartzmann, and F.Y. Liew. 2002. Treatment with soluable interleukin-15α exacerbates intracellular parasitic infection by blocking the development of memory CD8⁺ T cell response. Journal of Experimental Medicine **195**: 1463-1470.

Kuniyoshi, J.S., C.J. Kuniyoshi, A.M. Lim, F.Y. Wang, E.R. Bade, R. Lau, E.K. Thomas and J.S. Weber. 1999. Dendritic cell secretion of IL-15 is induced by recombinant huCD40LT and augments the stimulation of antigen-specific cytolytic T cells. Cell Immunology **193**: 48-58.

Lindberg, R. E. and J. K. Frenkel. 1977. Toxoplasmosis in nude mice. Journal of Parasitology **63**: 219-221.

Lowin, B., M. Hahne, C. Mattamnn and J. Tschopp. 1994. Cytolytic T-cell cytotoxicity is mediated through perforin and Fas lytic pathways. Nature **370**: 650-652.

Luder, C.G.K. and U. Gross. 1998. Toxoplasmosis: from clinics to basic science. Parasitology Today **14**: 43-45.

Luft, B.J. and J.S. Remington. 1992. AIDS commentary: toxoplasmic encephalitis. Clinical Infectious Diseases **15**: 211-222.

Maggi, E., M. Mazzetti, A. Ravina, F. Annunziato, M.D. Carli, M.P. Piccini, R. Manetti, M. Carbonari, A.M. Pesced, G. Del Prete and S. Romagrani. 1994. Ability of HIV to promote a Th1 to Th0 shift and to replicate preferentially in Th2 and Th0 cells. Science **265**: 244-248.

Montoya, J.G., K.E. Lowe, C. Clayberger et al. 1996. Human CD4 and CD8 T lymphocytes are both cytotoxic to *Toxoplasma gondii*-infected cells. Infection and Immunity **64**: 176-181.

Nakano, Y., H. Hisaeda, T. Sakai, M. Zhang, Y. Maekawa, T. Zhang, M. N. H. Ishikawa and K. Himeno. 2001. Granule-dependent killing of *Toxoplasma gondii* by CD8 T cells. Immunology **104**: 289-98

Neumann, H., H. Schmidt, E. Wilharm, L. Behrens, and H. Wekerle. 1997. Interferon γ gene expression in gene expression in sensory neurons: evidence for autocrine gene regulation. Journal of Experimmental Medicine **186**: 2023-2031.

Neyer, L.E., G. Grunig, M. Fort, J.S. Remington, D. Rennick, and C.A. Hunter. 1997. Role of interleukin-10 in regulation of T-cell-dependent and T-cell-independent mechanisms of resistanced to *Toxoplasma gondii*. Infection and Immunity **65**: 1675-1682.

Nickdel, M.B., F. Roberts, F. Brombacher, J. Alexander and C.W. Roberts. 2001. Counter-productive role for interleukin-5 during acute *Toxoplasma gondii* infection. Infection and Immunity **69**: 1044-1052.

Ohteki, T., T. Kukao, K. Suzue, C. Maki, M. Ito, M. Nakamura and S. Koyasu. 1999. Interleukin 12-dependent interferon γ production by CD8α[+] lymphoid dendritic cells. Journal of Experimental Medicine **189**: 1981-1986.

Parker, S.J., C.W. Roberts, and J. Alexander. 1991. CD8[+] T cells are the major lymphocyte population involved in the protective immune response to *Toxoplasma gondii* in mice. Clinical and Experimental Immunology **84**: 207-212.

Peterson, P.K., G. Gekker, S. Hu, and C.C. Chao. 1995. Human astrocytes inhibit intracellular multiplication of *Toxoplasma* by a nitric oxide – mediated mechanism. Journal of Infectious Diseases **171**: 516-518.

Reis e Sousa, C.S. Hieny, T. Scharton-Kerston, D. Jankovic, H. Charest, R.N. Germain and A. Sher. 1997. In vivo microbial stimulation induces rapid CD40 ligand-independent production of interleukin 12 by dendritic cells and their redistribution to T cell areas. Journal of Experimental Medicine **186**: 1819-1829.

Roberts, C.W., D.J.P. Ferguson, H. Jebbari, A. Satoskar, H. Bluethmann, and J. Alexander. 1996. Different roles for interleukin-4 during the course of *Toxoplasma gondii* infection. Infection and Immunity **64**: 897-904.

Sarcicron, M.E. and A. Gherardi. 2000. Cytokines involved in Toxoplasmic encephalitis. Scand. Journal of Immunology **52**: 534-543.

Scharton-Kerston, T. M., T.A. Wynn, E.Y. Denkers, S. Bara, L. Shoue, E. Grunvald, E. Hieny, R.T. Gazzinelli and A. Sher. 1996. In the absence of endogenous IFNγ mice develop unimpaired IL12 responses to *Toxoplasma gondii* while failing to control acute infection. Journal of Immunology **157**: 4045-4054.

Schluter, D., T. Meyer, A. Strack, S. Reiter, M. Kretschmar, C.D. Wiestler, H.Hof, and M. Deckert. 2001. Regulation of microglia by CD4+ and CD8+ T cells: selective analysis in CD45-congenic normal and *Toxoplasma gondii*-infected bone marrow chimeras. Brain Pathology **11**: 44-55.

Sedgwick, J.D., R. Mossner, S. Schwender and V. ter Meulen . 1991. Major histocompatibility complex expressing nonhematopoietic astroglial cells prime only CD8+ T lymphocytes: astroglial cells as perpetuators but not initiators of CD4+ T cell responses in the central nervous system. Journal of Experimental Medicine **173**: 1235-1246.

Stenger, S., R.J. Mazzaccaro, K. Uyemura, S. Cho, P.F. Barnes, J.P. Rosat, A. Sette, M.B. Brenner, S.A. Porcelli, B. Bloom and R.L. Modlin. 1997. Differential effects of cytolytic T cell subsets on intracellular infection. Science **276**: 1684-1687.

_____, J. P. Rosat, B.P. Bloom, A,M, Krensky, and R.L. Modlin. 1999. Granulysin: a lethal weapon of cytolytic T cells. Immunology Today **20**: 390-394.

Subauste, C.S., A.H. Koniaris and J.S. Remington. 1991. Murine CD8[+] cytotoxic T lymphocytes lyse *Toxoplasma gondii*-infected cells. Journal of Immunology **147**: 3955-3959.

_____, and M. Wessendarp. 2000. Human dendritic cells discriminate between viable and killed *Toxoplasma gondii* tachyzoites: dendritic cell activation after infection with viable parasites results in CD28 and CD40 ligand signaling that controls IL-12-dependent and –independent T cell production of IFN-γ. Journal of Immunology **165**: 14998-1505.

Suzuki, Y. 2002. Host Resistance in the Brain against *Toxoplasma gondii*. Journal of Infectious Diseases **185**: S58-S65.

_____, and J.S. Remington. 1990a. The effect of anti-IFN-γ antibody on the protective effect of Lyt-2[+] immune T cells against toxoplasmosis in mice. Journal of Immunology **144**: 1954-5196.

_____, and _____. 1990b. Dual regulation of resistance against *Toxoplasma gondii* infection by Lyt-2[+] and Lyt-1[+] T cells in mice. Journal of Immunology **141**: 1954-1956.

_____, F.K. Conley and J.S. Remington. 1989. Importance of endogenous IFN-gamma for prevention of toxoplasmic encephalitis in mice. Journal of Immunology **143**: 2045-2050.

_____, Y., K. Joh, S. and A. Kobayashi. 1991. Tumor necrosis factor-independent protective effect of recombinant IFN-gamma against acute toxoplasmosis in T cell deficient mice. Journal of Immunology **147**: 2728-2733.

_____, M.A. Orellana, R.D. Schreiiber and J.S. Remington. 1988. Interferon-γ: the major mediator of resistance against *Toxoplasma gondii*. Science **240**: 516-518.

_____, Q. Yang, S. Yang, N. Nguyen, S. Lim, O. Liesenfeld, T. Kojima and J.S. Remington. 1996. Il-4 is protective against development of Toxoplasmic Encephalitis. Journal of Immunology **157**: 2564-2569.

_____, S. Rani, O. Liesenfeld, T. Kojima, S. Lim, T.A. Nguyen, S.A. Dalrymple, R. Murray and J.S. Remington. 1997. Impaired resistance to the development of toxoplasmic encephalitis in interleukin-6-deficient mice. Infection and Immunity **65**: 2339-2345.

Taylor, G.A., C.M. Collazo, G.S. Yap, K. Nguyen, T. Gregorio, L.S. Taylor, B. Eagleson, L. Secrest, E. A. Southon, S.W. Reid, L. Tessarollo, M. Bray, D.W. McVicar, K.L. Komschlies, H.A. Young, C.A. Biron, A. Sher, and G.F. Vande Woude. 2000. Pathogen specific loss of host resistance in mice lacking the interferon-γ-inducible gene IGTP. Proceedings of the National Academy of Sciences USA **97**: 751-755.

Yamashita, K., K. Yui, M. Ueda and A. Yano. 1998. Cytotoxic T-lymphocyte-mediated lysis of *Toxoplasma gondii*-infected target cells does not lead to death of intracellular parasites. Infection and Immunity **66**: 4651-4655.

Yap, G.S. and A. Sher. 1999. Effector cells of both nonhemopoietic and hemopoietic origin are required for interferon (IFN)-γ- and tumor necrosis factor (TNF)-α-dependent host resistance to the intracellular pathogen, *Toxoplasma gondii*. Journal of Experimental Medicine **189**: 1083-1091.

_____, M. Pesin and A. Sher. 2000. IL-12 is required for the maintenance of IFN-γ production in T cells mediating chronic resistance to the intracellular pathogen, *Toxoplasma gondii*. Journal of Immunology **165**: 628-631.

Zenner, L., F. Darcy, A. Capron and M.F. Cesbron-Delauw. 1998. *Toxoplasma gondii*: kinetics of the dissemination in the host tissues during the acute phase of infection of mice and rats. Experimental Parasitology **90**: 86-94.

Zhang, Y. and E.Y. Denkers. 1999. Protective role for interleukin-5 during chronic *Toxoplasma gondii* infection. Infection and Immunity **67**: 4383-4392.

# DEVELOPMENTAL STAGE CONVERSION: INSIGHTS AND POSSIBILITIES

Kami Kim[1,2,*] and Louis M. Weiss[1,3]

[1]Departments of Medicine, [2]Microbiology and Immunology, and [3]Pathology Albert Einstein College of Medicine, 1300 Morris Park Avenue, Bronx, New York, 10461. USA

## ABSTRACT

*Toxoplasma gondii* infection is acquired by oral ingestion. After dissemination, tachyzoites differentiate into bradyzoites within cysts that remain latent. The ability of bradyzoites to transform back into tachyzoites is a central factor in the ability of *T. gondii* to reactivate and cause disease in immunosuppressed individuals. Both tachyzoites and bradyzoites develop in tissue culture. Recent advances in the genetic manipulation of *T. gondii* have expanded the molecular tools that can be applied to studies on bradyzoite differentiation. Evidence is accumulating that this differentiation event is stress mediated and may share common pathways with stress-induced differentiation events in other eukaryotic organisms. Bradyzoite differentiation defective mutants have been studied using microarray analysis, allowing identification of genes that may be coordinately regulated during this stage transition. Characterization of unique bradyzoite-specific structures, such as the cyst wall, and stage-specific metabolic pathways are other approaches being used to study developmental stage conversion.

**Key words:** differentiation, tachyzoite, bradyzoite, stress, *T. gondii,* stage-specific antigens

## INTRODUCTION

*Toxoplasma gondii* is a ubiquitous Apicomplexan protozoan parasite of mammals and birds (McLeod et al., 1991; Luft and Remington, 1988; Wong and Remington, 1993). There are three infectious stages: tachyzoites (asexual), bradyzoites (in tissue cysts, asexual) and sporozoites (in oocysts, sexual reproduction). Infection is often acquired by ingestion of undercooked meat, such as rare lamb, harboring tissue cysts (which contain bradyzoites). Infection can also be acquired by ingestion of food contaminated with oocysts (which contain sporozoites) or by exposure to cat feces containing oocysts. Upon ingestion, sporozoites or bradyzoites will invade the intestinal epithelium, differentiate into the rapidly growing tachyzoite form, and disseminate throughout the body. After oral ingestion, in the intestine of definitive hosts (i.e. cats), parasites can continue to differentiate into merozoites, and sexual reproduction can occur with development of oocysts containing sporozoites. In both definitive and intermediate hosts, tachyzoites, after dissemination, differentiate into bradyzoites that remain latent. Bradyzoites contained in cysts are refractory to most chemotherapeutic agents used for treatment of toxoplasmosis.

In immunocompromised hosts, e.g. AIDS patients, the development of *Toxoplasmic* encephalitis is believed to be due to the transition of the resting or bradyzoite stage to the active and rapidly replicating tachyzoite stage. It is thought that in chronic toxoplasmosis, bradyzoites in tissue cysts regularly transform to tachyzoites but that the tachyzoites are controlled by the immune system. Although a significant opportunistic infection in the early AIDS era, the incidence of clinically apparent toxoplasmosis has waned with the development of highly active antiretroviral treatment (HAART). Although less prevalent in AIDS patients, *T. gondii* has been implicated in waterborne outbreaks and deaths in marine wildlife due to water pollution, leading to its listing as a Category B bioterrorism agent ((Bowie et al., 1997; Miller et al., 2002; Aramini et al., 1999)). Chorioretinitis, a late manifestation of congenital infection, is also ascribed to bradyzoite reactivation although it is becoming increasingly clear that chorioretinitis may also be a manifestation of acute infection with highly virulent strains of *T. gondii* ((Bowie et al., 1997; Grigg et al., 2001)). Due to its central importance in disease pathogenesis, the biology of stage differentiation has been an active area of research. While progress has been made, bradyzoites to tachyzoite interconversion is not well understood. Many initial seminal observations have been reviewed elsewhere (Weiss and Kim, 2000).

## THE MORPHOLOGY AND BIOLOGY OF BRADYZOITES AND CYSTS

*T. gondii* is an obligate intracellular parasite that replicates within a parasitophorous vacuole within the host cell. Compared to tachyzoites (tachy = fast), bradyzoites (brady = slow; or cystozoite) replicate slowly. Like tachyzoites, bradyzoites remain intracellular and divide by a unique binary fission termed endodyogeny. The size of tissue cysts is variable, but on average a mature cyst is 50 to 70 μm and contains from 1000-2000 crescent-shaped 7 by 1.5 μm bradyzoites. Tissue cyst size is dependent on cyst age, the host cell parasitized, the strain of *T. gondii* and the cytological method used for measurement. Young and old cysts can be distinguished readily by their ultrastructural features (Dubey et al., 1998; Fortier et al., 1996; Sims et al., 1989; Scholytyseck et al., 1974), and it is unclear whether bradyzoites within mature cysts continue to replicate or are quiescent.

Bradyzoites differ ultrastructurally from tachyzoites in that they have a posteriorly located nucleus, solid rhoptries, numerous micronemes and polysaccharide (amylopectin) granules (Ferguson and Hutchison, 1987; Dubey, 1997). Lipid bodies are not seen in bradyzoites, but are numerous in sporozoites and occasionally seen in tachyzoites. Bradyzoites stain with periodic acid-Schiff (PAS) whereas tachyzoites do not. Bradyzoites are more resistant to acid pepsin (1 to 2 hour survival in pepsin-HCl) than tachyzoites (10 min survival) (Jacobs and Remington, 1960; Popiel et al., 1996). The prepatent period (time to oocyst sheeding) in cats following feeding of bradyzoites is shorter (3 to 7 days) than that following feeding of tachyzoites (over 14 days). This is the most sensitive biologic marker of mature functional tissue cysts (Dubey et al., 1997; Dubey, 1997)

## THE DEVELOPMENT OF CYSTS AND BRADYZOITES *IN VITRO*

The bradyzoite specific mAbs developed in the 1990s facilitated studies of bradyzoite development *in vitro* and the development of techniques for the induction of differentiation. In tissue culture studies, it is evident that bradyzoites spontaneously convert to tachyzoites and that tachyzoites spontaneously convert to bradyzoites (Bohne et al., 1993; 1994; De Champs et al., 1997; Jones et al., 1986; Lane et al., 1996; Lindsay et al., 1991; Lindsay et al., 1993a; Lindsay et al., 1993b; McHugh et al., 1993; Popiel et al., 1994; Popiel et al., 1996; Soete et al., 1993; Soete et al., 1994; Soete and Dubremetz, 1996; Weiss et al., 1994; Parmley et al., 1995). The rate of conversion appears to be strain dependent. Thus, low virulence strains, i.e. high cyst forming strains in mice, such as ME49, have a higher spontaneous rate of cyst formation in culture than do virulent strains such as RH (Soete et al., 1994). The rate of spontaneous differentiation also diminishes with prolonged tissue culture.

Stress conditions are associated with an increased frequency of bradyzoite development. Conditions that have been reported to induce bradyzoite formation are temperature stress (43°C , Soete et al., 1994)), pH stress (pH 6.6-6.8 or 8.0-8.2 Weiss et al., 1995; Soete et al., 1994; Soete and Dubremetz, 1996) or chemical stress (Na arsenite, Soete et al., 1994). IFN-γ increased bradyzoite antigen expression in murine bone marrow macrophage lines and this appears related to nitric oxide (NO) induction (Bohne et al., 1994). Bradyzoite induction is also enhanced by sodium nitroprusside (SNP), an exogenous NO donor (Bohne et al., 1994). Similarly, both oligomycin, an inhibitor of mitochondrial ATP synthetase function, and antimycin A, an inhibitor of the electron transport of the respiratory chain, increased bradyzoite antigen expression (Bohne et al., 1994; Tomavo and Boothroyd, 1995).

## Table I: Common Bradyzoite Markers

| Name of Antigen | Monoclonal Antibody | Size on Immunoblot (kDa) | Location by IFA | Proposed Function | Cloned? |
|---|---|---|---|---|---|
| BAG1/hsp30 aka BAG5 | 7E5 74.1.8 | 28 | cytoplasm | Heat shock protein | Yes |
| BSR4 (p36) | T84A12 | 36 | surface | SAG1 family antigen | Yes by promoter trap |
| SAG4A (p18) | T83B1 | 18 | surface | surface antigen | Yes |
| None | DC11 | not reactive | surface | ? | one of SAG family? |
| p21 | T84G11 | 21 | surface | surface antigen | one of SAG family? |
| p34 | T82C2 | 34 | surface | ? | one of SAG family? |
| MAG1 | None | 65 | matrix | | yes, revised studies indicate also expressed in tachyzoites |
| none | E7B2 | 29 | matrix | | No |
| none | 93.2 [Weiss unpublished] | not reactive | matrix | | No |
| none | 1.23.29 [Weiss unpublished] | 19 | matrix | | No |
| CST1 | 73.18; also recognized by DBA lectin | 116 | cyst wall | | No |
| CST1? | CC2 | 115 (bradyzoite) 40 (tachyzoite) | cyst wall | same as CST1? | No |
| LDH2 | polyclonal sera weakly cross reacts to LDH1 | 35 kDa (33kDa for LDH1) | cytosolic | glycolysis | Yes, tachyzoite isoform is LDH1 |
| ENO1 | polyclonal sera to ENO2 and ENO1 do not cross react | | nuclear and cytosolic | glycolysis | Yes, tachyzoite isoform ENO2 |

The contribution of the host cell to stage conversion remains to be elucidated. Two groups have reported (Weiss et al., 1998; Yahiaoui et al., 1999) that exposure of extracellular tachyzoites to stress conditions will result in an increase in bradyzoite differentiation, consistent with a direct effect of stress on the parasite. When bradyzoite differentiation occurs in cell culture following infection with tachyzoites all of the currently available markers for bradyzoite formation, with the exception of p21 (mAb T84G10), can be detected within 24 hours of infection (Gross et al., 1996; Lane et al., 1996). This includes markers of bradyzoite surface antigens as well as those related to cyst wall formation. There may be differences in antigen expression and replication in early bradyzoites and late bradyzoites, but a detailed chronology of marker expression has not yet been developed. A summary of the bradyzoite markers is in Table I.

## THE IDENTIFICATION OF BRADYZOITE-SPECIFIC GENES

Several bradyzoite specific genes have been identified and cloned. Random sequencing of cDNA libraries (EST projects) from bradyzoites and tachyzoites has led to the identification of Apicomplexa-specific genes. Genes that are induced on bradyzoite differentiation or unique to the bradyzoite libraries have also been identified, but many of these encode proteins of unknown function (Ajioka, 1998; Ajioka et al., 1998; Manger et al., 1998a). Further efforts with sporozoite libraries as well as other tachyzoite and bradyzoite libraries are underway and data are accessible at http://toxodb.org (Li et al., 2003). The *T. gondii* genome is being sequenced with 6x coverage currently available at http://toxodb.org and up to 8x coverage with annotation expected by the end of 2003.

*T. gondii* microarrays enriched in bradyzoite-specific cDNAs (from sequencing a bradyzoite library) have been used to identify groups of genes that are coordinately regulated during bradyzoite differentiation (Manger et al., 1998a; Cleary et al., 2002). These analyses have confirmed induction of known bradyzoite genes and identified other potential bradyzoite-specific genes. In addition, genes not known to be regulated during differentiation were shown to have altered mRNA expression (Cleary et al., 2002). The data from these studies and genetic studies suggest that bradyzoite differentiation is a complex pathway.

In a complementary approach, bradyzoite-specific genes have also been identified using a subtractive cDNA library approach (Yahiaoui et al., 1999). Sixty-five cDNA clones were analyzed from a bradyzoite subtractive cDNA library. Of these, many were identified that were exclusively or preferentially transcribed in bradyzoites. This included homologues of chaperones (mitochondria heat shock protein 60 and T complex protein 1), nitrogen fixation protein, DNA damage repair protein, KE2 protein, phophatidylinosoitol synthase, glucose-6-phosphate isomerase and enolase.

A bradyzoite-specific P type ATPase has been characterized and localizes in punctate pattern to the region of the plasma membrane (Holpert et al., 2001). Its mRNA and protein are preferentially expressed in bradyzoites. A second P-type ATPase also may be preferentially expressed in bradyzoites as judged by RT-PCR of steady-state mRNA although mRNA can also be detected in tachyzoites. These findings underscore the many metabolic differences likely to be present between tachyzoites and bradyzoites.

Despite identification of many bradyzoite specific genes and mutants that are unable to differentiate (see below), a unified model for bradyzoite differentiation cannot yet been proposed. Several themes, however, are emerging: (1) tachyzoites and bradyzoites express related genes encoding structural homologues in a mutually exclusive way; (2) metabolic genes that are stage specific exist suggesting these stages are metabolically distinct; and (3) stress related differentiation pathways and stress proteins are associated with these stage transitions.

## CYST WALL AND MATRIX ANTIGENS

The formation of the cyst wall and parasitophorous vacuole matrix are early events accompanying bradyzoite differentiation. The cyst wall and matrix probably protect bradyzoites from harsh environmental conditions such as dehydration and also provide a physical barrier to host immune defenses. Much of this may be due to carbohydrates that make up the cyst wall. The cyst wall is formed from the parasitophorous vacuole and has a ruffled appearance with precipitation of underlying material. The cyst wall is formed by the parasite and is enclosed in host cell membrane (Ferguson and Hutchison, 1987; Scholytyseck et al., 1974). It is periodic acid-Schiff (PAS) positive, stains with some silver stains, and binds the lectins *Dolichos biflorus* (DBA) and succinylated-wheat germ agglutinin (S-WGA), suggesting that specific polysaccharides are present in the cyst wall (Boothroyd et al., 1997). Lectin binding can be inhibited by competition with the sugar haptens N-acetylgalactosamine (GalNAc) for DBA and N-acetylglucosamine (GlcNAc) for S-WGA (Boothroyd et al., 1997).

CST1 is a 116kDa glycoprotein recognized by DBA as well as monoclonal 73.18 (Weiss et al., 1992; Zhang et al., 2001). It is likely to be identical to the antigen recognized by serum of animals with chronic infection (Zhang and Smith, 1995; Smith et al., 1996; Smith, 1993) and by a rat monoclonal antibody CC2 (Gross et al., 1995). Data that the cyst wall binds S-WGA and is disrupted by chitinase are consistent with the presence of chitin in this structure (Boothroyd et al., 1997). Lectin overlay experiments of 2 dimensional gels suggest that the DBA and S-WGA recognize different antigens, with S-WGA recognizing a 48 kDa antigen (Zhang et al., 2001). Protocols for purification of the cyst wall have now been developed and should facilitate identification of compenent proteins using a proteomic approach (Zhang et al., 2001, Zhang and Weiss, unpublished).   Several glycosyl transferase genes, including a polypeptide N-acetylgalactosaminyltransferase that may be involved in cyst wall formation, have been identified, expressed and characterized in *T. gondii* (Stwora-Wojczyk et al., 2004; Wojczyk et al., 2003).

MAG1 is a 65 kDa protein expressed in the matrix of the cyst between bradyzoites and is also seen in the cyst wall (Parmley et al., 1994). Initially it was thought that MAG1 was not expressed in tachyzoites (Parmley et al., 1994). Further experiments revealed that MAG1 is expressed in tachyzoites and secreted into the parasitophorous vacuole, albeit less abundantly than in bradyzoites (Ferguson and Parmley, 2002). Antibody to recombinant MAG1 reacts with extracellular material in the cyst matrix and to a lesser extent with the cyst wall, but not with the surface or cytoplasm of bradyzoites.   RT-PCR data indicates that mRNA for MAG1 is present in both tachyzoites and bradyzoites (Parmley et al., 1994).

The dense granule protein GRA5 (Lecordier et al., 1993) is found in both tachyzoites and bradyzoites.   In bradyzoites GRA5 has been demonstrated to be in the cyst wall and appears at the same time as other early bradyzoite markers (Lane et al., 1996).   At present no other proteins that are present in the cyst wall have been identified, but given the documented localization of a number of dense granule proteins (GRA3, GRA5) to the parasitophorous vacuole, it seems possible that many of the protein components of the cyst wall will be known dense granule proteins with new carbohydrate modifications or bradyzoite-specific glycoproteins secreted from dense granules. In support of this idea is that rat monoclonal antibody CC2 reacts with a 115kDa antigen similar, if not identical, to CST1 in bradyzoites, but recognizes a 40 kDa protein in tachyzoites (Gross et al., 1995).

## SURFACE ANTIGENS

The EST sequencing project and cloning of *T. gondii* surface antigens has shown that many *T. gondii* surface antigens are members of a gene family with similar structure to SAG1 including conserved placement of 12 cysteines (Boothroyd et al., 1998; Manger et al., 1998b; Lekutis et al., 2001).   These antigens include SAG1, SAG3 (p43), BSR4 (p36) and SRS 1-4 (SAG-related sequences), SAG5, SAG5.1 and SAG 5.2.   A second family of surface antigens is separate but related to SAG1 with less consistent conservation of cysteine spacing. This family includes SAG2A (SAG2 or p22) and SAG2B-SAG2D.   It is not clear why so many family members exist (at least 150 genes, although some are pseudogenes) since antigenic variation as described for trypanosomes or malaria has not been described. Some of these surface antigens are stage-specific (SAG1 and BRS4) while others (SAG3) are not. Bradyzoite specific family members include SAG2C, SAG2D, BSR4, and SAG5A. Some of the family members such as SAG2C and SAG2D are only detected on *in vivo* bradyzoites not *in vitro* bradyzoites suggesting they are expressed later in bradyzoite differentiation (Lekutis et al., 2001). BRS4, or a cross reacting antigen, has also been reported in merozoites. Tachyzoite specific family members include SAG1, SRS1-SRS3, SAG2A and SAG2B.

The reason for stage-specific expression of these surface proteins is not known. All of these SAGs appear to be attached to the plasma membrane by a similar glycolipid anchor. Roles for these surface antigens in evasion of host immune defenses and host cell invasion have been hypothesized. Both SAG3 (p43) and SAG1 (p30) have been implicated in adhesion to host cells. Disruption of SAG3 leads to two-fold decreased adhesion of parasites and parasites are less virulent (Dzierszinski et al., 2000). Adhesion of parasites can be blocked by SAG1 antibodies or antisera, but disruption of SAG1 in RH strain results in parasites that are more invasive but less virulent in animals (Mineo et al., 1993). In contrast a PLK SAG1 mutant with a nonsense mutation (the "B" mutant) does not appear to have impaired invasion or virulence (Mineo et al., 1993; Kim, Buelow and Boothroyd, unpublished).

BSR4/p36 (AF015290) is a surface protein specific to bradyzoites that reacts with mAb T84A12 (Knoll and Boothroyd, 1998a). A related protein may be present in merozoites. BSR4 was isolated from pH 8.0 treated *T. gondii* cultures in human fibroblasts using a promoter trap strategy. This gene demonstrates a restriction fragment length polymorphism between ME49 (PLK; Type II) and CEP (Type III) strains, which correlates with the lack of mAb T84A12 binding to CEP strain. Although expression of the protein appears to be bradyzoite-specific, levels of RNA do not show significant differences suggesting that post transcriptional regulation of BSR4/p36 occurs. The mechanisms for this are unknown. Disruption of BSR4 did not lead to any obvious phenotypic differences (Knoll and Boothroyd, 1998a).

SAG4 (p18; now SAG4A) (Z69373) is an 18 kDa surface protein of bradyzoites (Odberg-Ferragut et al., 1996). A SAG4 homologue SAG4.2 (AF015715; SAG4B) has also been identified and this family is separate from the SAG1 related genes. No expression of SAG4A is seen in tachyzoites. This gene similar to *BAG1* appears to be transcriptionally regulated during bradyzoite development.

## METABOLIC DIFFERENCES BETWEEN BRADYZOITES AND TACHYZOITES

Bradyzoite differentiation from tachyzoites is likely a programmed response related to a slowing of replication and lengthening of the cell cycle (Jerome et al., 1998). The tachyzoite cell cycle has features similar to that of higher eukaryotic cells and is characterized by major G1 and S phases and a relatively short G2+M. When *T. gondii* replication slows there is an increase in duration of the G1 phase of the cell cycle. Although grossly similar to cell cycle of higher eukaryotic cells, checkpoints within the cell cycle may differ from those observed in yeast and mammalian cells (Radke et al., 2001; Khan et al., 2002). When VEG strain sporozoites are used to infect cells, they transform to rapidly growing tachyzoites with a half-life of 6 hours. After 20 divisions these organisms shift to a slower growth rate with a half-life of 15 hours. Bradyzoite differentiation is not seen in the rapidly growing stages, but occurs spontaneously when the population shifts to a slower growth rate. This analogous to the programmed expansion and differentiation reported in other coccida. Flow cytometry measurements of DNA content of mature bradyzoites reveals them all to have 1N DNA content consistent with their being in a quiescent $G_o$ state (Radke et al., 2003).

When cells are infected by bradyzoites (from tissue cysts) differentiation to tachyzoites and the appearance of tachyzoite specific antigens (SAG1) occurs within 15 hours. Thus SAG1 appears before any cell division has occurred (Soete and Dubremetz, 1996). Vacuoles containing organisms expressing only tachyzoite antigens are clearly evident within 48 hours of infection. Thus, conversion between these developmental stages is a rapid event and commitment to differentiation may be occurring at the time of or shortly after invasion. Although a small proportion of replicating parasites (<10%) have 2N DNA content (i.e. are in a G2 premitotic state), parasites that co-express bradyzoite marker BAG1 and tachyzoite marker SAG1 are much more likely to have a G2 premitotic DNA content (approximately 50%; Radke et al., 2003). These data suggest that commitment to differentiation may occur at a particular point in the cell cycle and that transit through the cell cycle is required for differentiation.

Spontaneous differentiation occurs less readily in rapidly dividing organisms, and stress conditions that slow growth induce differentiation (Bohne et al., 1994; Jerome et al., 1998; Weiss and Kim, 2000). Stress conditions also induce a higher percentage of the organisms to differentiate into bradyzoites. Bradyzoite differentiation cannot be uncoupled from slowing of the cell cycle and may be a stochastic event that occurs at a specific point in the cell cycle when replication has slowed sufficiently. It appears that all conditions that slow, but do not arrest the cell cycle, result in bradyzoite differentiation, whereas conditions that block cell cycle progression do not result in appreciable differentiation (Fox and Bzik, 2002; Khan et al., 2002).

Given the different rate of growth and location of bradyzoites in a thick walled vacuole it is likely that the energy metabolism of tachyzoites and bradyzoites is different. Tachyzoites utilize the glycolytic pathway with the production of lactate as their major source of energy. Mitochondria with a functional TCA cycle exist in tachyzoites and are thought to contribute to energy production. While both tachyzoites and bradyzoites utilize the glycolytic pathway for energy, data suggests that bradyzoites lack a functional TCA cycle and respiratory chain (Denton et al., 1996).

An unusual and unexplained feature of *T. gondii* differentiation is the presence of stage specific differences in the activity and isoforms of glycolytic enzymes. Lactate dehydrogenase (LDH) and pyruvate kinase activity is higher in bradyzoites than tachyzoites while $PP_i$-phosphofructokinase activity is higher in tachyzoites than bradyzoites (Denton et al., 1996). In addition, the bradyzoite enzymes are resistant to acidic pH. These data are consistent with bradyzoite energy metabolism being dependent on the catabolism of amylopectin (see below) to lactate and suggest that bradyzoites are resistant to acidification resulting from the accumulation of these glycolytic products. The pyruvate kinase of *T. gondii* has been cloned and characterized (Maeda et al., 2003). Regulation and activation of glycolysis appear to be different in *T. gondii* than many other eukaryotes (Saito et al., 2002; Maeda et al., 2003; Denton et al., 1996).

Lactate dehydrogenase 2 (LDH2) is a 35 kDa cytoplasmic antigen that is expressed in bradyzoites but not in tachyzoites (Yang and Parmley, 1997; 1995). Its expression appears to be transcriptionally regulated because *LDH2* mRNA is detectable by RT-PCR only in bradyzoites. Antiserum to recombinant LDH2 reacts strongly with a 35 kDa antigen with a PI of 7.0 in bradyzoites and also weakly with a 33 kDa antigen with a PI of 6.0 in tachyzoites. The tachyzoite isoform is identified as LDH1 and has also been cloned. LDH1 and LDH2 are 71.4% identical and despite its apparent stage-specificity, attempts to disrupt *LDH2* have failed (Singh et al., 2002). The LDH activity of tachyzoite and bradyzoite extracts is different. It is believed that these two stage specific LDH isoforms account for this observation although no significant differences in recombinant enzyme activities were reported.

Stage specific enolases have also been cloned and characterized. Enolase catalyzes the conversion of 2-phosphoglycerate to phosphoenolpyruvate (Yahiaoui et al., 1999; Manger et al., 1998a). This is consistent with the hypothesis that utilization of the glycolytic pathway is different in tachyzoites compared to bradyzoites. The ENO1 and ENO2 have similar Michaelis constants (Km) but ENO2, the tachyzoite form, has 3 fold higher specific activity than ENO1, the bradyzoite enzyme (Dzierszinski et al., 2001). Although ENO1 and ENO2 are very similar, polyclonal antisera to each isoform do not cross react ((Dzierszinski et al., 1999; Dzierszinski et al., 2001)). Interestingly the localization of both forms is nuclear in dividing cells, but in late bradyzoites, which are quiescent, ENO1 is cytoplasmic (Ferguson et al., 2002). The significance of these observations is unknown but it is possible that glycolytic enzymes may have alternate functions that are not yet fully understood.

Another obvious difference between tachyzoites and bradyzoites is the presence of cytosolic granules of amylopectin that are composed of glucose polymers. Amylopectin is also present in the sexual cycle in the cat intestine in macrogametes, persists during oocyst formation, and is seen in sporozoites. Merozoites lack amylopectin. Amylopectin granules disappear from bradyzoites when they transform into tachyzoites during cell culture (Coppin et al., 2003). Amylopectin is hypothesized to serve as a carboyhydrate store for the bradyzoite or sporozoite during long periods of quiescence and nutrient deprivation. Structural studies of amylopectin have revealed it to be plant-like amylopectin with predominantly ($\alpha$1-4) linkages. Candidate genes for enzymes involved in amylopectin breakdown and synthesis have also been identified (Coppin et al., 2003).

What remains to be determined is if the metabolic changes observed cause bradyzoite differentiation or are a consequence of the differentiation process. One could envision a scenario where monitoring nutrient deprivation might serve as the sensor for differentiation in *T. gondii,* however, no such sensors have been identified.

## ROLE OF STRESS PROTEINS IN DIFFERENTIATION

It is now accepted that bradyzoite differentiation is a stress-related response of *T. gondii* to environmental conditions such as the inflammatory response of the host to the tachyzoite stage. Temperature, pH, mitochondrial inhibitors, sodium arsenite and many of the other stressors associated with bradyzoite development *in vitro* are also associated with the induction of heat shock proteins. Bradyzoite differentiation probably shares features common to other stress induced differentiation systems such as glucose starvation and hyphae formation in fungi or spore formation in *Dicyostelium* (Thomason et al., 1999; Soderbom and Loomis, 1998). These systems have demonstrated unique proteins related to specific differentiation structures in each organism as well as the utilization of phylogenetically conserved pathways. Many of these signaling pathways involve cyclic nucleotides and kinases as part of the regulatory system in differentiation. Heat shock proteins (hsps) are also involved in these pathways as chaperones for both regulatory and stage-specific proteins involved in differentiation.

It is likely that bradyzoite differentiation involves a number of heat shock proteins and other proteins classically associated with the stress response. In the *T. gondii* dbEST, BAG1 (also known as BAG5) is the most abundant bradyzoite specific gene representing ≈3% of all clones that are bradyzoite specific. BAG1 has homology to small heat shock proteins (smHsps) with strongest homology to smHsps of plants (Bohne et al., 1995; Parmley et al., 1995). It is a 28 kDa antigen expressed in the cytoplasm of bradyzoites, and there is no expression of this antigen in tachyzoites or sporozoites. Both *BAG1* mRNA and protein are upregulated during bradyzoite formation, suggesting transcriptional regulation of its expression. Expression of BAG1 is seen early in differentiation *in vitro*, and cells expressing BAG1 are seen within 24 hours of exposure to pH 8.0 or other stress conditions. The carboxyl-terminal region of BAG1 has a small heat shock motif most similar to the small heat shock proteins of plants (carrot hsp 17.7). Near the N-terminus is a synapsin Ia like domain that may be involved in the association of this small heat shock protein with proteins during development. Yeast 2 hybrid screening with both the N-terminus and C-terminus domains has resulted in potential BAG1 interacting proteins that are currently being confirmed (Ma and Weiss, unpublished).

Further evidence for a role of stress-induced genes in bradyzoite development is the finding that *T. gondii* HSP70 homologues (AF045559, U85649, U85648) are induced during bradyzoite formation (Weiss *et al.*, 1998; Lyons and Johnson, 1995; 1998; Miller et al., 1999). Induction of HSP70 can be demonstrated at both the protein and RNA level. Quercetin, an inhibitor of HSP synthesis, was able to suppress HSP70 levels whereas indomethacin, an inducer of HSP transcription, was able to enhance the production HSP levels (Weiss et al., 1998; Weiss et al., 1996). Extracellular *T. gondii* treated with a one-hour exposure to pH 8.1 versus pH 7.1 expressed a 72kDa inducible hsp70 (detected with mAb C92F3A-5; Stressgen) (Weiss et al., 1998). The heat shock protein is induced by a pre-treatment protocol that also induced bradyzoite formation. *T. gondii* infected cultures treated with pH 8.1 showed 4 fold induction of the hsp70 levels compared to *T. gondii* grown in pH 7.1 treated cells (Weiss et al., 1998; Weiss et al., 1996). The 3 to 4 fold change demonstrated in *T. gondii* with stress as well as differentiation is comparable with the magnitude of the hsp70 response demonstrated in *Trypanosoma cruzi*, *Theileria annulata* and *Plasmodium falciparum* (Shiels et al., 1997). Similar results were recently obtained with in vivo cysts during reactivation in a murine model induced by anti-γ-interferon (Silva et al., 1998), suggesting that hsp70 may be important in both tachyzoite to bradyzoite and bradyzoite to tachyzoite differentiation.

In recent studies, the promoter of the HSP70 region has been mapped. Expression of reporter genes driven by the HSP70 promoter is also responsive to conditions that induce bradyzoite formation including SNP and pH shock. Transcription factors responsible for regulation of HSP70 have not yet been identified although electromobility shift assays (EMSA) suggest that there are specific proteins that recognize the HSP70 promoter region. Although there is an area of similarity between the *BAG1* promoter region and that of *HSP70,* oligonucleotides from this *BAG1* upstream region do not compete in EMSA (Ma and Weiss, in preparation).

## GENETIC STUDIES ON BRADYZOITE BIOLOGY

Knockout of bradyzoite specific genes should not be essential as growth could still occur in the tachyzoite stage. This strategy has been used by two groups to develop BAG1/hsp30 (BAG5) knockouts (Bohne et al., 1998; Zhang et al., 1999). One knockout was done using HGXPRT as a selectable marker in an HGXPRT[neg] PLK strain of *T. gondii* and the other using CAT as a selectable marker in a clone of PLK strain (passaged through mice to ensure it made cysts at the start of the study). Cyst formation both *in vitro* and *in vivo* occurs in both knockouts, indicating that BAG1 is not essential for cyst formation. Zhang et al. found that the number of cysts formed *in vivo* in CD1 mice was reduced in *bag1* knockouts and complementation of this knockout restored the production of similar numbers of cysts to that of the wild type PLK strain (Zhang et al., 1999). When parasites were grown in SNP, the *bag1* knockout grew faster than PLK. This may be a result of a difference in transition rate from the rapidly growing tachyzoite to the slowly growing bradyzoite stage in this *bag1* knockout. The decrease in cyst formation is a relatively subtle phenotype and was not observed by Bohne et al. (1998) when *BAG1* was disrupted in a different genetic background and cyst formation was tested in a highly susceptible mouse strain (C57BL/6 (Zhang et al., 1999) instead of the outbred CD1 mice used in the Zhang study (Zhang et al., 1999).

The capacity to convert from tachyzoite to bradyzoite is key for *T. gondii* persistence in the host, and thus it is likely that multiple genes with redundant functions are involved in this process. smHsps have been postulated to act as specialized chaperones for enzymes such as glutathione reductase during differentiation, but the exact function of small hsps in *T. gondii* remains an area of active investigation. BAG1 homologues that do not appear to be bradyzoite-specific are present in the EST sequencing project. It is possible that these homologues might be able to partially compensate for lack of BAG1.

Bradyzoite and tachyzoite specific promoter regions can be used to create reagents for the study of differentiation *in vitro* (Gross, 1996). Reporter genes like the chloramphenicol acetyltranferase (CAT) gene or beta-galactosidase have been used to map promoter activity and define minimal promoter sequences. These studies have confirmed the stage specific expression of SAG1 (as a tachyzoite marker), LDH2, and BAG1, but as yet stage-specific transcription factors or other regulatory molecules have not been identified (Yang and Parmley, 1997; Bohne et al., 1997). These constructs have been used in both transient and stable transfection assays. Another reporter gene that has been used for FACS identification of differentiation mutants is the green fluorescent protein (GFP).

Genetic methods have been used to identify genes that are induced during bradyzoite differentiation. Promoter trapping has been utilized to identify genes induced during bradyzoite differentiation using a promoterless hypoxanthine-xanthine-guanine phosphoribosyltranferase (HGXPRT) gene with 6-thoxanthine (6-TX) or 8-azaguanine (8-AzaH) as negative selection and mycophenolic acid with xanthine (MPA-X) as a positive selection. (Bohne et al., 1997; Knoll and Boothroyd, 1998a; Knoll and Boothroyd, 1998b). By growing transfected parasites at pH 7.0 in the presence of 6-TX all organisms that have the HGXPRT gene on a constitutive or tachyzoite promoter will be killed. This population of organisms is then exposed to pH 8.0 and MPA-X. Under these conditions only parasites which express HGXPRT (i.e. those with a bradyzoite or constitutive promoter in front of the HGXPRT gene) will survive. This strategy may be "leaky", depending on the concentrations of 6-TX and MPA-X used. Nonetheless, the 6-TX and MPA-X selections can be repeated several times to enrich the population for organisms with HGXPRT under the control of a bradyzoite specific gene promoter. Using this approach additional bradyzoite specific promoters (Donald and Roos, unpublished; cited in Matrajt et al., 2002) and 8 bradyzoite specific recombinant (BSR) strains were obtained (Knoll and Boothroyd, 1998a). One of these, BSR4, was later found to be the bradyzoite surface antigen known as p36 (Knoll and Boothroyd, 1998a).

Further genetic strategies have been performed to identify mutants unable to differentiate (Matrajt et al., 2002; Singh et al., 2002). Singh et al. (2002) generated point mutants in a *LDH2-GFP* Prugnaiud (Type II) background to obtain mutants with altered ability to transform into bradyzoite. Parasites unable to differentiate were identified by FACS enrichment of GFP-negative parasites exposed to bradyzoite inducing conditions. Matrajt et al used insertional mutagenesis of an engineered stable line expressing a bradyzoite-specific pT7-HGXPRT cassette in UPRT-deficient RH (TypeI) parasites (Matrajt et al., 2002). Earlier studies had shown that UPRT disruptants differentiate into bradyzoites under conditions of $CO_2$ starvation (Bohne and Roos, 1997). The pT7 stable line was obtained by rounds of negative and positive selection, alternating 6-TX (tachyzoite conditions) with MPA-X (bradyzoite conditions) selection. The result was a cell line where HGXPRT was highly regulated by differentiation conditions. Insertional mutagenesis was performed with the DHFR cassettes that earlier were shown to have a high frequency of nonhomologous insertion (Donald and Roos, 1993).

Both groups were able to demonstrate that the mutants they obtained were unable to differentiate and had global defects in expression of previously characterized markers as determined by immunofluorescence and microarray analysis. Due to technical difficulties the exact mutations could not be identified. The two insertional mutants also (like the *BAG1* disruptant described by Zhang et al. (1999) had more rapid growth under bradyzoite inducing conditions than seen in wild-type parasites (Matrajt et al., 2002). Microarray analysis of these mutants identified classes of genes whose expression was decreased in the differentiation mutants (Singh et al., 2002; Matrajt et al., 2002). Among these were a 14-3-3 homologue, a PISTLRE kinase and a probable vacuolar ATPase.

## SIGNALING PATHWAYS IN BRADYZOITE FORMATION

Although tachyzoites and bradyzoites are well defined morphologically, little is known about how interconversion from one to the other stage occurs or what signal(s) mediate this transformation. Studies of other microorganisms including the fungi and other protozoa suggest that conserved signaling pathways involved in response to stress or nutrient starvation are involved in differentiation. Similar pathways may be responsible for the tachyzoite-bradyzoite transition. Data from a number of model systems has implicated cyclic nucleotide signaling in stress-induced differentiation. The effect of cyclic nucleotides on bradyzoite differentiation was assessed by using non-metabolized analogues of cAMP and cGMP as well as forskolin (to stimulate a short pulse of cAMP) (Kirkman et al., 2001). These experiments demonstrated that forskolin and cGMP could induce bradyzoite formation. Experiments with isolated extracellular *T. gondii* tachyzoites demonstrated that conditions inducing bradyzoites (pH8.1, Forskolin or SNP exposure) lead to a 3 to 4 fold elevation in cAMP levels but no consistent changes in cGMP were seen. Within 30 minutes, the cAMP levels are comparable to those seen in control parasites incubated in pH 7.1 media.

Most of the effects of cAMP within cells can be attributed to regulation of cAMP-dependent protein kinase A activity. Effects of cGMP can be attributed to stimulation of a cGMP-dependent kinase, effects upon phosphodiesterases or other signaling molecules. Several kinases that are potentially involved in differentiation have been cloned. These include protein kinase A [Eaton, Tang and Kim, unpublished data], a glycogen synthase kinase (GSK-3) homologue (Qin et al., 1998) and a unique cGMP-dependent protein kinase (PKG) (Gurnett et al., 2002; Nare et al., 2002; Donald and Liberator, 2002; Donald et al., 2002). Inhibitors of PKA and

PKG induce bradyzoite differentiation (Eaton and Kim in preparation; Nare et al., 2002), suggesting that there are both inhibitory and stimulatory pathways affected by cyclic nucleotide signaling. Protein phosphorylation has proven to be a major mechanism of regulation of gene expression and integration and amplification of extracellular signals. PKA1 and PKA2 appear to have opposing effects upon parasite proliferation (Eaton and Kim, in preparation). The relevance to bradyzoite differentiation is currently being tested. The presence of highly conserved signaling molecules suggests that many of the pathways identified in other eukaryotes are likely to be preserved in *T. gondii,* however their exact role in differentiation awaits futher study. Yeast 2 hybrid analysis has been used to identify proteins that interact with PKA subunits.

## SUMMARY

Unique structural antigens as well as metabolic pathways exist in bradyzoites. Investigations into bradyzoite biology and the differentiation of tachyzoites into bradyzoites has been accelerated by the development of *in vitro* techniques to study and produce bradyzoites as well as by the genetic tools that exist for the manipulation of *T. gondii*. The *Toxoplasma* genome will be sequenced at 8x coverage by 2003 will be an invaluable aid in future studies on differentiation in this organism.

The triggers for differentiation have not yet been identified and bradyzoite differentiation has not yet been uncoupled from slowing of the cell cycle. It is likely that many signals can result in appropriate signals that induce bradyzoite formation. It is also possible that bradyzoite is the default pathway and that any situation that perturbs positive signals necessary for tachyzoite replication result in the formation of bradyzoites. The development of bradyzoite appears to be a stress mediated differentiation response that leads to metabolic adaptations. The mechanism by which development is triggered and coordinated may eventually lead to novel therapeutic strategies to control toxoplasmosis, perhaps allowing a radical cure of infections such is achieved with malarial chemotherapy. In addition, genetic strategies that prevent cyst formation may prove useful in the development of vaccine strains of this Apicomplexan protozoan pathogen.

## ACKNOWLEDGEMENT

This work was supported by the NIH AI39454 (LMW), AI01535 (KK), AI41058 (KK) and a Burroughs Wellcome Fund New Investigator in Molecular Parasitology Award (KK).

## REFERENCES.

Ajioka, J.W. 1998. *Toxoplasma gondii*: ESTs and gene discovery. International Journal for Parasitology **28** 1025-1031.

_____, J.C. Boothroyd, B.P. Brunk, A. Hehl, L. Hillier, I.D. Manger, M. Marra, G.C. Overton, D.S. Roos, K.L. Wan, R. Waterston, and L. D. Sibley. 1998. Gene discovery by EST sequencing in *Toxoplasma gondii* reveals sequences restricted to the Apicomplexa. Genome Research **8** 18-28.

Aramini, J.J., C. Stephen, J.P. Dubey, C. Engelstoft, H. Schwantje, and C.S. Ribble. 1999. Potential contamination of drinking water with *Toxoplasma gondii* oocysts. Epidemiology and Infection **122** 305-315.

Bohne, W., U. Gross, D.J. Ferguson, and J. Heesemann. 1995. Cloning and characterization of a bradyzoite-specifically expressed gene (hsp30/bag1) of *Toxoplasma gondii*, related to genes encoding small heat-shock proteins of plants. Molecular Microbiology **16** 1221-1230.

_____, J. Heesemann, and U. Gross. 1993. Induction of bradyzoite-specific *Toxoplasma gondii* antigens in gamma interferon-treated mouse macrophages. Infection and Immunity **61** 1141-1145.

_____, _____, and _____. 1994. Reduced replication of *Toxoplasma gondii* is necessary for induction of bradyzoite-specific antigens: a possible role for nitric oxide in triggering stage conversion. Infection and Immunity **62** 1761-1767.

_____, C.A. Hunter, M.W. White, D.J. Ferguson, U. Gross, and D.S. Roos. 1998. Targeted disruption of the bradyzoite-specific gene BAG1 does not prevent tissue cyst formation in *Toxoplasma gondii*. Molecular and Biochemical Parasitology **92::** 115-126.

_____, A. Wirsing, and U. Gross. 1997. Bradyzoite-specific gene expression in *Toxoplasma gondii* requires minimal genomic elements. Molecular and Biochemical Parasitology **85:** 1347-1354.

_____, A. Hehl, L.J. Knoll, and I.D. Manger. 1998. The surface of *Toxoplasma*: more and less. International Journal of Parasitology **28:** 3-9.

Bowie, W.R., A.S. King, D.H. Werker, J.L. Isaac-Renton, A. Bell, S.B. Eng, and S.A. Marion. 1997. Outbreak of toxoplasmosis associated with municipal drinking water. The BC *Toxoplasma* Investigation Team. Lancet **350:** 173-177.

Cleary, M.D., U. Singh, I.J. Blader, J.L. Brewer, and J.C. Boothroyd. 2002. *Toxoplasma gondii* asexual development: identification of developmentally regulated genes and distinct patterns of gene expression. Eukaryotic Cell **1:** 329-340.

Coppin, A., F. Dzierszinski, S. Legrand, M. Mortuaire, D. Ferguson, and S. Tomavo. 2003. Developmentally regulated biosynthesis of carbohydrate and storage polysaccharide during differentiation and tissue cyst formation in *Toxoplasma gondii*. Biochimie **85:** 353-361.

De Champs, C., C. Imbert-Bernard, A. Belmeguenai, J. Ricard, H. Pelloux, E. Brambilla, and P. Ambroise-Thomas. 1997. *Toxoplasma gondii*: in vivo and in vitro cystogenesis of the virulent RH strain. Journal of Parasitology **83:** 152-155.

Denton, H., C.W. Roberts, J. Alexander, K.W. Thong, and G.H. Coombs. 1996. Enzymes of energy metabolism in the bradyzoites and tachyzoites of *Toxoplasma gondii*. FEMS Microbiology Letters **137:** 103-108.

Donald, R.G., J. Allocco, S.B. Singh, B. Nare, S.P. Salowe, J. Wiltsie, and P.A. Liberator. 2002. *Toxoplasma gondii* cyclic GMP-dependent kinase: chemotherapeutic targeting of an essential parasite protein kinase. Eukaryotic Cell **1:** 317-328.

_____, and P.A. Liberator. 2002. Molecular characterization of a coccidian parasite cGMPdependent protein kinase. Molecular and Biochemical Parasitology **120:** 165-175.

_____, and D.S. Roos. 1993. Stable molecular transformation of *Toxoplasma gondii*: a selectable dihydrofolate reductase-thymidylate synthase marker based on drug- resistance mutations in malaria. Proceedings from the National Academy of Sciences USA **90:** 11703-11707.

Dubey, J.P. 1997. Bradyzoite-induced murine toxoplasmosis: stage conversion, pathogenesis, and tissue cyst formation in mice fed bradyzoites of different strains of *Toxoplasma gondii* [published erratum appears in Journal of Eukaryotic Microbiology 1998 May-Jun;45(3):367]. The Journal of Eukaryotic Microbiology **44:** 592-602.

_____, D.S. Lindsay, and C.A. Speer. 1998. Structures of *Toxoplasma gondii* tachyzoites, bradyzoites, and sporozoites and biology and development of tissue cysts. Clinical Microbiology Reviews **11**: 267-299.

_____, C.A. Speer, S.K. Shen, O.C. Kwok, and J.A. Blixt. 1997. Oocyst-induced murine toxoplasmosis: life cycle, pathogenicity, and stage conversion in mice fed *Toxoplasma gondii* oocysts. Journal of Parasitology **83**: 870-882.

Dzierszinski, F., M. Mortuaire, M.F. Cesbron-Delauw, and S. Tomavo. 2000. Targeted disruption of the glycosylphosphatidylinositol-anchored surface antigen SAG3 gene in *Toxoplasma gondii* decreases host cell adhesion and drastically reduces virulence in mice. Molecular Microbiology **37**: 574-582.

._____, _____, N. Dendouga, O. Popescu, and S. Tomavo. 2001. Differential expression of two plant-like enolases with distinct enzymatic and antigenic properties during stage conversion of the protozoan parasite *Toxoplasma gondii*. Journal of Molecular Biology **309**, 1017-1027.

_____, O. Popescu, C. Toursel, C. Slomianny, B. Yahiaoui, and S. Tomavo. 1999. The protozoan parasite *Toxoplasma gondii* expresses two functional plant-like glycolytic enzymes. Implications for evolutionary origin of apicomplexans. Journal of Biological Chemistry **274**, 24888-24895.

Ferguson, D.J., and S.F. Parmley. 2002. *Toxoplasma gondii* MAG1 protein expression. Trends in Parasitology **18**: 482.

_____, _____, and S. Tomavo. 2002. Evidence for nuclear localisation of two stage-specific isoenzymes of enolase in *Toxoplasma gondii* correlates with active parasite replication. International Journal of Parasitology **32**: 1399-1410.

_____, and W.M. Hutchison. 1987. An ultrastructural study of the early development and tissue cyst formation of *Toxoplasma gondii* in the brains of mice. Parasitology Research **73**: 483-491.

Fortier, B., C. Coignard-Chatain, M. Soete, and J.F. Dubremetz. 1996. [Structure and biology of *Toxoplasma gondii* bradyzoites]. Comptes Rendus des Seances de la Societe de Biologie et de ses Filiales **190**: 385-394.

Fox, B.A., and D.J. Bzik. 2002. De novo pyrimidine biosynthesis is required for virulence of *Toxoplasma gondii*. Nature **415**: 926-929.

Grigg, M.E., J. Ganatra, J.C. Boothroyd, and T.P. Margolis. 2001. Unusual abundance of atypical strains associated with human ocular toxoplasmosis. The Journal of Infectious Diseases **184**: 633-639.

Gross, U. 1996. *Toxoplasma gondii*. Berlin: Springer.

_____, W. Bohne, C.G. Luder, R. Lugert, F. Seeber, C. Dittrich, F. Pohl, and D.J. Ferguson. 1996. Regulation of developmental differentiation in the protozoan parasite *Toxoplasma gondii*. Journal of Eukaryotic Microbiology **43**: 114S-116S.

_____, H. Bormuth, C. Gaissmaier, C. Dittrich, V. Krenn, W. Bohne, and D.J. Ferguson. 1995. Monoclonal rat antibodies directed against *Toxoplasma gondii* suitable for studying tachyzoite-bradyzoite interconversion in vivo. Clinical and Diagnostic Laboratory Immunology **2**: 542-548.

Gurnett, A.M., P.A. Liberator, P.M. Dulski, S.P. Salowe, R.G. Donald, J.W. Anderson, J. Wiltsie, C.A. Diaz, G. Harris, B. Chang, S.J. Darkin-Rattray, B. Nare, T. Crumley, P.S. Blum, A.S. Misura, T. Tamas, M.K. Sardana, J. Yuan, T. Biftu, and D.M. Schmatz. 2002. Purification and molecular characterization of cGMP-dependent protein kinase from Apicomplexan parasites. A novel chemotherapeutic target. Journal of Biological Chemistry **277**: 15913-15922.

Holpert, M., C.G. Luder, U. Gross, and W. Bohne. 2001. Bradyzoite-specific expression of a P-type ATPase in *Toxoplasma gondii*. Molecular and Biochemical Parasitology **112**: 293-296.

Jacobs, L., and J. Remington. 1960. The resistance of the encysted form of *Toxoplasma gondii*. Journal of Parasitology **46**: 11-21.

Jerome, M.E., J.R. Radke, W. Bohne, D.S. Roos, and M.W. White. 1998. *Toxoplasma gondii* bradyzoites form spontaneously during sporozoite- initiated development. Infection and Immunity **66:** 4838-4844.

Jones, T.C., K.A. Bienz, and P. Erb. 1986. In vitro cultivation of *Toxoplasma gondii* cysts in astrocytes in the presence of gamma interferon. Infection and Immunity **51:** 147-156.

Khan, F., J. Tang, C.L. Qin, and K. Kim. 2002. Cyclin-dependent kinase TPK2 is a critical cell cycle regulator in *Toxoplasma gondii*. Molecular Microbiology **45:** 321-332.

Kirkman, L.A., L.M. Weiss, and K. Kim. 2001. Cyclic nucleotide signaling in *Toxoplasma gondii* bradyzoite differentiation. Infection and Immunity **69:** 148-153.

Knoll, L.J., and J.C. Boothroyd. 1998a. Isolation of developmentally regulated genes from *Toxoplasma gondii* by a gene trap with the positive and negative selectable marker hypoxanthine-xanthine-guanine phosphoribosyltransferase. Molecular and Cellular Biology **18:** 807-814.

_____, and _____. 1998b. Molecular biology's lessons about *Toxoplasma* development: stage-specific homologs. Parasitology Today **14:** 490-493.

Lane, A., M. Soete, J.F. Dubremetz, and J.E. Smith. 1996. *Toxoplasma gondii*: appearance of specific markers during the development of tissue cysts in vitro. Parasitology Research **82:** 340-346.

Lecordier, L., C. Mercier, G. Torpier, B. Tourvieille, F. Darcy, J.L. Liu, P. Maes, A. Tartar, A. Capron, and M.F. Cesbron-Delauw. 1993. Molecular structure of a *Toxoplasma gondii* dense granule antigen (GRA 5) associated with the parasitophorous vacuole membrane. Molecular and Biochemical Parasitology **59:** 143-153.

Lekutis, C., D.J. Ferguson, M.E. Grigg, M. Camps, and J.C. Boothroyd. 2001. Surface antigens of *Toxoplasma gondii*: variations on a theme. International Journal of Parasitology **31:** 1285-1292.

Li, L., B.P. Brunk, J.C. Kissinger, D. Pape, K. Tang, R.H. Cole, J. Martin, T. Wylie, M. Dante, S.J. Fogarty, D.K. Howe, P. Liberator, C. Diaz, J. Anderson, M. White, M.E. Jerome, E.A. Johnson, J.A. Radke, C.J. Stoeckert, Jr., R.H. Waterston, S.W. Clifton, D.S. Roos, and L.D. Sibley. 2003. Gene discovery in the apicomplexa as revealed by EST sequencing and assembly of a comparative gene database. Genome Research **13:** 443-454.

Lindsay, D.S., J.P. Dubey, B.L. Blagburn, and M. Toivio-Kinnucan. 1991. Examination of tissue cyst formation by *Toxoplasma gondii* in cell cultures using bradyzoites, tachyzoites, and sporozoites. Journal of Parasitology **77:** 126-132.

_____, R.R. Mitschler, M.A. Toivio-Kinnucan, S.J. Upton, J.P. Dubey, and B.L. Blagburn. 1993a. Association of host cell mitochondria with developing *Toxoplasma gondii* tissue cysts. American Journal of Veterinary Research **54:** 1663-1667.

_____, M.A. Toivio-Kinnucan, and B.L. Blagburn. 1993b. Ultrastructural determination of cystogenesis by various *Toxoplasma gondii* isolates in cell culture. Journal of Parasitology **79:** 289-292.

Luft, B.J., and J.S. Remington. 1988. AIDS commentary. Toxoplasmic encephalitis. Journal of Infectious Diseases **157:** 1-6.

Lyons, R.E., and A.M. Johnson. 1995. Heat shock proteins of *Toxoplasma gondii*. Parasite Immunology **17:** 353-359.

_____, and _____. 1998. Gene sequence and transcription differences in 70 kDa heat shock protein correlate with murine virulence of *Toxoplasma gondii*. International Journal for Parasitology **28:** 1041-1051.

Maeda, T., T. Saito, Y. Oguchi, M. Nakazawa, T. Takeuchi, and T. Asai. 2003. Expression and characterization of recombinant pyruvate kinase from *Toxoplasma gondii* tachyzoites. Parasitology Research **89:** 259-265.

Manger, I.D., A. Hehl, S. Parmley, L.D. Sibley, M. Marra, L. Hillier, R. Waterston, and J.C. Boothroyd. 1998a. Expressed sequence tag analysis of the bradyzoite stage of *Toxoplasma gondii*: identification of developmentally regulated genes. Infection and Immunity **66:** 1632-

1637.

_____, _____, and J.C. Boothroyd. 1998b. The surface of *Toxoplasma* tachyzoites is dominated by a family of glycosylphosphatidylinositol-anchored antigens related to SAG1. Infection and Immunity **66**: 2237-2244.

Matrajt, M., R.G. Donald, U. Singh, and D.S. Roos. 2002. Identification and characterization of differentiation mutants in the protozoan parasite *Toxoplasma gondii*. Molecular Microbiology **44**: 735-747.

McHugh, T.D., A. Gbewonyo, J.D. Johnson, R.E. Holliman, and P.D. Butcher. 1993. Development of an in vitro model of *Toxoplasma gondii* cyst formation. FEMS Microbiology Letters **114**: 325-332.

McLeod, R., D. Mack, and C. Brown. 1991. *Toxoplasma gondii*--new advances in cellular and molecular biology. Experimental Parasitology **72**: 109-121.

Miller, C.M., N.C. Smith, and A.M. Johnson. 1999. Cytokines, nitric oxide, heat shock proteins and virulence in *Toxoplasma*. Parasitology Today **15**: 418-422.

Miller, M.A., I.A. Gardner, C. Kreuder, D.M. Paradies, K.R. Worcester, D.A. Jessup, E. Dodd, M.D. Harris, J.A. Ames, A.E. Packham, and P.A. Conrad. 2002. Coastal freshwater runoff is a risk factor for *Toxoplasma gondii* infection of southern sea otters (*Enhydra lutris nereis*). International Journal of Parasitology **32**: 997-1006.

Mineo, J.R., R. McLeod, D. Mack, J. Smith, I.A. Khan, K.H. Ely, and L.H. Kasper. 1993. Antibodies to *Toxoplasma gondii* major surface protein (SAG-1, P30) inhibit infection of host cells and are produced in murine intestine after peroral infection. Journal of Immunology **150**: 3951-3964.

Nare, B., J.J. Allocco, P.A. Liberator, and R.G. Donald. 2002. Evaluation of a cyclic GMP-dependent protein kinase inhibitor in treatment of murine toxoplasmosis: gamma interferon is required for efficacy. Antimicrobial Agents and Chemotherapy **46**: 300-307.

Odberg-Ferragut, C., M. Soete, B. Engels, B. Samyn, A. Loyens, J. Van Beeumen, D. Camus, and J.F. Dubremetz. 1996. Molecular cloning of the *Toxoplasma gondii* sag4 gene encoding an 18 kDa bradyzoite specific surface protein. Molecular and Biochemical Parasitology **82**: 237-244.

Parmley, S.F., L.M. Weiss, and S. Yang. 1995. Cloning of a bradyzoite-specific gene of *Toxoplasma gondii* encoding a cytoplasmic antigen. Molecular and Biochemical Parasitology **73**: 253-257.

_____, S. Yang, G. Harth, L.D. Sibley, A. Sucharczuk, and J.S. Remington. 1994. Molecular characterization of a 65-kilodalton *Toxoplasma gondii* antigen expressed abundantly in the matrix of tissue cysts. Molecular and Biochemical Parasitology **66**: 283-296.

Popiel, I., M. Gold, and L. Choromanski. 1994. Tissue cyst formation of *Toxoplasma gondii* T-263 in cell culture. Journal of Eukaryotic Microbiology **41**: 17S.

_____, _____, and K.S. Booth. 1996. Quantification of *Toxoplasma gondii* bradyzoites. Journal of Parasitology **82**: 330-332.

Qin, C.L., J. Tang, and K. Kim. 1998. Cloning and in vitro expression of TPK3, a *Toxoplasma gondii* homologue of shaggy/glycogen synthase kinase-3 kinases. Molecular and Biochemical Parasitology **93**: 273-283.

Radke, J.R., M.N. Guerini, and M.W. White. 2003. A change in the premitotic period of the cell cycle is associated with bradyzoite differentiation in *Toxoplasma gondii*. Molecular and Biochemical Parasitology **131**: 119-127.

_____, B. Striepen, M.N. Guerini, M.E. Jerome, D.S. Roos, and M.W. White. 2001. Defining the cell cycle for the tachyzoite stage of *Toxoplasma gondii*. Molecular and Biochemical Parasitology **115**:, 165-175.

Saito, T., T. Maeda, M. Nakazawa, T. Takeuchi, T. Nozaki, and T. Asai. 2002. Characterisation of hexokinase in *Toxoplasma gondii* tachyzoites. International Journal for Parasitology **32**: 961-967.

Scholytyseck, E., H. Mehlhorn, and B.E. Muller. 1974. Fine structure of cyst and cyst wall of

*Sarcocystis tenella, Besnoitia jellisoni, Frenkelia* sp. and *Toxoplasma gondii*. Journal of Protozoology **21**: 284-294.

Shiels, B., N. Aslam, S. McKellar, A. Smyth, and J. Kinnaird. 1997. Modulation of protein synthesis relative to DNA synthesis alters the timing of differentiation in the protozoan parasite *Theileria annulata*. Journal of Cell Science **110**: 1441-1451.

Silva, N.M., R.T. Gazzinelli, D.A. Silva, E.A. Ferro, L.H. Kasper, and J.R. Mineo. 1998. Expression of *Toxoplasma gondii*-specific heat shock protein 70 during In vivo conversion of bradyzoites to tachyzoites. Infection and Immunity **66**: 3959-3963.

Sims, T.A., J. Hay, and I.C. Talbot. 1989. An electron microscope and immunohistochemical study of the intracellular location of *Toxoplasma* tissue cysts within the brains of mice with congenital toxoplasmosis. British Journal of Experimental Pathology **70**: 317-325.

Singh, U., J.L. Brewer, and J.C. Boothroyd. 2002. Genetic analysis of tachyzoite to bradyzoite differentiation mutants in *Toxoplasma gondii* reveals a hierarchy of gene induction. Molecular Microbiology **44**: 721-733.

Smith, J.E. 1993. Toxoplasmosis. Heidelberg: SpringeriVerlag.

_____, G. McNeil, Y.W. Zhang, S. Dutton, G. Biswas-Hughes, and P. Appleford. 1996. Serological recognition of *Toxoplasma gondii* cyst antigens. Current Topics in Microbiology and Immunology **219**: 67-73.

Soderbom, F., and W.F. Loomis. 1998. Cell-cell signaling during Dictyostelium development. Trends in Microbiology **6**: 402-406.

Soete, M., D. Camus, and J.F. Dubremetz. 1994. Experimental induction of bradyzoite-specific antigen expression and cyst formation by the RH strain of *Toxoplasma gondii* in vitro. Experimental Parasitology **78**: 361-370.

_____, and J.F. Dubremetz. 1996. *Toxoplasma gondii*: kinetics of stage-specific protein expression during tachyzoite bradyzoite conversion in vitro. Current Topics in Microbiology and Immunology **219**: 76-80.

_____, B. Fortier, D. Camus, and J.F. Dubremetz. 1993. *Toxoplasma gondii*: kinetics of bradyzoite-tachyzoite interconversion in vitro. Experimental Parasitology **76**: 259-264.

Stwora-Wojczyk, M., J. Kissinger, S. Spitalnik, and B. Wojczyk. 2004. Identification and analysis of a family of *Toxoplasma gondii* UDP-N-acetyl-D-galactosamine:polypeptide N-acetylgalactosaminyltransferases. International Journal for Parasitology **In press**.

Thomason, P., D. Traynor, and R. Kay. 1999. Taking the plunge. Terminal differentiation in *Dictyostelium*. Trends in Genetics **15**: 15-19.

Tomavo, S., and J.C. Boothroyd. 1995. Interconnection between organellar functions, development and drug resistance in the protozoan parasite, *Toxoplasma gondii*. International Journal of Parasitology **25**: 1293-1299.

Weiss, L.M., and K. Kim. 2000. The development and biology of bradyzoites of *Toxoplasma gondii*. Frontiers in Bioscience **5**: D391-405.

_____, D. Laplace, P. Takvorian, H.B. Tanowitz, and M. Wittner. 1996. The association of the stress response and *Toxoplasma gondii* bradyzoite development. Journal of Eukaryotic Microbiology **43**: 120S.

_____, _____, _____, A. Cali, H.B. Tanowitz, and M. Wittner. 1994. Development of bradyzoites of *Toxoplasma gondii* in vitro. Journal of Eukaryotic Microbiology **41**: 18S.

_____, _____, _____, _____, A. Cali, and M. Wittner. 1995. A cell culture system for study of the development of *Toxoplasma gondii* bradyzoites. Journal of Eukaryotic Microbiology **42**: 150-157.

_____, _____, H.B. Tanowitz, and M. Wittner. 1992. Identification of *Toxoplasma gondii* bradyzoite-specific monoclonal antibodies. Journal of Infectious Diseases **166**: 213-215.

_____, Y.F. Ma, P.M. Takvorian, H.B. Tanowitz, and M. Wittner. 1998. Bradyzoite development in *Toxoplasma gondii* and the hsp70 stress response. Infection and Immunity **66**: 3295-3302.

Wojczyk, B.S., F.K. Hagen, B. Striepen, H.C. Hang, C.R. Bertozzi, D.S. Roos, and S.L.

Spitalnik. 2003. cDNA Cloning and Expression of UDP-N-acetyl-D-galactosamine: Polypeptide N-Acetylgalactosaminyltransferase T1 from *Toxoplasma gondii*. Molecular and Biochemical Parasitology **131**: 93-107.

Wong, S.Y., and J.S. Remington. 1993. Biology of *Toxoplasma gondii* [editorial]. AIDS 7: 299-316.

Yahiaoui, B., F. Dzierszinski, A. Bernigaud, C. Slomianny, D. Camus, and S. Tomavo. 1999. Isolation and characterization of a subtractive library enriched for developmentally regulated transcripts expressed during encystation of *Toxoplasma gondii*. Molecular and Biochemical Parasitology **99**: 223-235.

Yang, S., and S.F. Parmley. 1995. A bradyzoite stage-specifically expressed gene of *Toxoplasma gondii* encodes a polypeptide homologous to lactate dehydrogenase. Molecular and Biochemical Parasitology **73**: 291-294.

_____, and _____. 1997. *Toxoplasma gondii* expresses two distinct lactate dehydrogenase homologous genes during its life cycle in intermediate hosts. Gene **184**: 1-12.

Zhang, Y.W., S.K. Halonen, Y.F. Ma, M. Wittner, and L.M. Weiss. 2001. Initial characterization of CST1, a *Toxoplasma gondii* cyst wall glycoprotein. Infection and Immunity **69**: 501-507.

_____, K. Kim, Y.F. Ma, M. Wittner, H.B. Tanowitz, and L.M. Weiss. 1999. Disruption of the *Toxoplasma gondii* bradyzoite-specific gene BAG1 decreases in vivo cyst formation. Molecular Microbiology **31**: 691-701.

_____, and J.E. Smith. 1995. *Toxoplasma gondii*: reactivity of murine sera against tachyzoite and cyst antigens via FAST-ELISA. International Journal for Parasitology **25**: 637-640.

# *SARCOCYSTIS* OF HUMANS

Ronald Fayer

*United States Department of Agriculture, Agricultural Research Service, Beltsville, Maryland 2705, USA*

## ABSTRACT
Within the phylum Apicomplexa, protozoan parasites of the genus *Sarcocystis* require two hosts in a prey-predator relationship to complete their life cycle. Asexual stages develop only in intermediate hosts when, after ingesting sporocysts in food or water contaminated with animal feces, they become infected with stages that develop in blood vessels and muscles. Sexual stages develop only in definitive hosts when, after eating sarcocysts in animal muscles (meat), they become infected with stages that develop in the gastrointestinal tract. Humans can serve as both intermediate and definitive hosts for different species of *Sarcocystis*. Detection methods, means of identification, prevalence of infection, transmission routes, clinical signs, diagnostics, and methods of treatment and prevention are described in this chapter.

**Key words:** Sarcocystis, sarcocystosis, sarcosporidiosis, human, food animals, meat, life cycle, epidemiology, detection, treatment, control

## INTRODUCTION
*Sarcocystis* is neither a recently discovered nor newly emerged pathogen. It was first reported in 1843 by Miescher in striated muscles of a house mouse in Switzerland. The white threadlike cysts, unnamed for the next 20 years, were referred to as Meischer's tubules. Similar structures, reported in pig muscle in 1865 became the basis for naming the type species *Sarcocystis meischeriana* in 1899 (Dubey et al., 1989). For over 100 years after the discovery of these cysts, species were named simply after finding them in animal and human muscles. For many years scientists debated whether *Sarcocystis* species were protozoa or fungi. Electron microscopy demonstrated that the crescent-shaped bodies (bradyzoites) within the cysts had morphological features found in apicomplexan protozoa (Senaud, 1967).

Biological, biochemical, and molecular characteristics had not yet been used to further characterize organisms within the genus. The life cycle remained enigmatic until bradyzoites, liberated from *Sarcocystis* cysts in grackle (*Quiscalus quiscula*) muscles, developed within cultured mammalian cells into sexual stages and oocysts typically found in coccidian life cycles (Fayer,1970; Fayer,1972). Further clarification came with a study of *Sarcocystis fusiformis*, the name applied to all cysts found in cattle muscle. In a series of articles it was reported that *S. fusiformis* was actually 3 species, each with a morphologically distinct cyst and a different definitive host-dogs, cats, and humans (Rommel et al., 1972; Heydorn and Rommel, 1972; Rommel and Heydorn, 1972). These findings provided the basis for solving life cycles and for detecting, identifying, and naming species of *Sarcocystis*.

## LIFE CYCLES

All species of *Sarcocystis* are obligate intracellular parasites with a two-host life cycle described as a prey- predator, herbivore- carnivore, or intermediate-definitive host relationship. In the intermediate host, sarcocysts (sarco = muscle) are found in virtually all striated muscles of the body including the tongue, esophagus, and diaphragm, as well as cardiac muscle. To a lesser extent, sarcocysts have been found in neural tissue such as spinal cord and brain. Mature sarcocysts of each species vary in size from microscopic to macroscopic and develop morphologically unique sarcocyst walls. The wall surrounds nearly spherical metrocytes (mother cells) that give rise to large numbers of crescent-shaped bodies called bradyzoites (brady = slow, zoite = small animal). Bradyzoites appear to remain dormant until the sarcocyst is eaten by the definitive host. Following digestion of the sarcocyst wall, liberated bradyzoites become active and enter cells in the intestinal lamina propria. Each bradyzoite develops into a sexual stage: a microgametocyte from which several sperm-like microgametes emerge, or a macrogamete resembling an ovum. After fertilization, wall-forming bodies within the macrogamete (now a zygote) coalesce and the cytoplasm undergoes a series of changes (sporogony) until the zygote transforms into a mature oocyst containing two sporocysts. Oocysts released into the intestinal lumen pass from the body in the feces. The delicate oocyst wall often breaks, releasing individual sporocysts that can be observed microscopically (Figure 1). Most sporocysts measure about 10 x 15 μm, and contain 4 sporozoites and a discrete granular residual body. Sporocysts are immediately infectious for susceptible intermediate hosts. When sporocysts are ingested they pass through the stomach to the small intestine where they excyst, releasing the sporozoites. Sporozoites penetrate the gut epithelium and enter endothelial cells in mesenteric lymph node arteries where they undergo asexual multiplication (called schizogony or merogony) producing numerous merozoites (cells morphologically similar to sporozoites). In subsequent generations, merozoites pass downstream to arterioles, capillaries, venules and veins throughout the body (often most numerous in the kidney glomeruli) where they also develop in endothelial cells. Merozoites have been found in peripheral blood smears around the same time as mature second generation schizonts, some appear extracellular while others are in unidentified mononuclear cells. The number of asexual generations and their primary sites of development varies with each species of *Sarcocystis*. Merozoites from the terminal generation of schizogony enter muscle cells, round up to form metrocytes, and initiate sarcocyst formation. While the sarcocyst wall develops, metrocytes multiply and give rise to bradyzoites. Maturation varies with the species and takes 2 months or more until bradyzoites form and sarcocysts become infectious for the definitive host. Sarcocysts may persist for months or years.

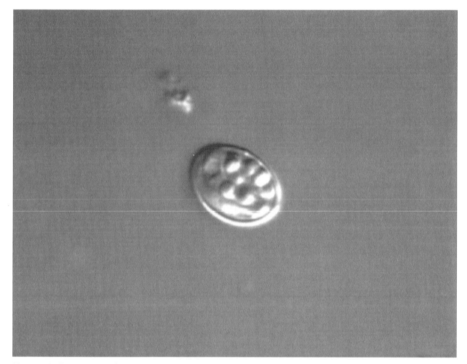

*Figure 1. Sporocyst of Sarcocystis species. The sporocyst is 11 μm in length.*

Humans (and some nonhuman primates) become infected after ingesting undercooked or raw meat containing mature cysts. Humans serve as definitive hosts for 2 named species of *Sarcocystis*: *Sarcocystis hominis* acquired from eating sarcocysts in the muscles of cattle, and *Sarcocystis suihominis* acquired from eating sarcocysts in the muscles of swine. Chimpanzees and macaques can also serve as definitive hosts for *S. suihominis* (Fayer et al., 1979).

Humans can also serve as intermediate hosts with sarcocysts developing in striated muscles. In such cases humans apparently are accidental hosts (see Detection, Identification, and Host Specificity).

## DETECTION, IDENTIFICATION, AND HOST SPECIFICITY

Oocysts with two sporocysts or, more frequently, individual sporocysts in human feces are diagnostic for intestinal infection. When shed, sporocysts contain 4 sporozoites and a granular residual body. Sporocysts of *S. hominis* (avg = 9.3 x 14.7 μm) and *S. suihominis* (avg = 10.5 x 13.5 μm) cannot be distinguished from one another or from sporocysts shed by other hosts.

Sarcocysts of *S. hominis* are microscopic in the muscles of cattle, whereas those of *S. suihominis* are macroscopic in muscles of swine. Sarcocysts can be examined in the muscles of these intermediate hosts by microscopy of histological sections. Sarcocysts have distinctive physical features that aid in identification of species such as overall size, presence or absence of septae, and ultrastructural morphology of the wall. However, within each species these features vary depending on the age of the sarcocyst, the host cell type, and methods of fixation. As many as 24 wall types have been identified for 62 species, e.g. the walls of *S. hominis* and *S. suihominis* sarcocysts are both type 10 (Dubey et al., 1989). The wall of *S. hominis* is up to 6 μm thick and radially striated from villar protrusions up to 7 μm long; bradyzoites, arranged in packets, are 7 to 9 μm long. The wall of *S. suihominis* is 4 to 9 μm thick with villar protrusions up to 13 μm long; bradyzoites are 15 μm long.

Molecular methods have been used to differentiate species of *Sarcocystis*. *S. hirsuta*, *S. hominis* and *S. cruzi* from cattle and bison were identified by directly sequencing polymerase chain reaction (PCR) products of 18s ribosomal DNA (Fischer and Odening, 1998). Also based on 18s rRNA gene sequences, *Sarcocystis* from a water buffalo was found nearly identical (0.1% difference) to *S. hominis*, indicating that multiple ruminant species serve as intermediate hosts and potential sources of human infection for this parasite (Yang et al., 2001).

Like most other species of *Sarcocystis*, *S. hominis* and *S. suihominis* are genetically programmed to complete their life cycles in specific intermediate hosts or within closely related host species. For example, sporocysts of *S. hominis* infect cattle and not pigs whereas those of *S. suihominis* infect pigs but not cattle. However, *S. cruzi* from dogs can infect cattle (*Bos taurus*), water buffalo (*Bubalus bubalis*), and bison (*Bison bison*). Similarly, humans appear to serve as intermediate hosts for several unidentified species of *Sarcocystis* perhaps acquiring infections by ingesting sporocysts excreted by predators of nonhuman primates. To determine the species responsible for acute fulminant infection in a captive-born rhesus monkey with immature and mature schizonts in endothelial cells throughout the body and mature sarcocysts in muscle, 18s rRNA gene sequences were examined (Lane et al., 1998). Homology of 95-96% was found with several species of *Sarcocystis* but complete identity was lacking. That report indicates the susceptibility of a primate to life-threatening infection with unknown species of *Sarcocystis* even in the apparent absence of a typical definitive host.

## PREVALENCE

Prevalence data for all *Sarcocystis* infections must be interpreted carefully. They often reflect the findings of physicians, public health workers, veterinarians, or scientists with specific interests, or just those who took the time to publish their findings. Much data are unreported and no truly large scale population surveys have been conducted.

Based on examination of tissues from abattoirs, a high percentage of cattle worldwide have been found infected with *Sarcocystis*, *S. cruzi* (infectious from cattle to canines) being the most prevalent and easiest to identify histologically (van Knapen et al., 1987). Most studies have not attempted to differentiate species, e.g., Ono and Ohsumi (1999) reported on the prevalence of *Sarcocystis* in Japanese and imported beef but did not identify species. Because *S. hominis* (infectious from cattle to humans) and *S. hirsuta* (infectious from cattle to felines) are difficult to distinguish except by electron microscopy, some prevalence data may be erroneous. *S. hominis* has not been detected in the United States, whereas up to 63% of cattle in Germany have been reported to be infected. *S. suihominis* was found more prevalent in Germany than Austria but little information is available from other countries. In Brazil all 50 samples of raw kibbe from 25 Arabian restaurants in Sao Paulo contained sarcocysts (Pena et al., 2001). Based on wall structure, *S. hominis*, *S. hirsuta*, and *S. cruzi* were found in 94, 70, and 92% of the samples, respectively. The overall prevalence of *Sarcocystis* in pigs appears low; 3%-36% worldwide. *S. suihominis* and *S. hominis* have been reported in slaughtered pigs and cattle raised in Japan (Saito et al., 1998, 1999).

Based on limited, somewhat focal surveys, intestinal sarcocystosis in humans was reported as more prevalent in Europe than any other continent (Dubey et al., 1989). A prevalence of 10.4% of fecal specimens was found in children in Poland and 7.3% of samples from Germany. Of 1228 apprentices from the Hanoi-Haiphong area of Viet Nam who worked in Central Slovakia in 1987-1989, 14 (1.1%) had sporocysts of *Sarcocystis* spp. detected in their stool (Straka et al., 1991). Kibbe positive for *S. hominis* was fed to 7 human volunteers; 6 excreted sporocysts, 2 developed diarrhea (Pena et al., 2001). After eating raw beef, a patient in Spain with abdominal discomfort, loose stools, and sporulated oocysts in the feces, was diagnosed with *S. hominis* (Clavel et al., 2001). In Tibet, where *Sarcocystis* was detected in 42.9% of beef specimens examined from the marketplace, *S. hominis* and *S. suihominis* were found in stools from 21.8% and 0 to 7% of 926 persons, respectively (Yu, 1991).

Muscular sarcocystosis in humans is rarely reported. Of approximately 46 cases reported by 1990 (Dubey et al., 1989) most were from Asia and Southeast Asia. In India, sarcocysts were found in biopsies from 4 cases with lumps or pain in limbs (Mehrotra et al., 1996). Additional cases, supported by histologic evidence, include 13 from Southeast Asia, 11 from India, 2 from Malaysians of Indian origin, 5 from Central and South America, 4 each from Africa, Europe, and the United States, one from China, and 2 of undetermined origin. An outbreak in 7 persons of a 15 member military team in Malaysia is the largest group occurrence on record (Arness et al., 1999).

## TRANSMISSION AND RESERVOIR HOSTS

Humans acquire gastrointestinal sarcocystosis only by ingesting raw or undercooked meat from cattle or pigs harboring mature cysts of *S. hominis or S. suihominis*, respectively. Other species of *Sarcocystis* such as *S. cruzi* in cattle muscle are not known to infect humans. Based on cases in Thailand of intestinal lesions in persons having eaten undercooked meat from *Bos indicus* cattle (Bunyaratvej et al., 1982) and possibly other animals (authors comment), there may be other unnamed species of *Sarcocystis* infectious for humans. Other meat animals that harbor *Sarcocystis* with unknown life cycles include camels, llamas, water buffalo, yaks, and species of pigs other than the domesticated *Sus scrofa*. Many reptiles, birds, and mammals that harbor sarcocysts serve as food animals in various parts of the world.

*Sarcocystis* causing muscular infection has been reported for less than 100 humans. In such cases, humans harbor the sarcocyst stage and therefore serve as the intermediate host. Based on all other *Sarcocystis* life cycles, infected human tissues must be eaten by a carnivore to complete the life cycle. Because there is no known predatory or scavenging cycle in nature in which human tissues are eaten regularly by carnivores, humans most likely become infected accidentally by ingestion of food or water contaminated with feces from a carnivore that participates in a primate-carnivore cycle involving an unknown species of *Sarcocystis*.

## CLINICAL SIGNS

Human volunteers that ate raw beef containing *S. hominis* became infected and shed oocysts in their feces (Aryeetey and Piekarski, 1976; Rommel and Heydorn, 1972). One person became ill. Symptoms that appeared 3 to 6 hr after eating the beef included nausea, stomachache, and diarrhea. Other volunteers ate raw pork containing *S. suihominis*, became infected, shed oocysts, and had dramatic signs that appeared 6 to 48 hr after eating the pork (Rommel and Heydorn, 1972; Heydorn, 1977). These signs included bloat, nausea, loss of appetite, stomachache, vomiting, diarrhea, difficulty breathing, and rapid pulse. In contrast, volunteers who ate well-cooked meat from the same pigs had no clinical signs (Heydorn, 1977). Six persons in Thailand who reportedly ate beef from zebu cattle uncooked in hot-chili dishes (but who may have eaten a variety of other animal products; authors note) developed segmental necrotizing enteritis with sexual stages attributed to *Sarcocystis* and Gram-positive bacilli (Bunyaratvej et al., 1982).

About 2 weeks after cattle and sheep are fed sporocysts of *Sarcocystis* collected from the feces of dogs fed raw beef or lamb, respectively, the first asexual generation develops in arteries with little more clinical signs than a slightly elevated body temperature for a day. When the second asexual generation matures approximately 4 weeks after ingestion of sporocysts a severe vasculitis may result. Depending on the number of parasites ingested, acute infection may be characterized by massive perivascular infiltration of mononuclear cells and multi-organ petechial hemorrhaging associated with weakness, fever, abortion in pregnant animals, and sometimes death (Johnson, et al., 1974, 1975; Proctor et al., 1976; Leek et al., 1977). Chronic infection in livestock has been characterized by inappetence, weight loss, loss of hair or wool breakage, poor or stunted growth, muscle atrophy, lethargy, and weakness. Histologically, such animals have widespread myositis, including glossitis and inflammation of cardiac muscle. As late as 1980, only 8 of 19 human cases of *Sarcocystis* were reported with evident vasculitis and/or myositis (reviewed by McLeod et al., 1980). They also reported the case of a 40-year-old male in California who, 4 years earlier, had traveled extensively in Asia. For over a year, intermittently, he had lesions on his arms, legs, soles of his feet, and trunk beginning as subcutaneous masses associated with overlying erythema and subsiding spontaneously about 2 weeks later. Biopsies of nodules revealed vasculitis in capillaries, venules, and arterioles consisting primarily of perivascular lymphocytes and/ or neutrophils. There were scattered clusters of microorganisms surrounded by thin-walled cyst in striated muscle fibers without significant myositis.    It was not unequivocally determined if *Sarcocystis* was responsible for the vasculitis, but the occurrence of both in other humans and animals appears more than coincidental. Because the patient felt well except for the nodules, no treatment was attempted.

Seven of 15 U.S. military personnel in Malaysia developed acute fever, myalgias, bronchospasm, pruritic rashes, lymphadenopathy,and subcutaneous nodules associated with eosinophilia, elevated erythtocyte sedimentation rate, and elevated creatinine kinase levels (Arness et al., 1999). Sarcocysts were found in biopsies from the index case whose symptoms were ameliorated with albendazole but which lasted for more than 5 years. Symptoms in 5 others were mild to moderate and self-limited, and one person with abnormal blood chemistries was asymptomatic.

## DIAGNOSIS

Presumptive diagnosis of human intestinal sarcocystosis is based on clinical signs and a history of having eaten raw or undercooked meat. Definitive diagnosis requires identification of sporocysts in feces. Sporocysts of *S. hominis* are shed 14 to 18 days after ingesting beef and those of *S. suihominis* are shed 11 to 13 days after ingestion of pork. Sporocysts are seen by bright-field microscopy at the uppermost optical plane of a sucrose-fecal flotation wet mount just beneath the cover slip. Because sporocysts overlap in size and shape, species cannot be distinguished from one another when both are present in the same specimen. (For average sizes see Detection, Identification and Host Specificity section).

*Sarcocystis* sarcocysts in muscle biopsies can be identified by microscopic examination of hematoxylin and eosin stained histologic sections. Inflammatory cells are sometimes, but not always, associated with infection.

*Sarcocystis* can be detected in meat by direct observation of macroscopic sarcocysts or microscopic examination of histologic sections. Larger quantities of meat also can be examined by grinding meat and then artificially digesting it in a mixture of pepsin and hydrochloric acid. After centrifugation, the pellet is examined microscopically for the presence of bradyzoites. For many years eosinophilic myositis, observed macroscopically as a blue-green tint on the surface of a fresh carcass, was thought to be associated with *Sarcocystis* infection because sarcocysts were usually found in muscles in the carcass. However, many cattle harbor sarcocysts without an inflammatory cell or eosinophil response to sarcocysts, making the linkage unsubstantiated. Numerous experimental infections of livestock also failed to result in eosinophilic myositis (Proctor et al., 1976; Stahlheim et al., 1976).

## TREATMENT

There is no known treatment for intestinal sarcocystosis. It is self-limiting and usually of short duration. The effectiveness of treatment with cotrimoxazole (Croft, 1994) or furazolidone (Mensa et al., 1999) has not been clearly demonstrated. For 6 persons with segmental necrotizing enteritis diagnosed as *Sarcocystis* and with Gram-positive bacilli, drastic treatment consisted of surgical resection of the small intestine followed by antibiotics (Bunyaratvej et al., 1982).

Treatment for myositis, vasculitis, or related lesions in humans has not been developed. Guidance may come from animal studies in which prophylaxis was achieved under experimental conditions (see Prevention) but data are lacking for treatment of established infections. Whether immunosuppressive therapy to treat vasculitis or myositis might enable greater development of the parasite is unknown. The usefulness of pyrimethamine, which is active against related coccidia such as *Toxoplasma*, is also unknown.

## PREVENTION

To prevent human intestinal sarcocystosis, meat should be cooked thoroughly or frozen to kill bradyzoites in cysts. *S. meisheriana* sarcocysts in pig thigh muscles were rendered noninfectious for puppies after cooking small pieces of meat at 60, 70, and 100 C for 20, 15 and 5 minutes, respectively (Saleque et al., 1990). Likewise exposure to -4 and -20 C for 48 and 24 hours, respectively, rendered meat noninfectious (Saleque et al., 1990).

Under experimental conditions, chemoprophylaxis using the anticoccidial drugs amprolium and salinomycin was effective in preventing severe illness and death in calves and lambs (Fayer and Johnson, 1975; Leek and Fayer, 1980 and 1983). There has been no occasion to attempt prophylaxis in humans.

To interrupt the life cycle and prevent infection of food animals they must be prevented from ingesting the sporocyst stage from the feces of carnivores, including humans. That means that fecally-contaminated water, feed, and bedding must not be present in areas where livestock are raised. Where such preventative measures cannot be followed and meat might be harboring cysts, it must be thoroughly frozen for two or more days, or thoroughly cooked to kill infectious bradyzoites. The aforementioned measures will prevent development of intestinal stages where humans might serve as definitive hosts. Where humans might serve as intermediate hosts with stages that develop in blood vessels and muscles, the ingestion of sporocysts must be prevented. The most likely source of such sporocysts would be water contaminated with feces from a carnivore or omnivore. Where contaminated water is suspected, boiling is the best method to ensure disinfection of all infectious agents.

## REFERENCES

Arness, M.K., J.D. Brown, J.P. Dubey, R.C. Neafie, and D.E. Granstrom. 1999. An outbreak of acute eosinophilic myositis due to human *Sarcocystis* parasitism. American Journal of Tropical Medicine and Hygiene **61**: 548-553.

Aryeetey, M.E., and G. Piekarski. 1976. Serologische *Sarcocystis*-studien an Menschen und Ratten. Zeitschrift fur Parasitenkunde **50**: 109-124.

Bunyaratvej, S., P. Bunyawongwiroj, and P. Nitiyanant. 1982. Human intestinal sarcosporidiosis: report of six cases. American Journal of Tropical Medicine and Hygiene **31**: 36-41.

Clavel, A., O. Doiz, M. Varea, S. Morales, F.J. Castillo, M.C. Rubio, and R. Gomez-Lus. 2001. Molestias abdominales y heces blandas en consumidor habitual de carne de vacuno poco cocinada. Enfermedades Infecciosas Microbiologia Clinica **19**: 29-30.

Croft, J.C. 1994. Nonamebic protozoal enteridities. 1994. *In* Infectious Processes, 5th edition. P.D. Hoeprich,, Jordan, M.C. and A.R. Ronald (eds.). Philadelphia, Lippincott 769-774.

Dubey, J.P., C.A. Speer, and R. Fayer. 1989. Sarcocystis of Animals and Man. Boca Raton, Florida: CRC Press, 215 p.

Fayer, R. 1970. *Sarcocystis*: development in cultured avian and mammalian cells. Science **168**: 1104-1105.

_____, 1972. Gametogony of *Sarcocystis* sp. in cell culture. Science **175**: 65-67.

_____, R., A. O. Heydorn, A.J. Johnson, and R.G. Leek.1979. Transmission of *Sarcocystis suihominis* from humans to swine to nonhuman primates (*Pan troglodytes, Macaca mulatta, Macaca irus*). Zeitschrift fur Parasitenkunde **59**: 15-20.

_____, and A. J. Johnson. 1975. Effect of amprolium on acute sarcocystosis in experimentally infected calves. Joutnal of Parasitology **61**: 932-936

Fischer, S. and K. Odening.1998. Characterization of bovine *Sarcocystis* species by analysis of their 18S ribosomal DNA sequences. Journal of Parasitology **84**: 50-54.

Heydorn, A. O. 1977. Sarkosporidien enfiziertes Fleisch als mogliche Krankheitsurache fur den Menschen. Archives fur Lebensmittelhygiene **28**: 27-31.

_____, and M. Rommel. 1972. Beitrage zum Lebenszyklus der Sarkosporidien. II. Hund und Katze als Ubertrager der Sarkosporidien des Rindes. Berlin Muenchen Tieraerztliche Wochenschrift **85**: 121-123.

Johnson, A.J., R. Fayer, and P.K. Hildebrandt. 1974. The pathology of experimental sarcosporidiosis in the bovine. Laboratory Investigation **30**: 377-378.

_____, P.K. Hildebrandt, and R. Fayer. 1975. Experimentally induced *Sarcocystis* infection in calves: Pathology. American Journal of Veterinary Research **3**: 995-999.

Lane, J.H., K.G. Mansfield, L.R. Jackso,, R.W. Diters, K.C. Lin, J.J. MacKey, and V.G. Sassevelle. 1998. Acute fulminant sarcocystosis in a captive-born rhesus macaque. Veterinary Pathology **35**: 499-505.

Leek, R. G. and R. Fayer. 1980. Amprolium for prophylaxis of ovine *Sarcocystis*. Journal of Parasitology **66**: 100-106.

_____, , and _____.1983. Experimental *Sarcocystis ovicanis* infection in lambs: Salinomycin chemoprophylaxis and protective immunity. Journal of Parasitology **69**: 271-276.

_____, _____, and A. J. Johnson. 1977. Sheep experimentally infected with *Sarcocystis* from dogs. Disease in young lambs. Journal of Parasitology **63**: 642-650.

McLeod, R., Hirabayashi, R.N., Rothman, W. and Remington, J.R. 1980. Necrotizing vasculitis and *Sarcocystis*: a cause and effect relationship? Southern Medical Journal **73**: 1380-1383.

Mehrotra, R., D. Bisht, P.A. Singh, S.C. Gupta, and R.K. Gupta. 1996. Diagnosis of human sarcocystis infection from biopsies of the skeletal muscle. Pathology **28**: 281-282.

Mensa, J., J.M. Gatell, Jimenez de Anta, and G. Prats.1999. Guia e terapeutica antimicrobiana, 9th edition. Barcelona, Masson, S.A., 219.

Ono, M. and Ohsumi, T. 1999. Prevalence of *Sarcocystis* spp. cysts in Japanese and imported beef (Loin: Musculus longissimus). Parasitology International **48**: 91-94.

Pena, H. F., S. Ogassawara, and I.L. Sinhorini. 2001. Occurrence of Cattle *Sarcocystis* species in raw kibbe from Arabian food establishments in the city of Sao Paolo, Brazil, and experimental transmission to humans. Journal of Parasitology **87**: 1459-1465.

Proctor, S.J., D. Barnett, O. H. V. Stalheim, and R. Fayer. 1976. Pathology of *Sarcocystis fusiformis* in cattle. 19th Annual Proceedings of Veterinary Laboratory Diagnosticians 329-336.

Rommel, M. and A. O. Heydorn. 1972. Beitrage zum Lebenszyklus der Sarkosporidien. III. *Isospora hominis* (Railiet und Lucet, 1891) Wenyon, 1923, eine Dauerform des Sarkosporidien des Rindes und des Schweins. Berlin Muenchen Tieraerztliche Wochenschrift **85**: 143-145.

_____, _____, and F. Gruber. 1972. Beitrage zum Lebenszyklus der Sarkosporidien. I. Die Sporozyste von S. tenella in den Fazes der Katze. Berlin Muenchen Tieraerztliche Wochenschrift **85**: 101-105.

Saito, M., Y. Shibata, A. Ohno, M. Kubo, K. Shimura, and H. Itagaki. 1998. *Sarcocystis suihominis* detected for the first time from pigs in Japanese Journal of Veterinary Medical Science **60**: 307-309.

_____, _____, M. Kubo, I. Sakakibara, A. Yamada, and H. Itagaki. 1999. First isolation of *Sarcocystis hominis* from cattle in Japan. Japanese Journal of Veterinary Medical Science **61**: 307-309.

Seleque, A., P.D. Juyal, and B.B. Bhatia. 1990. Effect of temperature on the infectivity of *Sarcocystis meischeriana* cysts in pork. Veterinary Parasitology **36**: 343-346.

Senaud, J. 1967. Contribution a l'etude des sarcosporidies et des toxoplasmes (Toxoplasmea). Protistologica **3**: 169-232.

Stalheim, O. H., S.J. Proctor, R. Fayer, and M. Lunde. 1976. Death and abortion in cows experimentally infected with *Sarcocystis* from dogs. 19th Annual Proceedings of Veterinary Laboratory Diagnosticians 317-327.

Straka, S., J. Skracikova, I. Konvit, M. Szilagyiova, and L. Michal. 1991. *Sarcocystis* species in Vietnamese workers. Cesk. Epidemiol. Mikrobiol. Immunol. **40**: 204-208.

Van Knapen, F., D. Bouwmann, and E. Greve. 1987. Study on the incidence of *Sarcocystis* spp. in Dutch cattle using various methods. Tijdschr Diergeneeskd. **112**: 1095-1100.

Yang, Z.Q., Y.X. Zuo, B. Ding, X.W. Chen, J. Luo, and Y.P. Zhang. 2001. Identification of *Sarcocystis hominis* -like (Protozoa:Sarcocystidae) cyst in water buffalo (*Bubalus bubalis*) based on 18s rRNA gene sequences. Journal of Parasitology **87**: 934-937.

Yu, S. 1991. Field survey of *Sarcocystis* infection in the Tibet autonomous region. Zhongguo Yi Xue Ke Xue Yuan Xue Bao **13**: 29-32. (In Chinese)

# ZOONOTIC MICROSPORIDIA FROM ANIMALS AND ARTHROPODS WITH A DISCUSSION OF HUMAN INFECTIONS

K.F. Snowden

*Texas A&M University, College of Veterinary Medicine, College Station, Texas, USA.*

## ABSTRACT

Members of the phylum Microspora are intracellular spore-forming eukaryotic organisms that have recently been recognized as opportunistic pathogens, especially in immunocompromised humans. Several new species have been described in human and animal hosts, but the biology and transmission of these organisms are not well known. Morphologic and molecular methods have been developed to clarify the species identity of these parasites and to understand the epidemiologic patterns of infection. The zoonotic potential of various microsporidia species and the role of animals as reservoirs for potential human exposure are topics of current research.

**Key words:** microsporidia, zoonosis, parasite, *Enterocytozoon*, *Encephalitozoon*

## INTRODUCTION

Microsporidia is a non-taxonomic term describing organisms belonging to the phylum Microspora. They are obligate intracellular spore-forming eukaryotes that were recognized more than 100 years ago as parasites of insects and fish. Subsequently, more than 1200 species have been identified in a wide variety of invertebrate and vertebrate hosts. Within the past 15 years a number of new microsporidial species have been recognized as "emerging opportunistic" pathogens, especially in immunocompromised humans. In a few cases, those parasites were previously known in animal hosts, but the biology and host range of most of these new parasites are not well known.

The 2 most frequently diagnosed microsporidian species in humans are *Enterocytozoon bieneusi* (Desportes et al., 1985) and *Encephalitozoon*

*(Septata) intestinalis* (Cali et al., 1993) as diarrhea-associated pathogens. A third microsporidian, *Encephalitozoon hellem*, is identified primarily as a cause of ocular or respiratory disease (Didier et al., 1991). These 3 organisms were first described in human infections, but an expanding range of non-human hosts have been identified as improved microscopic and molecular diagnostic methods have become widely available.

In contrast, a fourth microsporidial pathogen, *E. cuniculi* is widely recognized as a pathogen in rabbits and rodents (Snowden and Shadduck, 1999). This organism has also been reported in dog, fox and a few additional mammalian hosts (Snowden and Shadduck, 1999). The organism has also been identified in a small number of human cases (Fournier et al., 2000).

A number of additional microsporidial species have been identified in small numbers of human case reports. Some of these species were only characterized using morphological descriptions, while others have been more fully characterized by tissue culture, molecular analyses and/or in experimental animal infections. Little is known about infections with these parasites in non-human hosts with the exception of *Brachiola (Nosema) algerae*.

To date, no direct evidence of human-animal transmission for any microsporidial species has been reported, but based on an increasing body of evidence, a zoonotic potential exists. The role of animals as reservoir hosts or as direct or indirect sources of human exposure is likely, and further investigation is required for each of these parasites.

## *ENTEROCYTOZOON BIENEUSI*

The most frequently diagnosed species, *E. bieneusi,* was first described in 1985 as a enteric pathogen (Desportes et al., 1985). The organism has been identified on a worldwide basis, primarily in severely immunocompromised patients. A number of clinical studies have reported prevalence rates ranging from 1 to 50% depending on the study population, the clinical presentation, the HIV status and the diagnostic detection method employed (Deplazes et al., 2000). While originally associated with diarrhea and enteric disease in AIDS patients, the parasite has occasionally been identified in the biliary tract, in respiratory infections and in disseminated infections (Weber et al., 2000). *E. bieneusi* has also been identified in a small number of transplant patients and in HIV negative humans, especially in developing countries (Gainzarain et al., 1998; Ligoury et al., 2001; Sing et al., 2001).

Although *E. bieneusi* has been found primarily in humans, natural infections have been reported in several mammals and in one avian species. In a retrospective study, infections were detected in 3 species of simian-immunodeficiency-virus-infected macaque monkeys (*Macaca mulatta, M.*

*nemestrina, M. cyclopis*) (Mansfield et al., 1997). In a subsequent study at the same facility, 33.8% (18 of 53) SIV infected rhesus macaques (*M. mulatta*) and 16.7% (22 of 131) normal rhesus macaques were PCR positive for *Enterocytozoon* in fecal samples (Mansfield et al., 1998). In both reports, parasites localized in the biliary tree rather than in small intestinal mucosa. Domestic pigs (*Sus scrufa*) have been shown to be naturally infected with *E. bieneusi*. *Enterocytozoon*-like spores were first identified in fecal samples of swine in Switzerland (Deplazes et al., 1996a). In a subsequent study, analyses of fecal samples from 109 pigs in Switzerland showed a prevalence of 35% infection (Breitenmoser et al., 1999). In a larger study of swine in the northeastern USA, molecular analyses of various specimens showed that 32% of 202 pigs were positive for the parasite, suggesting a prevalence similar to that in the Swiss study (Buckholt et al., 2002).

   *E. bieneusi* infections have also been described in additional domestic animals including 2 reports in dogs, 2 reports in domestic cats, and single reports in domestic rabbits, in cattle, in a llama (reviewed by Dengjel et al., 2001) and in domestic chickens (Reetz et al., 2002). Additionally, the experimental infection of macaque monkeys (Tzipori et al., 1997) and gnotobiotic pigs (Kondova et al., 1998) with human isolates showed that strict host species barriers do not exist.

   Marked genetic variation of human and animal isolates of *E. bieneusi* has been described using molecular analyses. Comparisons of the internal transcribed spacer (ITS) region of the ribosomal RNA genes from eight human isolates showed three genotypes, designated A, B, and C (Rinder et al., 1997). In a larger study evaluating 78 human isolates, 4 genotypes were identified (Ligoury et al., 1998). More recently, 100 isolates were classified into 5 genotypes using similar methodology (Ligoury et al., 2001). These genotypes segregated into 2 groups of isolates from HIV positive and HIV negative patients. This distribution of genotypes may suggest differences in the epidemiology of infection of these patient populations and deserves further investigation.

   Molecular studies of animal isolates showed that the ITS sequences from 3 macaque isolates had 97% to 100% identity with isolates of human origin (Mansfield et al., 1997). Molecular analysis of 28 swine isolates showed 4 genotypes different from those described in human isolates (Breitenmoser et al., 1999). In an additional study of swine samples, several isolates were identical to previously reported human and swine genotypes, but 9 new genotypes were also identified (Buckholt et al., 2002). Two recent reports summarize data on the highly polymorphic ITS gene region of *E. bieneusi* by describing 23 (Dengjel et al., 2001) or 30 genotypes (Buckholt et al., 2002). Most of these genotypes segregate by host species on phylogenetic analysis (Dengjel et al., 2001), and the biological significance of the genetic diversity of *E. bieneusi* is still unknown.

In summary, *E. bieneusi* has been found primarily in humans and swine and in a few additional domestic animals. The high prevalence of natural infections in swine suggests that these animals may be under-recognized hosts for *E. bieneusi*. However, the genetic diversity indicated by ITS polymorphism argues against the role of swine as important reservoir hosts for human infection. Further research is needed to clarify the zoonotic potential of this organism and the role of animals as reservoirs for potential human exposure.

## *ENCEPHALITOZOON INTESTINALIS*

The second most common microsporidian identified in humans is *E. intestinalis*. This organism was first identified in small intestinal biopsy material from several AIDS patients in 1993 (Cali et al., 1993) and was renamed 2 years later following further molecular and immunological characterization (Hartskeerl et al., 1995). Most human infections are associated with chronic diarrhea and enteritis in AIDS patients, and a small number of extra-intestinal cases have been reported including biliary, respiratory and renal infections (Weber et al., 2000).

The most frequent descriptions of this organism are case reports, but a few epidemiologic studies show prevalences ranging from 0.9% to 7.3% in AIDS patients (Deplazes et al., 2000). *E. intestinalis* has also been associated with traveler's diarrhea, and infections have been reported in immunocompetent humans in a Mexican survey (Enriquez et al, 1998; Weber et al., 2000).

*E. intestinalis* has also been reported in a limited number of animal hosts. In one study, spores were identified in fecal samples from 2 pigs, 1 dog, 1 cow, 1 goat and 1 donkey using microscopic staining methods (Bornay-Llinares et al., 1998). In another study, 3 of 100 fecal samples from 43 mountain gorillas (*Gorilla gorilla beringei)* were positive for *E. intestinalis* based on PCR amplification of a portion of the SSU rRNA gene (Graczyk et al., 2002). Two of 62 samples from humans from that region were also positive for the parasite. The authors of that study suggested that based on the molecular similarity of ape and human isolates, anthropozoonotic transmission of *E. intestinalis* probably occurred. However, unlike other *Encephalitozoon* parasites, heterogeneity has not yet been reported in the SSU or ITS ribosomal gene regions of *E. intestinalis* (Liguory et al., 2000). Since only one genotype has been identified, the suggestion of human to ape transmission might be an over-interpretation of the limited data available. Evaluation of additional isolates from non-human hosts are needed to clarify the zoonotic potential of this microsporidian species.

# ENCEPHALITOZOON HELLEM

*E. hellem* was first differentiated from *E. cuniculi* in 1991 through biochemical and immunological comparisons of 3 parasite isolates from human eye lesions (Didier et al., 1991). This species has subsequently been identified as a cause of ocular and respiratory infections in AIDS patients, and occasionally causes disease in other organ systems or disseminates as a systemic infection (Franzen and Müller, 2001).

To date, *E. hellem* has not been reported in mammalian hosts other than man. However, an increasing number of reports describe *E. hellem* infections in avian hosts. Between 1975 and 1989 microsporidial infections were described in several species of lovebirds (*Agapornis* spp.) in at least 7 case reports (reviewed by Snowden et al., 2000). Microsporidia were also reported in other psittacine birds including budgerigars and Amazon parrots (Deplazes et al., 2000; Snowden et al., 2000). The organisms were sometimes described as *Encephalitozoon* sp. or *E. cuniculi*- like organisms, but species identity of those organisms was not established since *E. cuniculi* and *E. hellem* are morphologically indistinguishable at the light microscopic level, and *E. hellem* was not described until 1991.

In 1997, *E. hellem* was identified in a flock of budgerigars (*Melopsitticus undulatus*) based on histological findings and southern blotting analysis of the SSU rRNA gene (Black et al., 1997). Subsequently, morphologic and molecular sequencing of portions of the ribosomal genes have been used to identify *E. hellem* in the tissues of additional psittacine hosts including eclectus parrots (*Eclectus roratus*), a yellow-streaked lory (*Chalcopsitta scintillata*), a blue-fronted Amazon parrot (*Amazona aestiva*) (reviewed by Snowden et al., 2000) and an umbrella cockatoo (*Cacatua alba*) (D. Phalen, Texas A&M University, personal communication). Additionally, *E. hellem* has been identified in several non-psittacine avian hosts including an ostrich (*Struthio camelus*) (Snowden and Logan, 1999), 3 species of hummingbirds (Snowden et al., 2001) and Lady Gouldian finches (*Erythrura gouldiae*) (Carlisle et al., 2002). Thus, *E. hellem* has been identified in an expanding number of avian hosts.

To further explore microsporidial infection in avian hosts, an epidemiologic survey was conducted in 3 species of clinically normal lovebirds, where 25% of 198 birds from 8 flocks shed parasite spores (Barton et al., in press). Interestingly, there was a significant correlation between spore shedding and infection with the immunosuppressive avian circovirus, Psittacine Beak and Feather Disease Virus (PBFDV). The role of immunosuppression in avian microsporidial infections should be further explored.

No epidemiologic study has directly linked bird exposure with *E. hellem* infections in humans. Anecdotally, in several reports, patients with ocular microsporidiosis owned or were exposed to pet birds (Friedberg et al., 1990;

Orenstein et al., 1990; Yee et al., 1991).

Molecular analysis of the SSU rRNA and ITS gene showed heterogeneity among human isolates, and *E. hellem* was subsequently divided into 3 genotypes (Mathis et al., 1999). Genotype 1 was shown as the predominant strain in human isolates. In additional molecular studies, DNA sequencing of the polar tube protein gene from a number of *E. hellem* isolates supported the genotyping system established using the ITS region, and analysis of 24 isolates revealed 4 genotypes (Xiao et al., 2001a). A single avian isolate of *E. hellem* has been established in culture from the droppings of a peach-faced lovebird (*A. rosiecollis*) (Snowden et al., 2000). Molecular analysis of the SSU rRNA and ITS gene regions showed that this isolate was *E. hellem*, genotype 1.

In summary, there is increasing evidence that birds are under-recognized hosts for *E. hellem*. Naturally occurring infections have been identified in increasing numbers of avian species, and epidemiologic data suggest that unrecognized, asymptomatic infections are common in lovebirds. To date, genetic variation has been defined in human isolates of *E. hellem*, and limited molecular analyses of avian isolates of the parasite show no distinct differences from the human isolates.

## *ENCEPHALITOZOON CUNICULI*

In contrast to the 3 previous microsporidian organisms that are found primarily in human hosts, *E. cuniculi* is widely recognized as a parasite of rodents and rabbits. It was first described in rabbits with neurological disease in 1922 (Wright and Craighead, 1922). Natural infections have subsequently been described in a number of rodent hosts and rabbits including both laboratory animals and wild-caught animals (reviewed by Snowden and Shadduck, 1999; Deplazes et al., 2000). In some cases, the animals showed clinical abnormalities with neurologic, renal or systemic lesions; however, in most cases the infections were asymptomatic.

Historically, this parasite was problematic in laboratory animal colonies as unapparent infections that sometimes caused misinterpretation of experimental results (Snowden and Shadduck, 1999). Now the parasite has been virtually eliminated in laboratory rodent colonies used in biomedical research.

*E. cuniculi* has been identified in a number of other mammalian hosts. Natural and experimental infections have been documented in domestic dogs in the USA, Europe and Africa, and in farmed blue fox (*Alopex lagopus*) in Scandinavia (reviewed by Snowden and Shadduck, 1999). A variety of additional domestic and wild animal hosts have been identified primarily as case reports in single or small groups of animals. These include domestic cats, squirrel monkeys (*Saimiri sciureus*), a horse, a goat, and several wild carnivores including wild dogs (*Lycaon pictus*), meerkats

(*Suricata suricatta*), clouded leopards (*Neofelis nebulosa*), Siberian polecats (*Mustela eversmanii satunini*), mink (*Mustela vison*) and red fox (*Vulpes vulpes*) (reviewed by Snowden and Shadduck, 1999; Deplazes et al., 2000). Generally, clinical presentations included neurologic, renal or systemic disease in these cases, often associated with reproductive failures or neonatal deaths.

Although *E. cuniculi* is typically viewed as a parasite of rabbits and rodents, at least 12 infections have been reported in human case studies. Most of these patients were immunodeficient AIDS patients with neurological or multi-organ disseminated infections (Fournier et al., 2000). Localized infections have also been reported, and at least one infection has been documented in an HIV negative patient (Weber et al., 2000).

Direct transmission between animals and humans has not been documented, but several lines of evidence suggest that zoonotic transmission of *E. cuniculi* is likely. Molecular studies have shown multiple genotypes of *E. cuniculi* in isolates of animal and human origin. Molecular sequencing of the ITS region of ribosomal RNA genes showed three patterns, Type 1, 2 and 3 that segregated by their host origin as rabbit, mouse and dog strains respectively (Didier et al., 1995). Subsequent analyses of additional genes including the polar tube protein (PTP) and spore wall protein I (SWP01) supported this grouping of *E. cuniculi* into three genotypes (Xiao et al., 2001b). Human isolates have most frequently been characterized as genotype III, the dog strain (Didier et al., 1996; Snowden et al., 1999). Human *E. cuniculi* infections have also been genotyped as the rabbit strain, genotype I, in Europe where rabbits are commonly maintained as pets (Deplazes et al., 1996b; Rossi et al., 1998). The identification of the mouse strain II in foxes also supports the lack of host specificity for *E. cuniculi* (Mathis et al., 1996). Experimental infections have also shown that the various strains of *E. cuniculi* do not show strict host specificity. Rabbits have been infected with isolates representing all three genotypes (Mathis et al., 1997; Deplazes et al., 2000). Therefore, *E. cuniculi* appears to be a microsporidian infecting a variety of mammals, occasionally including humans.

## *BRACHIOLA (NOSEMA) ALGERAE*

Microsporidian organisms associated with invertebrate hosts have also been shown to cause occasional opportunistic infections in humans. The organism, *Nosema algerae* was first identified as a parasite of Anopheline mosquitoes (Vávra and Undeen, 1970). It was evaluated extensively in the 1970's and early 1980's as a potential biological control agent, and multiple reports described parasite infection in non-target hosts (Van Essen and Anthony, 1976). Experimental infections were established in mice confirming the possibility of mammalian infection (Undeen and Alger,

1976). Morphologic characteristics, thermophilic in vitro culture characteristics and molecular analyses of ribosomal RNA genes suggested that *N. algerae* was significantly different from the type-species of the genus, *Nosema bombycis*, a parasite of silkworms (*Bombyx mori*) (Baker et al., 1995). However, the taxonomic reclassification was not suggested until recently.

In 1999, the first human infection with *N. algerae* was identified from an eye lesion in an immunocompetent patient (Visvesvara et al., 1999). At approximately the same time, a new microsporidian species, *Brachiola vesicularum*, was described from an AIDS patient and established the type species for the new genus in the family Nosematidae (Cali et al., 1998). Morphological and thermophilic characteristics of this new species showed close similarities to *N. algerae* resulting in the renaming of *N. algerae* to *Brachiola algerae* (Lowman et al., 2000).

In summary, an insect microsporidian organism, *Brachiola (Nosema) algerae*, has shown zoonotic potential by naturally infecting a human. Medical parasitologists have suggested taxonomic reclassification of the parasite, while entomologists are not widely aware of that change. The appropriate name and taxonomic classification will be confirmed by future work and further evaluation using molecular methods.

## OTHER MICROSPORIDIA INFREQUENTLY REPORTED IN HUMANS

Several new microsporidian species have been named in recent years based on individual human cases. Examples of these organisms are *Vittaforma (Nosema) corneae*, *Brachiola vesicularum*, *Trachipleistophora hominis*, *T. anthropophthera* and others. The clinical presentations of these and several additional mammalian microsporidian species have been recently reviewed by Didier et al. (2000). To date, none of these parasite species have been associated with animal infections, and the origins, transmission patterns, or common hosts for these organisms have not yet been identified. Microsporidial parasites are truly an "emerging" infectious diseases in both human and veterinary fields that present exciting opportunities in biomedical research.

**REFERENCES**

Baker, M.D., C.R. Vossbrinck, E.S. Didier, J.V. Maddox, and J.A. Shadduck. 1995. Small subunit ribosomal DNA phylogeny of various microsporidia with emphasis on AIDS related forms. Journal of Eukaryotic Microbiology **42**: 564-570.

Barton, C.E., D.N. Phalen, and K.F. Snowden. 2003 in press. Prevalence of Microsporan spores shed by asymptomatic lovebirds: Evidence for a potential emerging zoonosis. Journal of Avian Medicine and Surgery.

Black, S.S., L.A. Steinohrt, D.C. Bertucci, L.B. Rogers, and E.S. Didier. 1997. *Encephalitozoon hellem* in Budgerigars (*Melopsittacus unodulatus*). Veterinary Pathology **34**: 189-198.

Bornay-Llinares, F.J., A.J. Da Silva, H. Moura, D.A. Schwartz, G.S. Visvesvara, N.J. Pieniazek, A. Cruz-Lopez, P. Hernandez-Jauregui, J. Guerrero, and F.J. Enriquez. 1998. Immunologic, microscopic, and molecular evidence of *Encephalitozoon intestinalis* (*Septata intestinalis*) infection in mammals other than humans. Journal of Infectious Diseases **178**: 820-826.

Breitenmoser, A.C., A. Mathis, E. Burgi, and R. Weber. 1999. High prevalence of *Enterocytozoon bieneusi* in swine with four genotypes that differ from those identified in humans. Parasitology **118**: 447-453.

Buckholt, M.A., J.H. Lee, and S. Tzipori. 2002. Prevalence of *Enterocytozoon bieneusi* in swine: an 18-month survey at the slaughterhouse in Massachusetts. Applied and Environmental Microbiology **68**: 2595-2599.

Cali, A., D.P. Kotler, and J.M. Orenstein. 1993. *Septata intestinalis* N.G., N.Sp., an intestinal microsporidian associated with chronic diarrhea and dissemination in AIDS patients. Journal of Eukaryotic Microbiology **40**: 101-112.

_____, P.M. Takvorian, S. Lewin, M. Rendel, C.S. Sian, M. Wittner, H.B. Tanowitz, E. Keohane, and L.M. Weiss. 1998. *Brachiola vesicularum*, N.G., N. Sp., a new microsporidium associated with AIDS and myositis. Journal of Eukaryotic Microbiology **45**: 240-251.

Carlisle, M.S., K. Snowden, J. Gill, M. Jones, P. O'Donoghue, and P. Prociv. 2002. Microsporidiosis in a Gouldian finch (*Erythrura [Chloebia] gouldiae*). Australian Veterinary Journal **80**: 41-44.

Dengjel, B., M. Zahler, W. Hermanns, K. Heinritzi, T. Spillman, A. Thomschke, and T. Loscher, R. Gothe, and H. Rinder. 2001. Zoonotic potential of *Enterocytozoon bieneusi*. Journal of Clinical Microbiology **39**: 4495-4499.

Deplazes, P., A. Mathis, C. Muller, and R. Weber. 1996a. Molecular epidemiology of *Encephalitozoon cuniculi* and first detection of *Enterocytozoon bieneusi* in faecal samples of pigs. Journal of Eukaryotic Microbiology **43**: 93S

_____, _____, R. Baumgartner, I. Tanner, and R. Weber. 1996b. Immunologic and molecular characteristics of *Encephalitozoon*-like microsporidia isolated from humans and rabbits indicate that *Encephalitozoon cuniculi* is a zoonotic parasite. Clinical Infectious Diseases **22**: 557-559.

_____, _____, and R. Weber. 2000. Epidemiology and zoonotic aspects of microsporidia of mammals and birds. Contributions in Microbiology. **6**: 236-260.

Desportes, I., Y. Lecharpentier , A. Galian, F. Bernard, B. Cochand-Priollet, A. Lavergne, P. Ravisse, and R. Modigliani. 1985. Occurrence of a new microsporidian: *Enterocytozoon bieneusi* n.g., n.sp., in the enterocytes of a human patient with AIDS. Journal of Protozoology **32**: 250-254.

Didier, E.S., P.J. Didier, D.N. Friedberg, S.M. Stenson, J.M. Orenstein, R.W. Yee, F.O. Tio, R.M. Davis, C. Vossbrinck, N. Millichamp, and J.A. Shadduck. 1991. Isolation and characterization of a new human microsporidian, *Encephalitozoon hellem* (n. sp.), from three AIDS patients with keratoconjunctivitis. Journal of Infectious Diseases **163**: 617-621.

_____, C.R. Vossbrinck, M.D. Baker, L.B. Rogers, D.C. Bertucci, and J.A. Shadduck. 1995. Identification and characterization of three *Encephalitozoon cuniculi* strains. Parasitology **111**: 411-421.

_____, G.S. Visvesvara, M.D. Baker, L.B. Rogers, D.C. Bertucci, M.A. Degroote, and C.R. Vossbrinck. 1996. A microsporidian isolated from an AIDS patient corresponds to *Encephalitozoon cuniculi* III, originally isolated from domestic dogs. Journal of Clinical Microbiology **34**: 2835-2837.

_____, P.J. Didier, K.F. Snowden, and J.A. Shadduck. 2000. Microsporidiosis in mammals. Microbes and Infection **2**: 709-720.

Enriquez, F.J., D. Taren, A. Cruz-Lopez, M. Muramoto, J.D. Palting, and P. Cruz. 1998. Prevalence of intestinal encephalitozoonosis in Mexico. Clinical Infectious Diseases **26**: 1227-1229.

Fournier, R, S., O. Liguory, C. Sarfati, F. David-Ouaknine, F. Derouin, J.M. Decazes, and J.M.

Molina. 2000. Disseminated infection due to *Encephalitozoon cuniculi* in a patient with AIDS: case report and review. HIV Medicine 1: 155-161.

Franzen, C., and A. Muller. 2001. Microsporidiosis: human diseases and diagnosis. Microbes and Infection 3: 389-400.

Friedberg, D.N., S.M. Stenson, J.M. Orenstein, P.M. Tierno, and N. C. Charles. 1990. Microsporidial keratoconjunctivitis in acquired immunodeficiency syndrome. Archives of Ophthalmology 108: 504-508.

Gainzarain, J. C., A. Canut, M. Lozano, A. Labora, F. Carrerass, S. Fenoy, R. Navajas, N.J. Pieniazek, A.J. Da Silva, and D.C. Aagula. 1998. Detection of *Enterocytozoon bieneusi* in two human immunodeficiency virus- negative patients with chronic diarrhea by polymerase chain reaction in duodenal biopsy specimens and review. Clinical Infectious Diseases 27: 394-398.

Graczyk, T. K., J. Boso-Nizeyi, A.J. Da Silva, I.N.S. Moura, N.J. Pieniazek, M.R. Cranfield, and A.H.D. Lindquist. 2002. A single genotype of *Encephalitozoon intestinalis* infects free-ranging gorillas and people sharing their habitats in Uganda. Parasitology Research 88: 926-931.

Hartskeerl, R. A., T. VanN Gool, A.R. Schuitema, E.S. Didier, and W.J. Terpstra. 1995. Genetic and immunological characterization of the microsporidian *Septata intestinalis* Cali, Kotler and Orenstein, 1993: reclassification to *Encephalitozoon intestinalis.*. Parasitology 110: 277-285.

Kondova, I., K. Mansfield, M.A. Buckholt, B. Stein, G. Widmer, A. Carville, A. Lackner, and S. Tzipori. 1998. Transmission and serial propagation of *Enterocytozoon bieneusi* from humans and rhesus macaques in gnotobiotic piglets. Infection and Immunity 66: 5515-5519.

Liguory, O., F. David, C. Sarfati, F. Derouin, and J. Molina. 1998. Determination of types of *Enterocytozoon bieneusi* strains isolated from patients with intestinal microsporidiosis. Journal of Clinical Microbiology 36: 1882-1885.

_____, S. Fournier, C. Sarfati, F. Derouin, and J. Molina. 2000. Genetic homology among thirteen *Encephalitozoon intestinalis* isolates obtained from human immunodeficiency virus-infected patients with intestinal microsporidiosis. Journal of Clinical Microbiology 38: 2389-2391.

_____, C. Sarfati, F. Derouin, and J. Molina. 2001. Evidence of different *Enterocytozoon bieneusi* genotypes in patients with and without human immunodeficiency virus infection. Journal of Clinical Microbiology 39: 2672-2674.

Lowman, P.M., P.M. Takvorian, and A. Cali. 2000. The effects of elevated temperatures and various time-temperature combinations on the development of *Brachiola (Nosema) algerae* N. Comb. in mammalian cell culture. Journal of Eukaryotic Microbiology 47: 221-234.

Mansfield, K.G., A. Carville D. Shvetz, J. Mackey, S. Tzipori, and A.A. Lackner. 1997. Identification of an *Enterocytozoon bieneusi*-like microsporidian parasite in simian-immunodeficiency-virus-inoculated macaques with hepatobiliary disease. American Journal of Pathology 150: 1395-1405.

_____, _____, D. Hebert , L. Chalifoux, D. Shvetz, K.C. Lin, S. Tzipori, and A.A. Lackner. 1998. Localization of persistent *Enterocytozoon bieneusi* infection in normal Rhesus Macaques (*Macaca mulatta*) to the hepatobiliary tree. Journal of Clinical Microbiology 36: 2336-2338.

Mathis, A., J. Akerstedt, J. Tharaldsen, O. Odegaard, and P. Deplazes. 1996. Isolates of *Encephalitozoon cuniculi* from farmed blue foxes (*Alopex lagopus*) from Norway differ from isolates from Swiss domestic rabbits (*Oryctolagus cuniculus*). Parasitology Research 82: 727-730.

_____, M. Michel, H. Kuster, C. Muller, R. Weber, and P. Deplazes. 1997. Two *Encephalitozoon cuniculi* strains of human origin are infectious to rabbits. Parasitology 114: 29-35.

_____, I. Tanner, R. Weber, and P. Deplazes. 1999. Genetic and phenotypic intraspecific variation in the microsporidian *Encephalitozoon hellem*. International Journal for Parasitology **29**: 767-770.

Orenstein, J.M., J. Seedor, D.N. Friedberg, S.M. Stenson, P.M. Tierno, N.C. Charles, D.M. Meisler, C.Y. Lowder, J.T. McMahon, D.L. Longworth, I. Rutherford, R.W. Yee, A. Martinez, F. Tio, and K. Held. 1990. Epidemiologic notes and reports: Microsporidian keratoconjunctivitis in patients with AIDS. Morbidity and Mortality Weekly Report **39**: 188-189.

Reetz, J., H. Rinder, A. Thomschke, H. Manke, M. Schwebs, and A. Bruderek. 2002. First detection of the microsporidium *Enterocytozoon bieneusi* in non-mammalian hosts (chickens). International Journal for Parasitology **32**: 785-787.

Rinder, H., S. Katzwinkel-Wladarsch, and T. Loscher. 1997. Evidence for the existence of genetically distinct strains of *Enterocytozoon bieneusi*. Parasitology Research **83**: 670-672.

Rossi, P., G. La Rosa, A. Ludovisi, A. Tamburrini, M.A. Gomez-Morales, and E. Pozio. 1998. Identification of a human isolate of *Encephalitozoon cuniculi* type I from Italy. International Journal for Parasitology **28**: 1361-1366.

Sing, A., K. Tybus, J. Heeseman, and A. Mathis. 2001. Molecular diagnosis of an *Enterocytozoon bieneusi* human genotype C infection in a moderately immunosuppressed human immunodeficiency virus seronegative liver-transplant recipient with severe chronic diarrhea. Journal of Clinical Microbiology **39**: 2371-2372.

Snowden, K. F., and K. Logan. 1999. Molecular identification of *Encephalitozoon hellem* in an ostrich. Avian Diseases **43**: 779-782.

_____, _____, and E.S. Didier. 1999. *Encephalitozoon cuniculi* Strain III Is a cause of encephalitozoonosis in both humans and dogs. Journal of Infectious Diseases **180**: 2086-2088.

_____, and J.A. Shadduck. 1999. Microsporidia in Higher Vertebrates; Wittner M, Weiss LM, editors. The Microsporidia and Microsporidiosis. Washington, D.C. American Society for Microbiology, 393-417.

_____, K. Logan, and D.N. Phalen. 2000. Isolation and characterization of an avian isolate of *Encephalitozoon hellem*. Parasitology **121**: 9-14.

_____, B. Daft, and R.W. Nordhausen. 2001. Morphological and molecular characterization of *Encephalitozoon hellem* in hummingbirds. Avian Pathology **30**: 251-255.

Tzipori, S., A. Carville, G. Widmer, D. Kotler, and K. Mansfield, and A. Lackner. 1997. Transmission and establishment of a persistent infection of *Enterocytozoon bieneusi*, derived from a human with AIDS, in simian immunodeficiency virus-infected Rhesus Monkeys. Journal of Infectious Diseases **175**: 1016-1020.

Undeen, A.H., and N.E. Alger. 1976. *Nosema algerae*: Infection of the white mouse by a mosquito parasite. Experimental Parasitology **40**: 86-88.

Van Essen, F.W., and D.W. Anthony. 1976. Susceptibility of nontarget organisms to *Nosema algerae* (Microsporidia: Nosematidae), a parasite of mosquitoes. Journal of Invertebrate Pathology **28**: 77-85.

Vávra, J., and A.H. Undeen. 1970. *Nosema algerae* n. sp. (Cnidospora, Microsporidia) a pathogen in a laboratory colony of *Anopheles stephensi* Liston (Diptera, Culicidae). Journal of Protozoology **147**: 293-306.

Visvesvara, G.S., M. Belloso, H. Moura, A.J. Da Silva, I.N.S. Moura, G.J. Leitch, D.A. Schwartz, P. Chevez-Barrios, S. Wallace, N.J. Pieniazek, and J.D. Goosey. 1999. Isolation of *Nosema algerae* from the cornea of an immunocompetent patient. Journal of Eukaryotic Microbiology **46**: 10S

Weber, R., P. Deplazes, and D. Schwartz. 2000. Diagnosis and clinical aspects of human microsporidiosis. Contributions in Microbiology **6**: 166-192.

Wright, J.H., and E.M. Craiggead. 1922. Infectious motor paralysis in young rabbits. Journal of Experimental Medicine **36**: 135-141.

Xiao, L., L. Lixia, H. Moura, I. Sulaiman A.A. Lal, S. Gatti, M. Scaglia, E.S. Didier, and G.S. Visvesvara. 2001a. Genotyping *Encephalitozoon hellem* isolates by analysis of the polar tube protein gene. Journal of Clinical Microbiology **39**: 2191-2196.

_____, _____, G.S. Visvesvara, H. Moura, E.S. Didier, and A.A. Lal. 2001b. Genotyping *Encephalitozoon cuniculi* by multilocus analyses of genes with repetitive sequences. Journal of Clinical Microbiology **39**: 2248-2253.

Yee, R.W., F.O. Tio, A. Martinez, K.S. Held, J.A. Shadduck, and E.S. Didier. 1991. Resolution of microsporidial epithelial keratopathy in a patient with AIDS. Ophthalmology **98**: 196-201.

# INSIGHTS INTO THE IMMUNE RESPONSES TO MICROSPORIDIA

Imtiaz A. Khan[1] and Elizabeth S. Didier[2]

[1]*Department of Microbiology, Immunology, and Parasitology, Louisiana State University Health Sciences Center, 1901 Perdido, New Orleans, LA 70112, U.S.A.,* [2]*Department of Microbiology and Immunology, Tulane National Primate Research Center, 18703 Three Rivers Road, Covington, LA 70433, U.S.A.*

**ABSTRACT:**
Microsporidiosis is a common infection of invertebrates and vertebrates, and has recently been recognized as an emerging and opportunistic infection in humans. Microsporidia infections in immune-competent mammals are often chronic and asymptomatic so that the host survives yet the parasite persists. Immune-deficient hosts such as AIDS patients or athymic and SCID mice infected with microsporidia, however, develop lethal disease. Hyperimmune responses leading to immune complex formation can develop into renal disease in carnivores infected with microsporidia. Well-regulated immune responses, therefore, are of key importance for preventing disease associated with microsporidiosis. Although humoral immune responses typically are generated against microsporidia in otherwise healthy mammalian hosts, cell-mediated immune responses are required for preventing morbidity and mortality in microsporidia-infected hosts. The purpose of this review is to relate what is currently known about the protective mammalian immune responses to microsporidia.

**Key words:** Microsporidia, opportunistic pathogen, emerging pathogen, immunology

## INTRODUCTION:

Microsporidia are ubiquitous, obligately intracellular, single-celled eukaryotic parasites which infect a wide range of invertebrates and vertebrates (Canning and Lom, 1986; Didier et al., 1998; Desportes-Livage, 2000). The microsporidia have long been considered protozoa, but were reclassified with the fungi in the phylum Microsporidia (Balbiani, 1882; Cavalier-Smith, 1998; Sprague and Becnel, 1998; Wittner, 1999). The infectious stage of the microsporidia is the spore which is oval in shape and, in species infecting mammals, is relatively small, measuring approximately 0.5-2.0 μM x 1.5-4.0 μM. The spore wall consists of a chitin endospore and glycoprotein exospore. A unique structure within the microsporidian spore is a coiled and relatively hollow polar filament which everts during germination to propel the spore contents into the host cell to initiate infection (Lom, 1972; Undeen, 1976; Weidner et al., 1984; Keohane and Weiss, 1999). Microsporidia contain Golgi-like membranes, endoplasmic reticulum and prokaryote-like ribosomes, but lack peroxisomes and have mitochondrial remnants suggesting a loss of mitochondria (Canning et al., 1986; Germot et al., 1997; Hirt et al., 1997; Didier et al., 1998; Williams et al., 2002).

Among mammals, microsporidiosis was first reported as a cause of motor paralysis in rabbits (Wright and Craighead, 1922; Levaditi et al., 1924). Since then, microsporidia have been identified as pathogens in a wide range of mammals including rodents, carnivores, and non-human primates (Canning et al., 1986; Didier et al., 1998; Snowden and Shadduck, 1999; Desportes-Livage, 2000). The earliest documented case of microsporidiosis in humans was in a 9-year-old child (Matsubayashi et al., 1959). Only a few cases of microsporidia were reported in humans until the beginning of the AIDS pandemic in the early 1980's, after which reports of human microsporidia infections dramatically increased. Microsporidiosis now has extended beyond the AIDS population to include organ transplant recipients, malnourished children, travelers, and the elderly (Sandfort et al., 1994; Sobottka et al., 1995; Bryan et al., 1996; Hautvast et al., 1997; Bryan and Schwartz, 1999; Lores et al., 2002).

Studies on the immune responses expressed against microsporidia after natural and experimental infections in laboratory animals have provided some understanding about the mechanisms of resistance versus pathogenesis, but little is known about the protective immune responses to microsporidia in humans. The purpose of this chapter is to present an overview about what is known about the immune responses generated against microsporidia in mammals.

## MICROSPORIDIAN SPECIES OF MAMMALS

Currently, 14 species of microsporidia have been identified as causing infections in mammals and all but one of these species were identified in humans (Table 1). Several genotypes within several of these microsporidian species also were identified by serology and molecular genetics methods (Didier et al., 1995; Rinder et al., 1997; Biderre et al., 1999; Breitenmoser et al., 1999; Deplazes et al., 2000; Delarbre et al.,

2001; Liguory et al., 2001, Xiao et al., 2001).

Table 1. Microsporidia species in mammals

| Species | Hosts | Sites of infection |
|---|---|---|
| Brachiola algerae [+] (Nosema algerae)* | Humans, mosquitoes | Cornea |
| Brachiola connori (Nosema connori)* | Humans | Disseminated |
| Brachiola vesicularum | Humans | Corneal stroma, skeletal muscle |
| Encephalitozoon cuniculi[+] | | |
|    Strain I [+] | Humans, rabbits, mice | Disseminated |
|    Strain II [+] | Mice, blue foxes, rats | Disseminated |
|    Strain III [+] | Humans, dogs, non-human primates | Disseminated |
|    Unclassified strains | Wide range of mammals | Disseminated |
| Encephalitozoon hellem [+] | Humans, birds | Cornea, disseminated |
| Encephalitozoon intestinalis [+] (Septata intestinalis)* | Humans, dogs, pigs, cows, donkies, goats | Disseminated |
| Enterocytozoon bieneusi | Humans, pigs, non-human primates, cats, farm dogs | Small intestine, biliary tract |
| Microsporidium africanum | Humans | Corneal stroma |
| Microsporidium ceylonensis | Humans | Corneal stroma |
| Nosema ocularum | Humans | Corneal stroma |
| Pleistophora sp. | Humans, fish | Skeletal muscle |
| Thelohania apodemi | Voles | Brain, skeletal muscle |
| Trachipleistophora anthropophthera | Humans | Disseminated |
| Trachipleistophora hominis [+] | Humans | Skeletal muscle |
| Vittaforma corneum [+] (Nosema corneae)* | Humans | Corneal stroma, Disseminated |

[+] Species of microsporidia which can be grown and harvested from cell culture.
* Previous names are noted in parentheses.

### Encephalitozoon species

Encephalitozoon cuniculi was the first microsporidian reported to infect mammals and was identified in the brain, spinal cord, and kidneys of a rabbit with motor paralysis (Wright and Craighead, 1922; Levaditi et al., 1924). E. cuniculi also was the first microsporidian successfully isolated from a mammalian host (a rabbit) for long-term culture which provided a source of organisms to study the basic biology of microsporidiosis and develop diagnostic methods (Shadduck, 1969). E. cuniculi has a wide host range among mammals, with infections having been reported in rodents, lagomorphs, ruminants, and carnivores, as well as human and non-human primates (Canning et al., 1986; Weber et al., 1994; Didier et al., 1998).

*Encephalitozoon hellem* was first identified in three AIDS patients with keratoconjunctivitis (Friedberg et al., 1990; Didier et al., 1991a; Yee et al., 1991). Since the morphology of *E. hellem* is nearly identical to *E. cuniculi*, some human cases attributed to *E. cuniculi* may have been due to *E. hellem* (Didier et al., 1991b). *E. hellem* infections are increasingly reported in psittacine birds (Black et al., 1997; Pulparampil et al., 1998), and because *E. hellem* replicates in tissue culture more efficiently at temperatures near 40°C, which is near the core body temperature of birds, speculation exists that *E. hellem* is a natural pathogen in birds and perhaps only incidentally infects humans (Snowden and Shadduck, 1999).

*Encephalitozoon (Septata) intestinalis* was first reported in AIDS patients with chronic diarrhea and is probably the second-most common microsporidian identified in humans after *E. bieneusi* (Blanshard et al., 1992; Orenstein et al., 1992a,b; Cali et al., 1993; Baker et al., 1995; Hartskeerl et al., 1995). *E. intestinalis* differs from the other *Encephalitozoon species* by secreting matrix material that "septates" the parasitophorous vacuole into compartments during its development. Like *E. cuniculi* and *E. hellem*, *E. intestinalis* can be grown in tissue culture (Van Gool et al.,1994; Didier et al., 1996; Visvesvara et al., 1999; Visvesvara, 2002). All three *Encephalitozoon* species can cause intestinal infections associated with diarrhea, but usually disseminate to cause clinical manifestations that can include conjunctivitis, sinusitis, myositis, peritonitis, hepatitis, nephritis, or encephalitis (Blanshard et al., 1992; Orenstein et al., 1992a; Kotler and Orenstein, 1998; 1999).

### Enterocytozoon bieneusi

The most commonly reported microsporidian which infects humans is *Enterocytozoon bieneusi* (Desportes et al. 1985; Orenstein et al., 1990; 1997; Orenstein, 1991; Weber et al., 1994). Additional natural hosts of *E. bieneusi* include pigs, cats, farm dogs, chickens, and non-human primates (Deplazes et al., 1996, 2000; Mansfield et al., 1997; Schwartz et al., 1998; Breitenmoser et al., 1999; Del Aguila et al., 1999; Mathis et al., 1999; Buckholt et al., 2002; Reetz et al., 2002). *E. bieneusi* infections usually remain localized to the small intestine to cause persistent diarrhea and weight loss, and some infections will spread to the gall bladder to cause cholangitis and cholescystitis (Chalifoux et al., 1998; Kotler and Orenstein, 1998; 1999). These organisms have not been grown in long-term culture and attempts to infect small laboratory rodents also have been unsuccessful, thereby hampering studies on this important microsporidian.

## Less-frequently encountered microsporidia

Several species of microsporidia have been identified less frequently as pathogens in humans and other mammals (Table 1). Of these species, *Brachiola algerae*, *Trachipleistophora hominis*, and *Vittaforma corneae*, can be grown in long-term culture and can be transmitted to infect mice and rats (Shadduck et al., 1990; Field et al., 1996; Hollister et al., 1996; Visvesvara, 2002).

## MICROSPORIDIA INFECTIONS IN MAMMALS

### Chronic/persistent infections in immune-competent hosts

Immunologically competent hosts infected with microsporidia usually develop chronic or persistent infections with few clinical signs of disease. Most of what is known about the immunology and pathology of microsporidiosis is based on *E. cuniculi* because it was the first mammalian microsporidian grown in culture making organisms readily available for experimental studies. Most *E. cuniculi* infections in mammals are chronic or persistent based on sporadic shedding of *E. cuniculi* spores with urine and the expression of high levels of microsporidia-specific serum antibody (Shadduck and Pakes, 1971). These infections often develop into balanced host-parasite relationships where the parasite persists in the face of a regulated immune response and the host survives with few if any clinical signs of disease (Nelson, 1967; Canning et al., 1986; Shadduck and Pakes, 1971; Didier et al., 1998; Snowden and Shadduck, 1999). It is less clear if immunologically competent humans infected with microsporidia eventually clear their infections or remain persistently infected. Survey data measuring microsporidia-specific antibody levels suggest that some populations of people are chronically infected with microsporidia (Canning et al., 1986; Van Gool et al., 1997). Studies were limited to using as antigen those species of microsporidia available from tissue culture, precluding use of *E. bieneusi*. A number of case reports have been published on HIV-seronegative individuals with microsporidia infections that resolved clinically after a few weeks and where parasite shedding was no longer detected. (Sandfort et al., 1994; Weber and Bryan, 1994; Sobottka et al., 1995; Silverstein et al. 1997; Gainzarain et al., 1998; Raynaud et al., 1998; Svenungsson et al., 1998; Lopez-Velez et al., 1999; Theng et al., 2001). As diagnostic methods improve and are more widely applied (eg. polymerase chain reaction), it will be possible to verify if humans clear their microsporidia infections or whether chronic infections persist. (Fedorko and Hijazi, 1996; Franzen and Müller, 1999; Garcia, 2002).

### Immunologically Compromised Hosts

In the absence of a competent immune system, microsporidia

infections cause morbidity and mortality of the host.  Immune-deficient athymic and SCID mice experimentally infected with *E. cuniculi, E. hellem*, or *V. corneae* developed lethal disease, usually associated with severe ascites that contained huge numbers of microsporidia organisms (Gannon, 1980b; Hermanek et al., 1993; Koudela et al., 1993; Schmidt and Shadduck, 1983; 1984; Silveira et al., 1993; Didier et al., 1994), and athymic mice inoculated with *T. hominis* developed severe skeletal muscle disease (Hollister et al., 1996).  Gamma-delta T cell-deficient mice and CD8$^{-/-}$ mice were susceptible to *E. cuniculi* infections and IFN-$\gamma$-deficient mice were susceptible to infection with *E. cuniculi* or *E. intestinalis* (Achbarou et al., 1996; Khan and Moretto, 1999; Khan et al., 2001; Moretto et al., 2001).  Also, hosts infected transplacentally with *E. cuniculi*, such as carnivores, squirrel monkeys, and horses often died as a consequence of their immature immune systems (Shadduck et al., 1978; Shadduck and Orenstein, 1993; Didier et al., 1998; Snowden and Shadduck, 1999).

In humans, microsporidiosis was first recognized in children with impaired immune systems (Matsubayashi et al., 1959; Margileth et al., 1973), and in persons with AIDS (Orenstein, 1991; Asmuth et al., 1994; Weber et al., 1994; Dascomb et al., 1999).  Microsporidiosis also is now recognized in immunologically naïve or compromised individuals such as travelers, organ transplant recipients undergoing immunosuppressive treatments, malnourished children, and the elderly (Sax et al., 1995; Bryan et al., 1996; Kelkar et al., 1997; Rabodonirina et al., 1996; Raynaud et al., 1998; López-Vélez et al., 1999; Lores et al., 2002).

**Immune-Mediated (Renal) Disease in Carnivores**
Carnivores such as domestic dogs and blue foxes, which become infected with *E. cuniculi* through transplacental transmission usually die due to encephalitis and nephritis.  Those animals that survive, however, remained infected and developed hypergammaglobulinemia, disseminated vasculitis, and eventual renal failure (Shadduck and Orenstein, 1993; Didier et al., 1998;  Snowden and Shadduck, 1999).

**HOST RESPONSES TO MICROSPORIDIA**

**Innate Resistance**
Innate barriers of resistance probably play a role in preventing or retarding microsporidia infections in mammals, particularly if one considers that of nearly 1200 species of microsporidia, only 14 species have been reported to infect mammals, and only four of these, *E. bieneusi* and three *Encephalitozoon* species, are commonly identified in mammals. Microsporidia are ubiquitous, so exposure to these organisms is likely to be commonplace.  In this context, a *Nosema* species of microsporidia was

detected in the feces of an AIDS patient who was not infected with this organism suggesting to McDougall and colleagues that incidental exposure to microsporidia does not always result in infection (McDougall et al., 1993). Host specificity exists among species of microsporidia that infect humans, as well. *E. bieneusi* and *E. intestinalis* infect a relatively wide range of mammals, but did not readily infect experimentally inoculated mice, while *E. cuniculi* commonly causes natural infections in rodents (Canning et al., 1986; Didier et al., 1998; Didier and Bessinger, 1999; Snowden and Shadduck, 1999). If microsporidia were to circumvent epithelial barriers, some organisms could replicate in macrophages, but non-specific serum factors such as opsonins and complement factors appeared to reduce their infectivity (Niederkorn and Shadduck, 1980; Schmidt and Shadduck, 1984). The number of microsporidia infecting a single macrophage may affect their survival, as well. Weidner and Sibley (1985) observed that macrophage phagosomes containing only one microsporidian failed to acidify thereby allowing these organisms to survive and replicate whereas phagosomes containing more than one organism tended to acidify leading to destruction of the microsporidia. *E. bieneusi* organisms, on the other hand, appear unable to replicate within macrophages, and are degraded if internalized (Orenstein et al., 1990; Orenstein, 1991; Kotler et al., 1993; Tzipori et al., 1997).

## Humoral immune responses

Both humoral and cell-mediated immune responses develop in mammals after infection with microsporidia, but resistance to lethal microsporidiosis depends upon intact T cell-mediated immune responses. Antibodies probably contribute to resistance, but under some circumstances, promote disease through immune complex-mediated hypersensitivity responses leading to renal disease.

## Expression of antibodies against microsporidia

Humoral immune responses most often have been measured against the *Encephalitozoon species* of microsporidia, particularly against *E. cuniculi*, because these species have a wide host range among mammals and can be grown in culture, making them readily available for experimental studies. Methods used to detect microsporidia-specific antibodies include India Ink reactivity, a complement fixation test, indirect immunofluorescence antibody (IFA) staining, ELISA, and western blot immunodetection (Pakes et al., 1984; Didier et al., 1993; Van Gool et al., 1997; Weber et al., 1999; Garcia, 2002).

Under experimental conditions, the time until specific antibodies were expressed in serum after exposure of immune-competent rodents, rabbits, and non-human primates to microsporidia generally varied with the route of inoculation (Didier et al., 1998; Didier and Bessinger, 1999). Specific IgM responses preceded IgG responses and then fell to baseline approximately one month after IgG levels peaked. Microsporidia-specific IgG responses developed earliest after inoculation by intravenous, intraperitoneal, or intracerebral routes and were first detected about one week later. Subcutaneous inoculations generally led to IgG detection approximately two weeks later, while oral inoculations resulted in detection of specific IgG approximately three weeks later. Specific IgG levels usually peaked about two or three month after exposure and would then plateau and persist if the hosts were chronically infected (Didier and Bessinger, 1999; Didier, 2000). Animals inoculated with dead organisms generated specific antibody responses but the antigen-specific antibodies eventually declined in the absence of continued antigen exposure (Liu and Shadduck, 1988; El Fakhry et al., 1998; Sobottka et al., 2001). The microsporidian-specific serum IgG responses expressed by microsporidia-infected mammals other than carnivores, remained consistently high, but did not continue to increase indefinitely, suggesting a level of immune regulation. Antigen-specific serum IgA responses have been detected in rabbits inoculated intrarectally and in mice inoculated orally (Wicher et al., 1991; El Fakhry et al., 1998), but were not monitored in most serological studies.

Antibody responses in immunologically naïve (ie. immature) and immunodeficient hosts varied depending on the status of the immune system at the time of exposure to microsporidia. Rabbits born to seropositive dams expressed maternal antibodies during the first two-to- four weeks which then waned after which the pups seroconverted to express specific IgG between eight and 14 weeks of age (Cox and Gallichio, 1978; Bywater and Kellett 1978a,1978b; 1979; Lyngset, 1980). Neonatal mice and kittens inoculated ip with *E. cuniculi* were slower to mount an antibody response but eventually generated antibody levels similar to that achieved by adult animals inoculated with microsporidia (Pang and Shadduck, 1985).

Immune-deficient rodents such as athymic and SCID mice, failed to express microsporidia-specific IgG responses unless reconstituted with lymphocytes from syngeneic immune-competent donors (Gannon, 1980b; Schmidt and Shadduck, 1983; 1984; Hermanek et al., 1993). SIV-infected, immunedeficient rhesus macaques (*Macaca mulatta*) also failed to express detectable specific antibody responses, but SIV-infected monkeys not yet immunedeficient did express specific antibodies but at a slower rate and at lower levels than expressed by microsporidia infected monkeys not infected with SIV (Didier et al., 1994; 1998).

As would be expected, HIV-infected individuals expressed variable antibody responses to microsporidia. In some cases, individuals with no history of microsporidiosis expressed relatively high ($\geq$ 1:800 ELISA titers) levels of microsporidia-specific antibodies yet some individuals from who microsporidia were detected and even isolated into culture, failed to express microsporidia-specific serum antibodies (Didier et al., 1991c; 1993). Based on the serological studies in SIV-infected monkeys and immunodeficient mice experimentally infected with microsporidia, the expression of antibody responses to microsporidia in HIV-infected individuals probably depends upon immune status of the individuals at the time infection occurs and raises the question about whether microsporidiosis can reactivate in persons with AIDS or whether new infections with microsporidia cause disease in these individuals.

Antibody responses in carnivores such as blue foxes and domestic dogs are particularly interesting because two extremes in immune responsiveness have been observed. Transplacentally infected carnivores, which are immune-deficient at the time of infection, usually developed acute renal disease and died (Mohn and Nordstoga, 1975; Nordstoga and Westbye, 1976; Mohn, 1982; Mohn et al., 1982; Shadduck and Orenstein, 1993; Snowden and Shadduck, 1999). Surviving animals however, expressed maternal passively-transferred specific IgG at three weeks of age, but usually remained infected. These animals subsequently developed renal disease associated with hypergammaglobulinemia. Carnivores infected experimentally with *E. cuniculi* at two days of age expressed specific IgM and IgG, and also developed hypergammaglobulinemia with total serum IgG levels that were three times that of uninfected animals (Shadduck et al., 1978; Mohn, 1982; Mohn et al., 1982; Stewart et al., 1986; 1988; Szabo and Shadduck, 1987; 1988). Adult domestic dogs inoculated orally, intravenously, or intraperitoneally with *E. cuniculi* expressed transient IgM responses from six-to- twelve weeks after inoculation. Microsporidia-specific IgG responses could be detected approximately six weeks after inoculations and the total serum IgG levels continued to increase over the following several months (Stewart et al., 1979; Szabo and Shadduck, 1988; Botha et al., 1979; 1986a,b; Stewart et al., 1986; 1988; Hollister et al., 1989).

### Functions of specific antibody in resistance

Passive transfer of hyperimmune serum failed to protect athymic BALB/c mice inoculated with *E. cuniculi* (Schmidt and Shadduck, 1983). In addition, some microsporidia-infected AIDS patients expressed relatively high levels of microsporidia-specific antibodies yet developed signs of disease associated with microsporidiosis (Didier et al., 1993). Data from *in vitro* studies, however, suggest that antibodies probably contribute to resistance against microsporidia. *E. cuniculi* is able to survive and replicate within parasitophorous vacuoles in macrophages due to an absence of phagosome-lysosome fusion , but pretreatment of spores with antiserum resulted in phagosume-lysosome fusion and reduced infectivity of surviving organisms (Weidner, 1975; Niederkorn and Shadduck, 1980; Schmidt and Shadduck, 1984). Complement fixing antibodies also have been detected against microsporidia and were the basis of a diagnostic test in rabbits (Wosu et al., 1977). The function of these complement fixing antibodies in resistance *in vivo* is unclear, since it is unlikely that the spore coat of a mature microsporidian can be penetrated by complement fixation, although less mature stages may be susceptible (Niederkorn and Shadduck, 1980; Schmidt and Shadduck, 1984; Didier and Bessinger, 1999). Finally, neutralizing antibodies to microsporidia have been reported to inhibit microsporidian infections of non-phagocytic cells, but their role *in vivo* also has not been established (Enriquez et al., 1998).

## Role of antibody in pathogenesis

In carnivores, antibody responses resulting from *E. cuniculi* infections appeared to contribute to the pathogenesis of microsporidiosis. Perivascular granulomatous lesions and granular deposits containing IgM and IgG in the glomerular basement membranes were typically observed in *E. cuniculi*-infected carnivores. In addition to renal lesions, vascular alterations were noted in heart, salivary gland, and prostate of *E. cuniculi*-infected blue foxes (Åkerstedt et al., 2002). Blue foxes and mink with clinical signs due to microsporidiosis expressed relatively high levels of specific serum antibodies while blue foxes that resolved their clinical symptoms also decreased their expression of serum antibodies to microsporidia (Mohn and Nordstoga, 1975; Zhou and Nordstoga, 1993; Snowden and Shadduck, 1999). It is still unclear, however, if the high expression of antibodies during *E. cuniculi* infection in these animals leads to autoimmunity or if trapped immunecomplexes promote inflammation.

## Serology as a diagnostics tool

Immune-competent laboratory animals infected with microsporidia continue to express high levels of specific antibodies which is a basis for serologic diagnosis of microsporidiosis (Bywater and Kellett, 1978a,b; 1979; Chalupsky et al., 1973; 1979; Gannon, 1980a; Lyngset, 1980; Shadduck and Baskin, 1989). Culling of seropositive laboratory animals resulted in clearance of microsporidiosis from these colonies (Cox et al., 1977; Bywater and Kellett, 1978b). Furthermore, inoculation of mice with killed *E. cuniculi* resulted in transient antibody responses and ELISA titers that soon fell to baseline levels (Liu and Shadduck, 1988) suggesting that antibody levels remained positive (ie. ELISA titer $\geq$ 1:800) if the parasites persisted. Immune-deficient lab animals with microsporidiosis often failed to produce high levels of specific antibodies.

Serology is problematic for diagnosing microsporidiosis in humans. Early serological surveys on humans utilized *E. cuniculi* as the antigen as it was the only mammalian microsporidian that could be cultured, but new species of microsporidia have now been identified and isolated from mammals for long-term culture. Unfortunately, the inability to culture *E. bieneusi* limits serology for detecting antibodies to this species. Numerous populations of individuals have tested positive for antibodies to *E. cuniculi*, but many of these individuals also carried other parasitic infections that may cause non-specific antibody expression (Bergquist et al., 1984; Canning et al., 1986; Hollister and Canning, 1987; Hollister and Willcox., 1991; Van Gool et al., 1997). Antibodies to microsporidia cross-react between species so that species-specific identification by serology is difficult (Niederkorn et al., 1980; Didier et al., 1991a,b,; 1993; Aldras et al., 1994; Weiss et al., 1992; Hartskeerl et al., 1995; Ombrouck et al., 1995). As PCR-based methods become available commercially, it will be possible to determine if positive serology correlates with true microsporidial infections in humans (Fedorko and Hijazi, 1996; Franzen and Mueller, 1999; Weiss, 2000; Garcia, 2002).

## Cell-mediated immune responses

Cellular immune mechanisms expressed during microsporidial infections have not been well studied, but the susceptibility of AIDS patients and other immune-deficient hosts demonstrates the importance of immune T cells in resistance to microsporidia (Schmidt and Shadduck, 1984; Hermanek et al., 1993; Dascomb et al., 1999). In experimental studies, adoptive transfer of immune T cell protected athymic and SCID mice from lethal *E.cuniculi* challenge, but transfer of naive lymphocytes or hyperimmune serum failed to protect or prolong the survival of the athymic animals (Schmidt and Shadduck, 1984; Didier and Bessinger, 1999; Khan et al., 2001). Similar results from human studies demonstrated that recovery of T cell levels via protease inhibitory therapy led to resolution of microsporidiosis in patients infected with HIV (Goguel et al., 1997; Carr et al., 1998; Conteas et al., 1998). Results from experimental studies demonstrated that T cell-mediated protection depended on cytolytic activity of CD8+ T cells and on cytokine production for activating macrophages to kill intracellular parasites by generation of reactive nitrogen intermediates (Didier, 1995; Khan et al., 2001). Other mechanisms must also play a role in controlling microsporidia, because iNOS-/- mice survived *E. cuniculi* infection (Khan et al., 2001). Nevertheless, these findings suggest that protective immune responses against *E. cuniculi* are likely dependent on macrophages as well as cytolytic and cytokine-producing immune T cells.

## Role of cytokines

Similar to other intracellular viral, bacterial and parasitic infections, protective immunity against *E. cuniculi* infection appears to depend on TH-1 cytokines such as IL-12 and IFN-γ. Based on *in vitro* observations, IFN-γ was observed to play an important role in the protective immunity against *E. cuniculi* infection (Didier, 1995; Braunfuchsova et al., 1999). Mice lacking the IFN-γ gene or given antibody to IFN-γ were unable to clear either *E. cuniculi* or *E. intestinalis* infections and mice treated with antibody to IL-12 likewise became susceptible to *E. cuniculi* (Achbarou et al., 1996; Khan and Moretto, 1999). The use of gene knock-out mice further validated the importance of TH-1 cytokines in immune responses against *E. cuniculi* since p40$^{-/-}$ mice (which are unable to produce IL-12) also succumbed to infection upon *E. cuniculi* challenge. Minimal TH-2 cytokine production was observed during *E. cuniculi* infection as the mRNA for IL-4, a prominent TH-2 cytokine, was undetectable in splenocytes of infected animals (Khan et al., 1999). Similarly, no circulating IL-4 was detected in the sera of infected mice. Expression of mRNA for IL-10, another TH-2 cytokine, however, increased in the splenocytes of *E. cuniculi*-infected animals. Since IL-10 has been reported to be involved in the regulation of TH-1 immune response in other infectious disease models (Gazzinelli et al., 1994; Trinchieri, 1997), it may play a similar role in *E. cuniculi* infection.

## Role of T cell subsets

Studies from our laboratory have demonstrated that protective immunity against i.p. *E. cuniculi* challenge in mice was predominantly associated with CD8$^{+}$ T cells with CD4$^{+}$ T cells playing a far less important role (Khan et al., 1999; Moretto et al., 2000). CD8$^{-/-}$ mice were susceptible to i.p. *E.cuniculi* infection and adoptive transfer of immune CD8$^{+}$ T cells to these knock out animals restored protection (Moretto et al., 2001). The protective immunity of CD8$^{+}$ T cells against *E.cuniculi* infection was due to their ability to produce cytokines and exhibit cytolytic activity against infected target cells (Khan et al., 1999; 2001). Mice lacking the gene for perforin, a granule that is important for cytotoxic function, also were highly susceptible to *E. cuniculi* infection. These observations indicate the importance of cytotoxic T lymphocyte (CTL) responses, along with macrophage-activating cytokines, in resistance to *E. cuniculi* infections.

Surprisingly, a lack of CD4$^{+}$ T cells did not affect the outcome of *E. cuniculi* infection in the knock out animals. Interestingly, normal antigen-specific CD8$^{+}$ T cell responses were detected in CD4$^{-/-}$ mice (Moretto et al., 2000). The lack of CD4$^{+}$ T cells also did not alter the magnitude of antigen-specific cytotoxic response. Although CD4$^{-/-}$ mice were able to resolve *E.cuniculi* infection, mice lacking the gamma-delta T cell population were susceptible to *E.cuniculi* when challenged with very high parasite doses. Unlike CD8$^{-/-}$ or gamma-delta T-cell-deficient mice, however, gamma-delta T-cell$^{-/-}$ animals were able to survive a comparatively low dose infection (Moretto et al., 2001). The susceptibility of gamma-delta T-cell[-/-] mice was attributed to down-regulation of CD8[+] T cell immunity, since a significant decrease in the antigen–specific CD8[+] T cell immune response was observed in these animals. Based on these observations, it appears that gamma-delta T cells play a prominent role in priming CD8[+] T cells during *E. cuniculi* infection. The induction of CD8[+] T cell immunity by gamma-delta T cells may be due to their ability to produce cytokines necessary from priming CD8[+] T cell response as reported in other models (Ferrick et al., 1995).

## CONCLUSIONS:

Reports of microsporidiosis in animals and humans have increased dramatically during the last decade and has led to an increased effort to better understand the immunology of this infection. Most of what is presently known is based on studies with *E. cuniculi* which seems to induce immune responses similar to those induced by related intracellular pathogens. This would support the hypothesis that *E. cuniculi* induces a strong burst of IL-12 production by host macrophages or dendritic cell which would lead to a polarization towards TH-1 cytokines manifested by high levels of IFN-γ in the circulation and tissues (Khan et al., 2001). The increased numbers of NK cells and gamma-delta T cells during the early stages of infection are likely sources of IFN-γ production, while TH-2 cytokines like IL-4 were not detectable throughout the course of infection (Niederkorn et al., 1983; Moretto et al., 2001). The increase in IFN-γ production is known to cause up-regulation of class I molecules on the infected cells which likely leads to antigen-specific CD8[+] T cell proliferation and generation of CTL (Boehm et al., 1997; Khan et al., 1999). It is unclear if CTLs, when killing the host cells, also kill the microsporidia within the host cells. Activated macrophages are the only cells currently known to be capable of killing microsporidia, and the role of CD4[+] T cells in protective immune response against *E.cuniculi* may be to release activating cytokines (eg. IFN-γ). While humans with decreased CD4[+] T cell levels become susceptible to microsporidiosis, CD4[+] T- cell-deficient mice are resistant to parasite challenge and CD8[+] T cell immune responses in these knock out mice are not compromised. As a result, the true function of CD4+ T cells in immunity to microsporidia remains unclear (Moretto et al., 2000). It appears, however, that the CD8[+] CTL response during *E. cuniculi* infection, which is critical for host protection in mice, can be launched independent of CD4[+] T cells. The presence of gamma-delta T cells appears to be crucial for the induction of optimal CD8[+] T cell immunity.

Important questions remain to be addressed. The interactions between IFN-γ and CD8[+] T cells during *E. cuniculi* infection and the mechanisms by which gamma-delta T cells induce the CD8[+] CTL response are unknown. Whether CTL kill microsporidia, and the mechanisms by which macrophages kill intracellular microsporidia still need to be addressed. The route of transmission is likely to affect the course of immunity against microsporidia, as well. Studies addressing these issues should results in major strides being made to better understand the immune responses that provide resistance to these opportunistic and emerging pathogens.

## REFERENCES

Achbarou, A., C. Ombrouck, T. Gneragbe, F. Charlotte, L. Renia, I. Desportes-Livage, and D. Mazier. 1996. Experimental model for human intestinal microsporidiosis in interferon gamma receptor knockout mice infected by *Encephalitozoon intestinalis*. Parasite Immunology 18: 387-392.

Åkerstedt, J., K. Nordstoga, A. Mathis, E. Smeds, and P. Deplazes. 2002. Fox encephalitozoonosis: isolation of the agent from an outbreak in farmed blue foxes (*Alopex lagopus*) in Finland and some hitherto unreported pathologic lesions. Journal of Veterinary Medicine 49: 400-405.

Aldras, A.M., J.M. Orenstein, D.P. Kotler, J.A. Shadduck, and E.S. Didier. 1994. Detection of microsporidia by indirect immunofluorescence antibody test using polyclonal and monoclonal antibodies. Journal of Clinical Microbiology 32: 608-612.

Asmuth, D., P.C. DeGirolami, M. Federman, C.R. Ezratty, D.K. Pleskow, G. Desai, and C.A. Wanke. 1994. Clinical features of microsporidiosis in patients with AIDS. Clinical Infectious Diseases 18: 819-825.

Balbiani, G. 1882. Sur les microsporidies ou sporogspermies des articules. Czech Republic Academy of Sciences 95: 1168-1171.

Baker, M.D., C.R. Vossbrinck, E.S. Didier, J.V. Maddox, and J.A. Shadduck.. 1995. Small subunit ribosomal DNA phylogeny of various microsporidia with emphasis on AIDS related forms. Journal of Eukaryotic Microbiology 42: 564-570.

Bergquist, R., L. Morfeldt-Mansson, P.O. Pehrson, B. Petrini, and J. Wasserman. 1984. Antibody against *Encephalitozoon cuniculi* in Swedish homosexual men. Scandavian Journal of Infectious Diseases 16: 389-391.

Biderre, C., A. Mathis, P. Deplazes, R. Weber, G. Metenier, and C.P. Vivares. 1999. Molecular karyotype diversity in the microsporidian *Encephalitozoon cuniculi*. Parasitology 118: 439-445.

Black, S.S., L.A. Steinohrt, D.C. Bertucci, L.B. Rogers, and E.S. Didier 1997. *Encephalitozoon hellem* in budgerigars (*Melopsittacus undulatus*). Veterinary Pathology 34: 189-198.

Blanshard, C., W.S. Hollister, C.S. Peacock, D.G. Tovey, D.S. Ellis, E.U. Canning, and B.G. Gazzard. 1992. Simultaneous infection with two types of intestinal microsporidia in a patient with AIDS. Gut 33: 418-420.

Boehm, U., T. Klamp, M. Groot, and J.C. Howard. 1997. Cellular responses to interferon-gamma. Annual Reviews in Immunology 15: 749-795.

Botha, W.S., A.F. van Dellen, and C.G. Stewart. 1979. Canine encephalitozoonosis in South Africa. Journal of the South African Veterinary Association 50: 135-144.

_____, I.C. Dormehl, and D.J. Goosen. 1986a. Evaluation of kidney function in dogs suffering from canine encephalitozoonosis by standard clinical pathological and radiopharmaceutical techniques. Journal of the South African Veterinary Association 57: 79-86.

_____, C.G. Stewart, and A.F. van Dellen. 1986b. Observations on the pathology of experimental encephalitozoonosis in dogs. Journal of the South African Veterinary Association 57: 17-24.

Braunfuchsova, P., J. Kopecky, O. Ditrich, and B. Koudela. 1999. Cytokine response to infection with the microsporidian, *Encephalitozoon cuniculi*. Folia Parasitologica **46**: 91-95.

Breitenmoser, C., A. Mathis, E. Bürgi, R. Weber, and P. Deplazes. 1999. High prevalence of *Enterocytozoon bieneusi* in swine with four genotypes that differ from those identified in humans. Parasitology **118**: 447-453.

Bryan, R.T., and D.A. Schwartz. 1999. "Epidemiology of Microsporidiosis" In The Microsporidia and Microsporidiosis, Murray Wittner, ed. American Society for Microbiology, Washington, DC. pp 502-516.

_____, R. Weber, and D.A. Schwartz. 1996. Microsporidiosis in patients who are not infected with human immunodeficiency virus. Clinical Infectious Diseases **23**: 114-117.

Buckholt, M.A., J.H. Lee, and S. Tzipori S. 2002. Prevalence of *Enterocytozoon bieneusi* in swine: an 18-month survey at a slaughterhouse in Massachusetts. Applied and Environmental Microbiology **68**: 2595-2599

Bywater, J.E., and B.S. Kellett. 1978a *Encephalitozoon cuniculi* antibodies in a specific-pathogen-free rabbit unit. Infection and Immunity **21**: 360-364.

_____, and _____. 1978b. The eradication of *Encephalitozoon cuniculi* from a specific pathogen-free rabbit colony. Laboratory Animal Science **28**: 402-404.

_____, and _____. 1979. Humoral immune response to natural infection with *Encephalitozoon cuniculi* in rabbits. Laboratory Animals **13**: 293-297.

Cali, A., D.P. Kotler, and J.M. Orenstein. 1993. *Septata intestinalis* N. G., N. Sp., an intestinal microsporidian associated with chronic diarrhea and dissemination in AIDS patients. Journal of Eukaryotic Microbiology **40**: 101-112.

Canning, E.U., J. Lom. 1986. The microsporidia of vertebrates. Academic Press, Inc., New York. 289 pp.

Carr, A., D. Marriott, A. Field, E. Vasak, and D.A. Cooper. 1998. Treatment of HIV-1-associated microsporidiosis and cryptosporidiosis with combination antiretroviral therapy. Lancet 1998; 351:256-261.

Cavalier-Smith, T. 1998. A revised six-kingdom system of life. Biological Reviews of the Cambridge Philosophical Society (London) **73**: 203-266.

Chalifoux, L.V., J. MacKey, A. Carville, D. Shvetz, K.C. Lin, A. Lackner, and K.G. Mansfield. 1998. Ultrastructural morphology of *Enterocytozoon bieneusi* in biliary epithelium of rhesus macaques (*Macaca mulatta*). Veterinary Pathology **35**: 292-296.

Chalupsky, J., I. Vavra, and P. Bedrnik. 1973. Detection of antibodies to *Encephalitozoon cuniculi* in rabbits by the indirect immunofluorescent antibody test. Folia Parasitologica (Praha) **20**: 281-284.

_____, _____, and _____. 1979. Encephalitozoonosis in laboratory animals-a serological survey. Folia. Parasitologia (Praha) **26**: 1-8.

Conteas, C.N., O.G. Berlin, C.E. Speck, S.S. Pandhumas, M.J. Lariviere, and C. Fu. 1998. Modification of the clinical course of intestinal microsporidiosis in acquired immunodeficiency syndrome patients by immune status and anti-human immunodeficiency virus therapy. American Journal for Tropical Medicine and Hygiene **58**: 555-558.

Cox, J.C., and H.A. Gallichio. 1978. Serological and histological studies on adult rabbits with recent, naturally acquired encephalitozoonosis. Research in Veterinary Science **24**: 260-261.

_____, _____, D. Pye, and N.B. Walden. 1977. Application of immunofluorescence to the establishment of an *Encephalitozoon cuniculi*-free rabbit colony. Laboratory Animal Science **27**: 204-209.

Dascomb, K., R. Clark, J. Aberg, J. Pulvirenti, R.G. Hewitt, P. Kissinger, and E.S. Didier. 1999. Natural history of intestinal microsporidiosis among patients infected with human immunodeficiency virus. Journal of Clinical Microbiology **37**: 3421-3422.

Delarbre, S., S. Gatti, M. Scaglia, and M. Drancourt. 2001. Genetic diversity in the microsporidian *Encephalitozoon hellem* demonstrated by pulsed-field gel electrophoresis. Journal of Eukaryotic Microbiology **48**: 471-474.

Del Aguila, C., F. Izquierdo, R. Navajas, N.J. Pieniazek,, G. Miró, A.I. Alonso, A.J. DaSilva, and S. Fenoy. 1999. *Enterocytozoon bieneusi* in animals: rabbits and dogs as new hosts. Journal of Eukaryotic Microbiology **46**: 8S-9S.

Deplazes, P., A. Mathis, C. Muller, and R. Weber. 1996. Molecular epidemiology of *Encephalitozoon cuniculi* and first detection of *Enterocytozoon bieneusi* in faecal samples of pigs. Journal of Eukaryotic Microbiology **43**: 93S.

_____, A. Mathis, and R. Weber. 2000. Epidemiology and zoonotic aspects of microsporidia of mammals and birds. Contributions to Microbiology **6**: 236-260.

Desportes, I., Y. Le Charpentier, A. Galian, F. Bernard, B. Cochand-Priollet, A. Lavergne, P. Ravisse, and R. Modigliani. 1985. Occurrence of a new microsporidan: *Enterocytozoon bieneusi* n.g., n. sp., in the enterocytes of a human patient with AIDS. Journal of Protozoology **32**: 250-254.

Desportes-Livage, I. 2000. Biology of microsporidia. Contributions to Microbiology **6**: 140-165.

_____, E. Doumbo, E. Pichard, I. Hilmarsdottir, H.A. Traore, I.I. Maiga, Y. El Fakhry, and A. Dolo. 1998. Microsporidiosis in HIV-seronegative patients in Mali. Transactions of the Royal Society for Tropical Medicine and Hygiene **92**: 423-424

Didier, E.S. 1995. Reactive nitrogen intermediates implicated in the inhibition of *Encephalitozoon cuniculi* (phylum microspora) replication in murine peritoneal macrophages. Parasite Immunology **17**: 405-412.

_____. 2000. Immunology of microsporidiosis. Contributions to Microbiology **6**: 193-208.

_____, and G.T. Bessinger. 1999. "Host-Parasite Relationships in Microsporidiosis animal Models and Immunology." In The Microsporidia and Microsporidiosis, Murray Wittner, ed. American Society for Microbiology, Washington, DC. pp. 225-257

_____, J.P. Didier, D.N. Friedberg, S.M. Stenson, J.M. Orenstein, R.W. Yee, F.O. Tio, R.M. Davis, C. Vossbrinck, N. Millichamp, and J.A. Shadduck. 1991a. Isolation and characterization of a new human microsporidian, *Encephalitozoon hellem* (n. sp.), from three AIDS patients with keratoconjunctivitis. Journal of Infectious Diseases **163**: 617-621.

_____, _____, J.M. Orenstein, and J.A. Shadduck. 1991b. Fine structure of a new human microsporidian, *Encephalitozoon hellem*, in culture. Journal of Protozoology **38**: 502-507.

_____, Kotler DP, Dieterich DT, Orenstein JM, Aldras AM, Davis R, Friedberg DN, Gourley WK, Lembach R, Lowder CY, Meisler DM, Rutherford I, R.W. Yee, and J.A. Shadduck. 1993. Serological Studies in Human Microsporidia Infections. AIDS **7**: S8-S11.

_____, L.B. Rogers, J.M. Orenstein, M.D. Baker, C.R. Vossbrinck, T. Van Gool, R. Hartskeerl, R. Soave, and L.M. Beaudet. 1996. Characterization of *Encephalitozoon* (*Septata*) *intestinalis* isolates cultured from nasal mucosa and bronchoalveolar lavage fluids of two AIDS patients. Journal Eukaryotic Microbiology **43**: 34-43.

_____, J.A. Shadduck, J.P. Didier, N. Millichamp, and C.R. Vossbrinck. 1991c. Studies on ocular microsporidia. Journal Protozoology **38**: 635-638.

_____, K.A. Snowden, and J.A. Shadduck. 1998. The biology of Microsporidian species infecting mammals. Advances in Parasitology **40**: 279-316.

_____, P.W. Varner, J.P. Didier, A.M. Aldras, N.J. Millichamp, M. Murphey-Corb, R. Bohm, and J.A. Shadduck 1994. Experimental microsporidiosis in immunocompetent and immunodeficient mice and monkeys. Folia Parasitolgica (Praha) **41**: 1-11.

_____, C.R. Vossbrinck, M.D. Baker, L.B. Roger, D.C. Bertucci, and J.A. Shadduck. 1995. Identification and characterization of three *Encephalitozoon cuniculi* strains. Parasitology **111**: 411-421.

El Fakhry Y, A. Achbarou, I. Desportes-Livage, and D. Mazier. 1998. *Encephalitozoon intestinalis*: humoral responses in interferon-gamma receptor knockout mice infected with a microsporidium pathogenic in AIDS patients. Experimental Parasitology **89**: 113-121.

Enriquez, F.J., G. Wagner, G. Fraguso, and O. Ditrich. 1998. Effects of an anti-exospore monoclonal antibody on microsporidial develoment in vitro. Parasitology **117**: 515-520.

Fedorko, D.P., and Y.M. Hijazi. 1996. Application of molecular techniques to the diagnosis of microsporidial infection. Emerging Infectious Diseases **2**: 183-191.

Ferrick, D.A., M.D. Schrenzel, T. Mulvania, B. Hsieh, W.G. Ferlin, and H. Lepper. 1995. Differential production of interferon-gamma and interleukin-4 in response to Th1- and Th2-stimulating pathogens by gamma delta T cells in vivo. Nature **373**: 255-257.

Field, A.S., D.J. Marriott, S.T. Milliken, B.J. Brew, E.U. Canning, J.G. Kench, P. Darveniza, and J.L. Harkness. 1996. Myositis associated with a newly described microsporidian, *Trachipleistophora hominis*, in a patient with AIDS. Journal of Clinical Microbiology **34**: 2803-2811.

Franzen, C., and A. Müller. 1999. Molecular Techniques for detection, species differentiation, and phylogenetic analysis of Microsporidia Clinical Microbiology Reviews **12**: 243-285.

Friedberg, D.N., S.M. Stenson, J.M. Orenstein, P.M. Tierno, and N.C. Charles. 1990. Microsporidial keratoconjunctivitis in acquired immunodeficiency syndrome. Archives of Ophthalmology **108**: 504-508.

Gainzarain, J.C., A. Canut, M. Lozano, A. Labora, F. Carreras, S. Fenoy, R. Navajas, N.J. Pieniazek, A.J. da Silva, and C. del Aguila. 1998. Detection of *Enterocytozoon bieneusi* in two human immunodeficiency virus-negative patients with chronic diarrhea by polymerase chain reaction in duodenal biopsy specimens and review. Clinical Infectious Diseases **27**: 394-398.

Gannon, J. 1980a. A survey of *Encephalitozoon cuniculi* in laboratory animal colonies in the United Kingdom. Laboratory Animal **14**: 91-94.

_____.1980b. The course of infection of *Encephalitozoon cuniculi* in immunodeficient and immunocompetent mice. Laboratory Animal **14**: 189-192.

Garcia, L..S. 2002. Laboratory identification of the microsporidia. Journal of Clinical Microbiology **40**: 1892-1901.

Gazzinelli, R.T., S. Hayashi, M. Wysocka, L. Carrera, R. Kuhn, W. Muller, F. Roberge, G. Trinchieri, and A. Sher. 1994. Role of IL-12 in the initiation of cell mediated immunity by *Toxoplasma gondii* and its regulation by IL-10 and nitric oxide. Journal Eukaryotic Microbiology **41**: 9S.

Germot, A., H. Philippe, and H. Le Guyader. 1997. Evidence for loss of mitochondria in microsporidia from a mitochondrial-type HSP70 in *Nosema locustae*. Molecular and Biochemical Parasitology **87**: 159-168.

Goguel, J., C. Katlama, C. Sarfati, C. Maslo, C. Leport, and J.-M. Molina. 1997. Remission of AIDS-associated intestinal microsporidiosis with highly active antiretroviral therapy. AIDS **11**: 1658-1659.

Hartskeerl, R.A., T. Van Gool, A.R. Schuitema, E.S. Didier, and W.J. Terpstra. 1995. Genetic and immunological characterization of the microsporidian *Septata intestinalis* Cali, Kotler and Orenstein, 1993: reclassification to *Encephalitozoon intestinalis*. Parasitology **110**: 277-285.

Hautvast, J.L., J.J. Tolboom, T.J. Derks, P. Beckers, and R.W. Sauerwein. 1997. Asymptomatic intestinal microsporidiosis in a human immunodeficiency virus-seronegative, immunocompetent Zambian child. Pediatric Infectious Disease Journal **16**: 415-416.

Hermanek, J., B. Koudela, Z. Kucerova, O. Ditrich, and J. Travnicek. 1993. Prophylactic and therapeutic immune reconstitution of SCID mice infected with *Encephalitozoon cuniculi*. Folia Parasitologica (Praha) **40**: 287-291.

Hirt, R.P., B. Healy, C.R. Vossbrinck, E.U. Canning, and T.M. Embley. 1997. A mitochondrial Hsp70 orthologue in *Vairimorpha necatrix*: molecular evidence that microsporidia once contained mitochondria. Current Biology **7**: 995-998.

Hollister, W.S., and E.U. Canning. 1987 An enzyme-linked immunosorbent assay (ELISA) for detection of antibodies to *Encephalitozoon cuniculi* and its use in determination of infections in man. Parasitology **94**: 209-219.

_____, _____, and M. Viney. 1989. Prevalence of antibodies to *Encephalitozoon cuniculi* in stray dogs as determined by an ELISA. Veterinary Record **124**: 332-336.

_____, _____, and A. Willcox. 1991. Evidence for widespread occurrence of antibodies to *Encephalitozoon cuniculi* (Microspora) in man provided by ELISA and other serological tests. Parasitology **102**: 33-43.

_____, _____, E. Weidner, A.S. Field, J. Kench, and D.J. Marriott. 1996. Development and ultrastructure of *Trachipleistophora hominis* n.g., n.sp. after *in vitro* isolation from an AIDS patient and inoculation into athymic mice. Parasitology **112**: 143-154.

Kelkar, R., P.S. Sastry, S.S. Kulkarni, T.K. Saikia, P.M. Parikh, and S.H. Advani. 1997. Pulmonary microsporidial infection in a patient with CML undergoing allogeneic marrow transplant. Bone Marrow Transplantation **19**: 179-182.

Keohane, E.M. and L.M. Weiss. 1999. "The Structure, Function, and Composition of the Microsporidian Polar Tube." In The Microsporidia and Microsporidiosis, Murray Wittner, ed. American Society for Microbiology, Washington, DC. Pp. 196-224.

Khan, I.A., and M. Moretto. 1999. Role of gamma interferon in cellular immune response against murine *Encephalitozoon cuniculi* infection. Infection and Immunity **67**: 1887-1893.

_____, _____, and L.M. Weiss. 2001. Immune response to *Encephalitozoon cuniculi* infection. Microbes and Infection **3**: 401-405.

_____, J.D. Schwartzman, L.H. Kasper, and M. Moretto. 1999. CD8+ CTLs are essential for protective immunity against *Encephalitozoon cuniculi* infection. Journal of Immunology **162**: 6086-6091.

Kotler, D.P., and J.M. Orenstein. 1998. Clinical syndromes associated with microsporidiosis. Advances in Parasitology **40**: 321-349.

_____, and _____. 1999. "Clinical Syndromes Associated with Microsporidiosis" In The Microsporidia and Microsporidiosis, Murray Wittner, ed., American Society for Microbiology, Washington, DC. Pp. 258-292.

_____, S. Reka, K. Chow, and J.M. Orenstein. 1993. Effects of enteric parasitoses and HIV infection upon small intestinal structure and function in patients with AIDS. Journal of Clinical Gastroenterology **16**: 10-15.

Koudela, B., J. Vitovec, Z. Kucerova, O. Ditrich, and J. Travnicek. 1993. The severe combined immunodeficient mouse as a model for *Encephalitozoon cuniculi* microsporidiosis. Folia Parasitologica (Praha) **40**: 279-286.

Levaditi, C., S. Nicolau, and R. Schoen. 1924 L'agent étiologique de l'encephalite épizootique du lapin (*Encephalitozoon cuniculi*). Comptes Rendus de l'Académie des Sciences **89**: 984-986

Liguory, O., C. Sarfati. F. Derouin, and J-M. Molina. 2001. Evidence of qifferent *Enterocytozoon bieneusi* genotypes in patients with and without human immunodeficiency virus infection. Journal of Clinical Microbiology **39**: 2672-2674.

Liu, J.J., and J.A. Shadduck. 1988. *Encephalitozoon cuniculi* infection in MRL/MPJ-LPR (lymphoproliferation) mice. Laboratory Animal. Science **38**: 685-688.

Lom, J. 1972. On the structure of the extruded microsporidian polar filament. Zeitschrift für Parasitenkunde **38**: 200-213.

López-Vélez, R.., M.C. Turrientes, C. Garrón, P. Montilla, R. Navajas, S. Fenoy, and C. del Aguila 1999. Microsporidiosis in travelers with diarrhea from the tropics. Journal of Travel Medicine **6**: 223-227.

Lores, B., I. López-Miragaya, C. Arias, S. Fenoy, J. Torres, and C. del Aguila 2002. Intestinal Microsporidiosis due to *Enterocytozoon bieneusi* in elderly human immunodeficiency virus-negative patients from Vigo, Spain. Clinical Infectious Diseases **34**: 918-921.

Lyngset, A. 1980. A survey of serum antibodies to *Encephalitozoon cuniculi* in breeding rabbits and their young. Laboratory Animal Science **30**: 558-561.

Mansfield, K.G., A. Carville, D. Shvetz, J. MacKey, S. Tzipori, and A.A. Lackner. 1997. Identification of an *Enterocytozoon bieneusi*-like microsporidian parasite in simian immunodeficiency- virus-inoculated macaques with hepatobiliary disease. American Journal of Pathology **150**: 1395-1405.

Margileth, A.M., A.J. Strano, R. Chandra, R. Neafie, M. Blum, and R.M. McCully. 1973. Disseminated nosematosis in an immunologically compromised infant. Archives of Pathology **95**: 145-150.

Mathis, A., A.C. Breitenmoser, and P. Deplazes. 1999. Detection of new *Enterocytozoon* genotypes in faecal samples of farm dogs and a cat. Parasite **6**: 189-193.

Matsubayashi, H., T. Koike, T. Mikata, and S. Hagiwara. 1959. A case of *Encephalitozoon*-like body infection in man. Archives of Pathology **67**: 181-187.

McDougall, R. J., M.W. Tandy, R.E. Boreham, D.J. Stenzel, and P.J. O'Donoghue. 1993. Incidental finding of a microsporidian parasite from an AIDS patient. Journal of Clinical Microbiology **31**: 436-439.

Mohn, S.F. 1982. Experimental encephalitozoonosis in the blue fox. Clinical and serological examinations of affected pups. Acta Veterinaria Scandinavia **23**: 503-514.

_____, and K. Nordstoga. 1975. Electrophoretic patterns of serum proteins in blue foxes with special reference to changes associated with nosematosis. Acta Veterinaria Scandinavia **16**: 297-306.

_____, _____, and I.W. Dishington. 1982. Experimental encephalitozoonosis in the blue fox. Clinical, serological and pathological examinations of vixens after oral and intrauterine inoculation. Acta Veterinaria Scandinavia **23**: 490-502.

Moretto, M., L. Casciotti, B. Durell, and I.A. Khan. 2000. Lack of CD4(+) T cells does not affect induction of CD8(+) T-cell immunity against *Encephalitozoon cuniculi* infection.. Infection and Immunity **68**: 6223-6232.

_____, B. Durell, J.D. Schwartzman, and I.A. Khan. 2001. Gamma delta T cell-deficient mice have a down-regulated CD8+ T cell immune response against *Encephalitozoon cuniculi* infection. Journal of Immunology **166**: 7389-7397.

Nelson, J.B. 1967. Experimental transmission of a murine microsporidian in Swiss mice. Journal of Bacteriology **94**: 1340-1345.

Niederkorn, J.Y., and J.A. Shadduck. 1980. Role of antibody and complement in the control of *Encephalitozoon cuniculi* infections by rabbit macrophages. Infection and Immunity **27**: 995-1002.

_____, _____, and E. Weidner. 1980. Antigenic cross-reactivity among different microsporidian spores as determined by immunofluorescence. Journal of Parasitology **66**: 675-677.

_____, J.K. Brieland, and E. Mayhew. 1983. Enhanced natural killer cell activity in experimental murine encephalitozoonosis. Infection and Immunity **41**: 302-307.

Nordstoga, K., and K. Westbye. 1976. Polyarteritis nodosa associated with nosematosis in blue foxes. Acta Pathologica and Microbiologica Scandinavia [A] **84**: 291-296.

Ombrouck, C., B. Romestand, J.M. da Costa, I. Desportes-Livage, A. Datry, F. Coste, G. Bouix, and M. Gentilini. 1995. Use of cross-reactive antigens of the microsporidian *Glugea atherinae* for the possible detection of *Enterocytozoon bieneusi* by western blot. American Journal of Tropical Medicine and Hygiene **52**: 89-93.

Orenstein, J.M. 1991. Microsporidiosis in the acquired immunodeficiency syndrome. Journal Parasitology **77**: 843-64.

_____, J. Chiang, W. Steinberg, P.D. Smith, H. Rotterdam, and D.P. Kotler. 1990. Intestinal microsporidiosis as a cause of diarrhea in human immunodeficiency virus-infected patients: a report of 20 cases. Human Pathology 21: 475-481.

_____, D.T. Dieterich, and D.P. Kotler. 1992a. Systemic dissemination by a newly recognized intestinal microsporidia species in AIDS. AIDS 6: 1143-1150.

_____, H.P. Gaetz, A.T. Yachnis, S.S. Frankel, R.B. Mertens, and E.S. Didier. 1997. Disseminated microsporidiosis in AIDS: are any organs spared? AIDS 11: 385-386.

_____, M. Tenner, A. Cali, and D.P. Kotler. 1992b. A microsporidian previously undescribed in humans, infecting enterocytes and macrophages, and associated with diarrhea in an acquired immunodeficiency syndrome patient. Human Pathology 23: 722-728.

Pakes, S.P., J.A. Shadduck, D.B. Feldman, and J.A. Moore. 1984. Comparison of tests for the diagnosis of spontaneous encephalitozoonosis in rabbits. Laboratory Animal Science 34: 356-359.

Pang, V.F., and J.A. Shadduck. 1985. Susceptibility of cats, sheep, and swine to a rabbit isolate of *Encephalitozoon cuniculi*. American Journal of Veterinary Research 46: 1071-1077.

Pulparampil, N., D. Graham, D. Phalen, and K, Snowden. 1998. *Encephalitozoon hellem* in two eclectus parrots *Eclectus roratus*: Identification from archival tissues. Journal of Eukaryotic Microbiology 45: 651-655.

Rabodonirina, M., M. Bertocchi, I. Desportes-Livage, L. Cotte, H. Levrey, M.A. Piens, G. Monneret, M. Celard, J.F. Mornex, and M. Mojon. 1996. *Enterocytozoon bieneusi* as a cause of chronic diarrhea in a heart-lung transplant recipient who was seronegative for human immunodeficiency virus. Clinical Infectious Diseases 23: 114-117.

Raynaud, L., F. Delbac, V. Broussolle, M. Rabodonirina, V. Girault, M. Wallon, G. Cozon, C.P. Vivares, and F. Peyron. 1998. Identification of *Encephalitozoon intestinalis* in travelers with chronic diarrhea by specific PCR amplification. Journal of Clinical Microbiology 36: 37-40.

Reetz, J., H. Rinder, A. Thomschke, H. Manke, M. Schwebs, and A. Bruderek. 2002 First detection of the microsporidium Enterocytozoon bieneusi in non-mammalian hosts (chickens). International Journal for Parasitology 32: 785-787.

Rinder, H., S. Katzwinkel-Wladarsch, and T. Loescher. 1997. Evidence for the existence of genetically distinct strains of *Enterocytozoon bieneusi*. Parasitology Research 83: 670-672.

Sandfort, J., A. Hannemann, H. Gelderblom, K. Stark, R.L. Owen, and B. Ruf. 1994. *Enterocytozoon bieneusi* infection in an immunocompetent patient who had acute diarrhea and who was not infected with the human immunodeficiency virus. Clinical Infectious Diseases 19: 514-516.

Sax, P.E., J.D. Rich, W.S. Pieciak, and Y.M. Trnka. 1995. Intestinal microsporidiosis occurring in a liver transplant recipient. Transplantation 60: 617-618.

Schmidt, E.C., and J.A. Shadduck JA. 1983. Murine encephalitozoonosis model for studying the host-parasite relationship of a chronic infection. Infection and Immunity 40: 936-942.

_____, and _____. 1984. Mechanisms of resistance to the intracellular protozoan *Encephalitozoon cuniculi* in mice. Journal of Immunology 133: 2712-2719.

Schwartz, D.A., D.C. Anderson, S.A. Klumpp, and H.M. McClure. 1998. Ultrastructure of atypical (teratoid) sporogonial stages of *Enterocytozoon bieneusi* (Microsporidia) in naturally infected rhesus monkeys *(Macacca mulatta)*. Archives of Pathology and Laboratory Medicine 122: 423-429.

Shadduck, J.A. 1969. *Nosema cuniculi: in vitro* isolation. Science 166: 516-517.

_____, and S.P. Pakes. 1971. Encephalitozoonosis (nosematosis) and toxoplasmosis. American Journal of Pathology 64: 657-671.

_____, and G.B. Baskin. 1989. Serologic evidence of *Encephalitozoon cuniculi* infection in a colony of squirrel monkeys (*Saimiri sciureus*). Laboratory Animal Science **39**: 328-330.

_____, and J.M. Orenstein. 1993. Comparative pathology of microsporidiosis. Archives of Pathology and Laboratory Medicine **117**: 1215-1219.

_____, R. Bendele, and G.T. Robinson. 1978. Isolation of the causative organism of canine encephalitozoonosis. Veteranary Pathology **15**: 449-460.

_____, R.A. Meccoli, R. Davis, and R.L. Font. 1990. Isolation of a microsporidian from a human patient. Journal of Infectious Diseases **162**: 773-776.

Silveira, H., E.U. Canning EU, and J.A. Shadduck. 1993. Experimental infection of athymic mice with the human microsporidian *Nosema corneum*. Parasitology **107**: 489-496.

Silverstein, B.E., E.T. Cunningham, T.P. Margolis, V. Cevallos, and I.G. Wong. 1997. Microsporidial keratoconjunctivitis in a patient without human immunodeficiency virus infection. American Journal of Ophthalmology **124**: 395-396.

Snowden, K.F., and J.A. Shadduck. 1999. "Microsporidia in Higher Vertebrates" In The Microsporidia and Microsporidiosis, Murray Wittner, ed., American Society for Microbiology, Washington, DC. Pp. 393-417.

Sobottka, I., H. Albrecht, J. Schottelius, C. Schmetz, M. Bentfeld, R. Laufs, and D.A. Schwartz. 1995. Self-limited traveller's diarrhea due to a dual infection with *Enterocytozoon bieneusi* and *Cryptosporidium parvum* in an immunocompetent HIV-negative child [letter]. European Journal of Clinical Microbiology and Infectious Diseases **14**: 919-920.

_____, F. Iglauer, T. Schüler, C. Schmetz, G.S. Visvesvara, H. Albrecht, D.A. Schwartz, N.J. Pieniazek, K. Bartscht, R. Laufs, and J. Schottelius. 2001. Acute and long-term humoral immunity following active immunization of rabbits with inactivated spores of various *Encephalitozoon* species. Parasitology Research **87**: 1-6.

Sprague, V, and J.J. Becnel. 1998. Note on the name-author-date combination for the taxon Microsporidies Balbiani, 1882, when ranked as a phylum. Journal of Invertebrate Pathology **71**: 91-94.

Stewart, C.G., W.S. Botha, and A.F. van Dellen. 1979. The prevalence of *Encephalitozoon* antibodies in dogs and an evaluation of the indirect fluorescent antibody test. Journal of the South African Veterinary Association **50**: 169-172.

_____, M.G. Collett, and H. Snyman. 1986. The immune response in a dog to *Encephalitozoon cuniculi* infection. Onderstepoort. Journal of Veterinary Research **53**: 35-37.

_____, F. Reyers, and H. Snyman. 1988. The relationship in dogs between primary renal disease and antibodies to *Encephalitozoon cuniculi* Journal of the South African Veterinary Association **59**: 19-21.

Svenungsson, B., T. Capraru, B, Evengårdi, R. Larsson, and M. Lebbad. 1998. Intestinal microsporidiosis in a HIV-seronegative patient. Scandinavian Journal of Infectious Diseases **30**: 314-316.

Szabo, J.R., and J.A. Shadduck. 1987. Experimental encephalitozoonosis in neonatal dogs. Veterinary Pathology **24**: 99-108.

_____, and _____. 1988. Immunologic and clinicopathologic evaluation of adult dogs inoculated with *Encephalitozoon cuniculi*. Journal of Clinical Microbiology **26**: 557-563.

Theng, J., C. Chan, M.L. Ling, and D. Tan. 2001. Microsporidial keratoconjunctivitis in a healthy contact lens wearer without human immunodeficiency virus infection. Ophthalmology **108**: 976-978.

Trinchieri, G. 1997. Cytokines acting on or secreted by macrophages during intracellular infection (IL-10, IL-12, IFN-gamma). Current Opinions in Immunology **9**: 17-23.

Tzipori, S., A. Carville, G. Widmer, D. Kotler, K. Mansfield, and A. Lackner. 1997. Transmission and establishment of a persistent infection of *Enterocytozoon bieneusi*, derived from a human with AIDS, in simian immunodeficiency virus-infected rhesus monkeys. Journal of Infectious Diseases **175**: 1016-1020.

Undeen, A.H. 1976. *In vivo* germination and host specificity of *Nosema algerae* in mosquitos. Journal of Invertebrate Pathology **27**: 343-347.

Van Gool, T., E.U. Canning, H. Gilis, M.A. Van Den Bergh Weerman, J.K. Eeftinck Schattenkerk, and J. Dankert. 1994. *Septata intestinalis* frequently isolated from stool of AIDS patients with a new cultivation method. Parasitology **109**: 81-289.

_____, V.C. Vetter, B. Weinmayr, A. Van Dam, F. Derouin, and J. Dankert. 1997. High seroprevalence of *Encephalitozoon* species in immunocompetent subjects. Journal of Infectious Diseases **175**: 1020-1024.

Visvesvara, G.S. 2002. In vitro cultivation of microsporidia of clinical importance. Clinical Microbiology Reviews **15**: 401-413.

_____, H. Moura, G.J. Leitch, and D.A. Schwartz. 1999. "Culture and Propagation of Microsporidia" In The Microsporidia and Microsporidiosis, Murray Wittner, ed., American Society for Microbiology, Washington, DC. Pp. 363-392.

Weber R, and R.T. Bryan. 1994. Microsporidial infections in immunodeficient and immunocompetent patients. Clinical Infectious Diseases **19**: 517-521.

_____, R.T. Bryan, D.A. Schwartz, and R.L. Owen. 1994. Human microsporidial infections. Clinical Microbiology Reviews **7**: 426-461.

Weber, R., D.A. Schwartz, and P. Deplazes. 1999. "Laboratory Diagnosis of Microsporidiosis." In The Microsporidia and Microsporidiosis, Murray Wittner, ed. American Society for Microbiology, Washington, DC. Pp. 315-362.

Weidner, E. 1975. Interactions between *Encephalitozoon cuniculi* and macrophages. parasitophorous vacuole growth and the absence of lysosomal fusion. Zeitschrift für Parasitenkunde **47**: 1-9.

_____, and L.D. Sibley. 1985. Phagocytized intracellular microsporidian blocks phagosome acidification and phagosome-lysosome fusion. Journal of Protozoology **32**: 311-317.

_____, W. Byrd, A. Scarborough, J. Pleshinger, and L.D. Sibley. 1984. Microsporidian discharge and the transfer of polaroplast organelle membrane into plasma membrane. Journal of Protozoology 31:195-198.

Weiss, L.M. 2000. "Molecular Phylogeny and Diagnostic Approaches to Microsporidia." In Cryptosporidiosis and Microsporidiosis, F. Petry ed., Contributions to Microbiolology, Basel, Karger. Pp. 209-235.

_____, A. Cali, A. Levee, D. LaPlace, H. Tanowitz, D. Simon, and M. Wittner. 1992. Diagnosis of *Encephalitozoon cuniculi* infection by western blot and the use of cross-reactive antigens for the possible detection of microsporidiosis in humans. American Journal of Tropical Medicine and Hygiene **47**: 456-462.

Wicher, V., R.E. Baughn, C. Fuentealba, J.A. Shadduck, F. Abbruscato, and K. Wicher. 1991 Enteric infection with an obligate intracellular parasite, *Encephalitozoon cuniculi*, in an experimental model. Infection and Immunity **59**: 2225-2231.

Williams, B.A.P., R.P. Hirt, J.M. Lucocq., and T.M. Embley. 2002. A mitochondrial remnant in the microsporidian *Trachipleistophora hominis*. Nature **418**: 865-869.

Wittner, M. "Historic Perspective on the Microsporidia: Expanding Horizons." In The Microsporidia and Microsporidiosis, Murray Wittner, ed. American Society for Microbiology, Washington, DC. 1999.

Wosu, N.J., R. Olsen, J.A. Shadduck, A. Koestner, and S.P. Pakes. 1977. Diagnosis of experimental encephalitozoonosis in rabbits by complement fixation. Journal of Infectious Diseases **135**: 944-948.

Wright, J.H., and E.M. Craighead. 1922. Infectious motor paralysis in young rabbits. Journal of Experimental Medicine **36**: 135-140.

Xiao, L., L. Li, G.S. Visvesvara, H. Moura, E. Didier, and A.A. Lal. 2001. Genotyping *Encephalitozoon cuniculi* by multilocus analyses of genes with repetitive sequences. Journal of Clinical Microbiology **39**: 2248-2253.

Yee, R.W., F.O. Tio, J.A. Martinez, K.S. Held, J.A. Shadduck, and E.S. Didier. 1991. Resolution of microsporidial epithelial keratopathy in a patient with AIDS. Ophthalmology **98**: 196-201.

Zhou, Z., and K. Nordstoga. 1993. Mesangioproliferative glomerulonephritis in mink with encephalitozoonosis. Acta Veterinaria Scandinavia **34**: 69-76.

# CHEMOTHERAPY OF MICROSPORIDIOSIS: BENZIMIDAZOLES, FUMAGILLIN AND POLYAMINE ANALOGUES

C.J. Bacchi[1], and L.M. Weiss[2]

[1]Department of Biology, Haskins Laboratory, Pace University, New York, New York [2]Departments of Pathology and Medicine, Albert Einstein College of Medicine, Bronx, New York. USA

**ABSTRACT:**
Microsporidia infection in humans is becoming increasingly recognized as a problem in immunocompromised as well as immune competent hosts. Human micropsoridian infections have been caused by several different microsporidian genera encompassing at least 12 species. Currently there are few controlled trials of the treatment of microsporidiosis in humans. Albendazole has been widely used for the treatment of microsporidiosis, however, it is clear from case reports that not all microsporidia are sensitive to this agent. Fumagillin has emerged as an alternative to albendazole treatment. This chapter discusses the recently identified enzyme target of fumagillin as well as the use of fumagilin in the treatment of microsporidiosis. Studies on polyamine metabolism of *Encephalitozoon cuniculi* indicate this organism relies primarily on uptake and interconversion of spermine to satisfy its polyamine requirements. Polyamine analogues having a bis aryl-3-7-3 configuration as well as pentamine and oligoamine analogues interfere with polyamine metabolism and cure model infections. These data suggest that the polyamine pathway should be a useful therapeutic target for the treatment of microsporidiosis. Polyamine metabolism in *Enc. cuniculi* is presented and discussed along with studies on the use of polyamine analogues in experimental models of infection.

**Key words:** Microsporidiosis, benzimidazole, albendazole, fumagillin, methionine aminopeptidases, polyamine metabolism.

## INTRODUCTION

Microsporidia are a phylum of ubiquitous obligate intracellular parasitic organisms consisting of over 1000 species distributed into over 140 genera. Speciation of these organisms is based on their ultrastructural morphology during their life cycles (for a review see Sprague, 1992). Molecular phylogenic techniques, based on rRNA and other genes, has also been used to provide information for speciation and classification of the microsporidia (for a review see Weiss and Vossbrinck, 1998). This molecular phylogenetic data suggests that Microsporidia are related to Fungi (Hirt et al., 1999; Weiss et al., 1999; Keeling et al., 2000).

Infections due to microsporidia have been described in both invertebrate and vertebrate hosts, including insects, fish, and mammals (Wittner and Weiss, 1999). Microsporidiosis was first recognized in 1857 with the identification of *Nosema bombycis* as the cause of disease in silkworms. While suspected of causing disease in humans in 1959, reports of clinical disease were rare in prior to 1985. With the advent of HIV infection (i.e. AIDS) and its attendant immune suppression there has been a significant increase in the number of reported human infections with the microsporidia. Microsporidiosis was first recognized in patients with chronic diarrhea, malabsorption and wasting (Modigliani et al., 1985; Desportes et al., 1985). Since then, microsporidia have been described in cases of hepatitis (Terada et al., 1987), peritonitis (Zender et al., 1989), keratoconjunctivitis, (Yee et al., 1991; Metcalfe et al., 1992; Rastreilli et al., 1994; Friedberg and Ritterband, 1999), over 200 cases of chronic diarrhea and wasting (Orenstein et al., 1990; Orenstein et al., 1992; Eeftinick et al., 1991; Leder et al., 1998; Dieterich et al., 1994), myositis (Chopp et al., 1993; Field et al., 1996; Cali et al.. 1996; Cali et al., 1998), cholangitis (Pol et al., 1993), dissemination to various organs (Cali et al., 1993; Asmuth et al., 1994; Gunnarsson et al., 1995), asymptomatic individuals (Rabeneck et al., 1993; Franzen et al., 1995), urethritis (Corcocan et al., 1996), sinusitis (Rossi et al., 1996), and the CNS in AIDS patients (Weber et al., 1997; Vavra et al., 1998), as well as in other non-HIV infected immune suppressed patients (Guerard et al., 1999). Microsporidia have also been described in immune competent patients as cases of self limited diarrhea, ocular infection and myositis. The microsporidia reported as pathogens in humans are listed in Table 1.

The most common presentation of microsporidiosis is still diarrhea and malabsorption (Kotler and Orenstein, 1998) due to infection with *Enterocytozoon bieneusi* and occasionally with *Encephalitozoon intestinalis*. Prevalence rates for microsporidiosis in case series of chronic diarrhea vary depending on the group and method of diagnosis (Kotler and Orenstein, 1998; Simon et al., 1991; Weber et al., 1992), but an average prevalence is 30%. Most of the AIDS patients who present with microsporidiosis as an enteric pathogen are severely immune suppressed, with CD4 counts below

100 (Kotler and Orenstein, 1998; Simon et al., 1991; Weber et al., 1992). *Enterocytozoon bieneusi* and *Enc. intestinalis* have also been associated with cholangitis including sclerosing cholangitis (Pol et al., 1993). Other gastrointestinal manifestations of microsporidiosis include hepatitis (Terada et al., 1987) and peritonitis (Zender et al., 1989). In addition, a patient with *Encephalitozoon intestinalis* has been described as presenting with an acute abdomen secondary to intestinal perforation that improved on treatment with albendazole (Soule et al., 1997).

**Table 1: Microsporidia identified as human pathogens**

*Brachiola*
        *Brachiola vesicularum*
        *Brachiola (Nosema) algerae*
        *Brachiola (Nosema) connori*
*Encephalitozoon*
        *Encephalitozoon cuniculi*
        *Encephalitzoon hellem*
        *Encephalitzooon intestinalis*
*Enterocytozoon*
        *Enterocytozoon bieneusi*
*Nosema*
        *Nosema ocularum*
        *Vittaforma (Nosema) cornea*
*Microsporidium*
        *Microsporidium africanus*
        *Microsporidium ceylonesis*
*Pleistophora*
        *Pleistophora sp.*
*Trachipleistophora*
        *Trachipleistophora hominis*
        *Trachipleistophora anthropopthera.*

Several AIDS patients presenting with refractory sinusitis have been demonstrated to have infection with *Encephalitozoon sp.* (Rossi et al., 1996; Rossi et al., 1999; Gritz et al., 1997). These patients were recognized by their failure to respond to antibacterial therapy and subsequently responded to albendazole. A case of reversible renal failure in an HIV positive patient due to *Enc. cuniculi* has been reported (Aarons et al., 1994). In this infected patient albendazole treatment led to resolution of the renal failure. Urethritis and prostatitis due to *Encephalitozoon sp.* that resolved with albendazole treatment has been described (Corcoran et al., 1996). Both *Encephalitozoon sp.* and *Brachiola algerae* have presented as cellulitis or skin nodules in immunocompromised hosts (leukemia and AIDS patients) (Wittner and Weiss, 1999). Both *Encephalitozoon cuniculi* (Weber et al., 1997) and *Trachipleistophora anthropopthera* (Vavra et al., 1998) have presented as encephalitis with mass lesions mimicking CNS Toxoplasmosis. *Pleistophora, Brachiola* and *Trachipleistophora* have presented as myositis (Chopp et al., 1993; Field et al., 1996; Cali et al., 1998) with fever, myalgia, weakness, muscle tenderness, and wasting.

Microsporidia have been described to cause stromal keratitis in immunocompetent patients with trauma, as well as superficial keratoconjunctivitis in patients with immunosuppression (Yee et al., 1991; Metcalf et al., 1992; Rastreilli et al., 1994; Friedberg and Ritterband, 1999; Diesenhouse et al., 1993; Gritz et al., 1997; Rossi et al., 1999; Font et al., 2000). In stromal keratitis the associated organisms have been *Nosema sp., Vittaforma corneae* (Silveira et al., 1995) or *Microsporidium sp.* and these organisms have penetrated and involved the deep stroma with spores located in corneal lamellae. In a case of stromal keratitis in an immunocompetent patient due to *Vittaforma corneae,* despite several treatment courses with topical fumagillin and systemic albendazole, the patient's stromal keratitis persisted, and eventually, a penetrating (full thickness) keratoplasty was needed in order to control the infection (Font et al., 2000). In cases of keratoconjuctivitis the associated organisms have been *Enc. hellem* and *Enc. cuniculi.* Infection is limited to the superficial cornea and conjunctiva, with keratoconjunctivitis characterized by punctate epithelial keratopathy. In such keratoconjuctivits, visual acuity can vary from 20/20 to light perception, and may fluctuate during the course of the illness.

Fumagillin and albendazole have demonstrated consistent activity against microsporidia both *in vitro* and *in vivo* (Molina et al., 1995; Molina et al., 2000; Gunnarsson et al., 1995; Weber et al., 1994; Dietrich et al., 1994; Dore et al., 1995; Rosberger et al., 1993; Haque et al., 1993; Beauvis et al., 1994; Ditrich et al., 1994; Diesenhouse et al., 1993; Blanshard et al., 1992; Franssen et al., 1995; Molina et al., 1998). Despite some reports of favorable treatment with metronidazole for intestinal infection with *Ent. bieneusi* this drug has not been effective in other studies (Pol et al., 1993; Eeftinck et al., 1991; Gunnarsson et al., 1995; Molina et al., 1998; Molina et al., 1997) and there is no *in vitro* activity of metronidazole against *Enc. cuniculi* (Beauvis et al., 1994). Other medications used without success in the treatment of gastrointestinal microsporidiosis are azithromycin, paromomycin (microsporidia lack the binding site for this drug) and quinacrine. Atovaquone has been anecdotally reported to have limited efficacy in microsporidiosis (Molina et al., 1997; Anwar-Bruni et al., 1996), however there is no *in vitro* activity (Beauvis et al., 1994). Sparfloxacin and chloroquine have demonstrated *in vitro* activity against microsporidia but have not been used clinically (Beauvis et al., 1994). Prophylaxis with trimethaprim-sulfamethoxazole is not effective for preventing microsporidiosis and this drug has no in vitro or in vivo activity against these organisms (Albrecht et al., 1995). Thalidomide (Sharpstone et al., 1995; Sharpstone et al., 1997) and octreotide have both been reported to decrease diarrhea in patients with microsporidiosis probably secondary to their effects on enterocytes. For a review of drugs used in microsporidiosis in humans and animals see Costa and Weiss (2000).

## BENZIMIDAZOLES

Benzimidazoles are widely used as antihelminthic agents in veterinary and human medicine, and as antifungal agents in agriculture. In addition, many of the benzimidazoles have been demonstrated to have activity against protists, such as *Giardia lamblia*. Albendazole, a benzimidazole, has been demonstrated to have activity against the microsporidia *Nosema bombycis* (Azizul Haque et al., 1993), *Encephalitozoon cuniculi* (Weiss et al., 1994; Colbourn et al., 1994) and *Encephalitozoon intestinalis* (Katiyar and Edlind, 1997). The primary mode of action of benzimidazoles involves their interaction with the cytoskeletal protein tubulin (Lacey, 1990; Edlind et al., 1996; Colbourn et al., 1994; Katiyar et al., 1994; Fayer and Fetter, 1995). In β-tubulin the aminoacids Glu-198, Phe-167, Arg-241 and Phe-200 correlate with benzimidazole susceptibility and are present in the β-tubulin sequence of *Enc. cuniculi* and *Enc. hellem* (Katiyar et al., 1994). The microsporidia *Vittaforma corneae*, related to *Ent. bieneusi*, is less sensitive to albendazole (Didier et al., 1998) and contains amino acid residues associated with albendazole resistance in other organisms [T.D. Edlind, personal communication].

Albendazole has demonstrated efficacy in human infections with *Encephalitozoon sp.* In *Enc. intestinalis* infection symptomatic improvement and clearance of parasites from post treatment stool examinations and intestinal biopsies was seen with albendazole treatment and in patients who died from other causes no organisms were seen on autopsy (Dore et al., 1995; Weber et al., 1994; Sbottka et al., 1995; Joste et al., 1996; Soule et al., 1997; Molina et al., 1998). Treatment with albendazole varied between 400 mg BID and TID, and lasted from two to four weeks with symptoms resolving within the first week of treatment. Severely immunocompromised patients relapsed after completion of treatment and thus maintenance therapy is indicated if the CD4 count is less than 200. *Encephalitozoon cuniculi* has been described as causing cerebral microsporidiosis (Weber et al., 1997) in an HIV-infected patient and treatment with albendazole appeared to be successful initially, with reduction in spore shedding and improvement in encephalitis, however, the patient continued to have urinary excretion of spores. In cases of chronic sinusitis and disseminated infection due to *Enc. hellem* treatment with 400 mg of albendazole twice daily resulted in resolution of symptoms and clearance of the organism (Lecuit et al., 1994; Visvesvara et al., 1994). In a patients with disseminated *Enc. cuniculi* infection involving the central nervous system, conjunctiva, sinuses, kidney and lungs clinical improvement was demonstrated with albendazole treatment (DeGroote et al., 1995; Weber et al., 1997). In a patient with disseminated infection with myositis due to *T. hominis* and in a patient with myositis due to a *Bachiola vesicularum* albendazole [400 mg BID] resulted in clinical improvement (Cali et al., 1998). Multiple studies on the utility of albendazole for the treatment of *Enerocytozoon bieneusi* infection have demonstrated a poor response to treatment with this agent (Leder et al., 1998; Dieterich et al., 1994; Dionisio et al., 1997; Blanshard et al., 1993). While morphologic changes could be seen in *Ent. bieneusi* during albendazole therapy the organism persisted and relapse occurred when the drug was stopped. There was, however, a decrease in bowel movements in about 50% of treated patients. Overall, while albendazole can be used for treatment in microsporidiosis due to *Ent. bieneusi*, it is clearly not an optimal agent.

## FUMAGILLIN

Fumagillin is a derivative of a natural product produced by *Aspergillus fumigatus*. In the 1950s this drug was used for the treatment of humans afflicted with amebiasis with *Entamoeba histolytica*. Fumidil B, (fumagillin bicyclohexammonium) is used in agriculture to treat honeybees infected with the microsporidium *Nosema apis* (Katznelson and Jamieson, 1953) and it has also had demonstrated activity against: *Nosema kingi*, a microsporidium of *Drosophila willistoni* (Armstrong, 1975), and *Octosporea muscaedomesticae*, a pathogen of the blowfly, *Phormia regina*. Fumagillin has been used to treat infections by microsporidia in various types of fish. Fumagillin and its derivatives are effective for the treatment of *Pleistophora anguillarum* in eels (Kano and Fukui, 1982; Kano et al., 1982; Molnar et al., 1987), *Sphaerospora renicola* in the common carp (Molnar et al., 1987), *Loma salmonae* in Chinook salmon (Kent and Dawe, 1994), and *Nucleospora (Enterocytozoon) salmonis* in Chinook salmon (Higgins et al., 1998). As *Nuc. salmonsis* is in the family Enterocytozoonidae, this suggested that fumagillin would have activity against *Ent. bieneusi.* Fumagillin, and its semisynthetic analogue TNP-470, have been studied *in vitro* and *in vivo* in microsporidia pathogenic for humans (Didier et al., 1997; Coyle et al., 1998; Molina et al., 2000; Molina et al., 1997; Diesenhouse et al., 1993) and have had activity demonstrated against *Enc. cuniculi, Enc. hellem, Enc. intestinalis* and *Vittaforma corneae.* TNP-470 was demonstrated to be more active (with an $ID_{50}$ of 0.001 ug/ml) than the parent compound fumagillin for these microsporidia (Didier et al., 1997; Coyle et al., 1998). TNP-470 was also active against microsporidiosis in a nude mouse model of *Enc. cuniculi* infection (Coyle et al., 1998). Fumidil B has been used topically as drops for the treatment of microsporidian keratoconjunctivitis due to *Enc. hellem* (Diesenhouse et al., 1993). Discontinuation of the drops resulted in recurrence of clinical symptoms and findings, such that the patients had to remain on continuous maintenance, albeit at a lesser dosage. Fumagillin at 20 mg TID given for three weeks was able to clear *Ent. bieneusi* spores from patients with *Ent. bieneusi* infection (Molina et al., 2000; Molina et al., 1997; Molina et al., 2002). A major limiting toxicity of this treatment was thrombocytopenia due to the direct effect of fumagillin on the bone marrow (Molina et al., 2000; Molina et al., 1997; Molina et al., 2002). This was reversible on the discontinuation of treatment.

Fumagillin has been demonstrated to bind irreversibly to a common bifunctional protein identified by mass spectrometry as methionine aminopeptidase type 2 (MetAP2) (Sin et al., 1997; Griffith et al., 1997; Griffith et al., 1998; Liu et al., 1998; Lowther et al., 1998). These drugs do not bind or inhibit the activity of methionine aminopeptidase type 1 (MetAP1). Both of these drugs selectively block the aminopeptidase, but not the elF-2 phosphorylation activity, of MetAP2. Methionine aminopeptidase-2 has been demonstrated to be the common target for other fumagillin analogues, e.g. TNP470/AGM-1470, as well as ovalicin. Crystallization studies have demonstrated that the specific binding site of the reactive epoxide of fumagillin and MetAP2 is a histidine residue at position 231 (Griffith et al., 1998). The ring epoxide in fumagillin is involved in covalent modification of the MetAP-2, whereas the side epoxide group of fumagillin appears to be dispensable. The specificity of fumagillin for MetAP2 is probably due to the three dimensional structure of the catalytic pocket of MetAP2 compared to methionine aminopeptidase type 1 (MetAP1). In yeast and higher eukaryotes two isoforms (type 1 and type 2) of MetAP exist (Sin et al., 1997). Yeast deficient in MetAP1 (ΔMetAP1 that only have MetAP2)

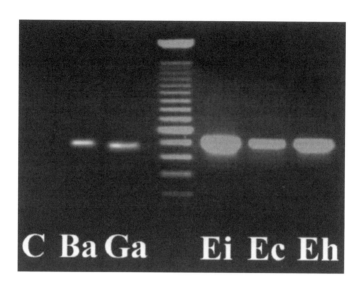

**Figure 1. PCR using primers M2 and M4 demonstrating the presence of MetAP2 in the microsporidia.** Center lane is a 100 bp ladder. Amplicons of 500 bp are demonstrated consistent with the predicted size of MetAP2 amplicon based on the location of these primers. C=control with no DNA; Ba = *Brachiola* (*Nosema*) *algerae*; Ga =*Glugea americanus* (*Spraguea lophi s*; Ei = *Encephalitozoon intestinalis* Ec: *Encephalitozoon cuniculi* Eh: *Encephalitozoon hellem*

were killed by fumagillin or ovalicin, but yeast deficient in MetAP2 (ΔMetAP2 that contain only MetAP1) were not (Sin et al., 1997). Deletion of both MetAP1 and MetAP2 is lethal to yeast. This confirms that fumagillin selectively targets MetAP2 and not MetAP1. Since fumagillin derivatives display tissue and species selectivity in their ability to inhibit MetAP2, this suggests that it should be possible to design selective MetAP2 inhibitors (Han et al., 2000; Taylor, 1993). It is not known why MetAP2 inhibition inhibits endothelial cell growth given the presence of both MetAP1 and MetAP2. However, it is likely that MetAP2 has an essential protein substrate in these cells, and that MetAP1 cannot process this protein. The presence of MetAP2 activity was demonstrated in *Enc. cuniculi* using L-Met-AFC and by antibody to Human MetAP2 (Weiss et al., 2001). Homology PCR cloning has demonstrated the presence of MetAP2 genes in the microsporidia *Enc. hellem, Enc. cuniculi, Enc.intestinalis, Ent. bieneusi, Brachiola algerae* and *Glugea americanus* (Figure one, Weiss LM unpublished data). These microsporidian MetAP2 genes lack the N-terminal region associated with the eIF-2 phosphorylation activity of other eukaryotic MetAP2 genes. No MetAP1 analogues were identified by this approach in any of the microsporidia examined. Confirmation is provided by analysis of the published *Enc. cuniculi* genome demonstrates the presence MetAP2 and the absence of MetAP1 in these organisms (Katinka et al., 2001). Thus, microsporidia are similar to the yeast MetAP1 knockout and inhibition of MetAP2 presents an attractive target for therapy. Recently, we have been able to express enzymatically active recombinant microsporidian MetAP2 (Weiss, L.M.; unpublished data), which should prove useful in the design of new drugs that are more specific for microsporidian MetAP2.

## POLYAMINES

The massive amount of information linking polyamines to biological functions has been summarized in a book by Seymour Cohen (Cohen, 1998). Polyamines are low molecular weight, multicationic molecules found universally in cells, whose intracellular content is closely regulated, and which participate in many interactions within the cell. The most commonly encountered polyamines are putrescine (1,4-diaminobutane) spermidine (1,8-diamino-4-azaoctane) and spermine (1,12-diamino-4,9-diazadodecane). Less commonly found are cadaverine (1, 5-diaminopentane) which is synthesized through lysine decarboxylase, and trypanothione (bis[glutathionyll]spermidine) found in parasitic protozoa of the Kinetoplastidae (Fairlamb and Cerami, 1992).

In mammalian cells and most protozoa, polyamine synthesis is initiated from the amino acid ornithine which is decarboxylated to putrescine via ornithine decarboxylase (ODC: Figure 2). Aminopropyl groups are added from decarboxylated S-adenosylmethionine (AdoMet) which then forms a purine nucleoside byproduct, 5'-methylthioadenosine (MTA). In most cells there are separate synthases catalyzing spermidine and spermine formation. In African trypanosomes and other kinetoplastids, spermine is not produced. Instead, two glutathione molecules are linked to a spermidine molecule to form trypanothione. This molecule, in turn, can only be reduced by a specific trypanothione reductase which gives trypanosomatids a unique way of handling oxidative stress. Glutathione reductase has not been detected in trypanosomatids (Fairlamb and Cerami,1992). Other protozoa have an alternative pathway to putrescine via arginine, arginine decarboxylase (ADC) and agmatine ureohydrolase to putrescine. This pathway is regarded as a plant pathway and *Cryptosporidium parvum* is the only protozoan demonstrated thus far to exhibit this route of putrescine synthesis (Keithly et. al., 1997).

Polyamines are also taken up by most cells through transporters whose mechanisms are not understood, although polyamine binding proteins on the surface of mammalian cells have been identified (Casero and Woster, 2001). Usually spermine is assimilated and interconverted to spermidine and putrescine via a retro-conversion pathway (Figure 2). In this pathway, spermine is converted to N'acetylspermine via spermidine/spermine acetyltransferse (SSAT). The acetylated spermine is then converted to spermidine and acetamidoproprionaldehyde (AAPA) by an FAD-dependent polyamine oxidase (PAO). The same set of enzymes converts spermidine to N'acetylspermidine and putrescine. A recent study has also demonstrated direct spermine to spermidine interconversion without the need for an acetylation step, producing aminopropanol as byproduct (Vujcic et. al., 2002). This enzyme, spermine oxidase, is thought to function in direct spermine to spermidine interconversion, while the SSAT pathway may function primarily to excrete excess polyamines in the acetylated state (Vujcic et. al., 2002).

Polyamines, as polycations, interact with many cell components and their functions are many. They are important in maintaining the three dimensional structure of nucleic acids and proteins, while metabolically acting to replace or assist metallic cations in enzyme catalysis, and are critical in ribosome structure and function. Thus, polyamines figure prominently in the synthesis of nucleic acids, proteins and other macromolecules. Cells mutated to diminish polyamine synthesis become auxotrophic for polyamines while cells treated with pharmacological agents blocking synthesis and/or transport do not divide and in many cases undergo apoptosis and cell death. Polyamine interactions with the cell membrane and signal transduction mechanisms and also, as hypusine, in cell cycle regulation are indications of their intracellular regulatory properties (Marton and Pegg, 1995; Cohen, 1998).

With many intracellular and intercellular functions attributed to polyamines, it is not surprising that their intracellular content is highly regulated through uptake, synthesis, interconversion and excretion. Major control points exist at ODC, AdoMetdc and SSAT. ODC is a pyridoxal-requiring amino-acid decarboxylase with a half-life of about 10 min. in mammalian cells. The content of ODC in the cell varies dramatically, depending on polyamine levels and the presence of another protein, antizyme, which specifically acts on the degradation of ODC and also acts to down-regulate polyamine transport (Murikami et. al., 1992; He et. al., 1994). Transport of polyamines is activated in some mammalian cells by insulin, estradiol and interleukins 3 and 4. Other factors, such as TNFα, IFNδ, P-glycoprotein efflux transporter MDR protein, all appear to reduce transport activity (Casero and Woster, 2001). The enzyme AdoMetdc, generating decarboxylated AdoMet, the aminopropyl group source for spermidine and spermine formation, is also highly regulated. AdoMetdc is synthesized as a proenzyme which is cleaved post-translationally to the active form. The mammalian enzyme is activated by putrescine, while the bacterial enzyme is $Mg^{++}$-activated. Increases of polyamine content lead to rapid decreases in ODC and AdoMetdc, due to a series of translational and post-translational controls (Porter et. al., 1992). A third regulatory point is SSAT, which attaches an acetyl group to the 3-carbon end of spermine and spermidine. This enzyme is the rate-limiting step of polyamine interconversion, and provides acetylated polyamines which can be excreted from the cell. SSAT is also highly inducible and has a short half-life. While a decrease in intracellular polyamine pools generally signals increase in ODC and AdoMetdc, SSAT levels increase with increasing internal polyamine concentrations. Changes in SSAT activity are accompanied by changes in mRNA levels indicating that SSAT gene expression is regulated at the level of transcription or stability of mRNA (Xiao and Casero, 1995). Since SSAT acetylates polyamines and neutralizes one amine positive charge, it also causes changes in binding of polyamines to target molecules in the cell. Since acetylated polyamines are also found outside of cells, SSAT may regulate polyamine pool sizes by encouraging excretion (Porter et. al., 1992). The overall effect of SSAT is that it increases in activity with increasing polyamine pool size, and converts polyamines to acetylated derivatives, encouraging excretion from the cell.

## POLYAMINE METABOLISM IN THE MICROSPORIDIA

Most of the studies on polyamines in Microsporidia have been carried out with *Encephalitozoon cuniculi* as model, since it can be cultured in vitro in RK-13 or MDCK cells (Visvesvara et. al., 1991). Infected cell monolayers can be maintained for several months allowing collection of mature spores in supernatant medium. Infected cell monolayers can also be harvested by scraping, and the collected material washed, homogenized and filtered through 12 and 5 μm filters removing most of the monolayer cells and debris. Utilizing a percoll gradient procedure (Weiss et. al., 1994; Green et. al., 1999; Bacchi et al., 2001), two bands are separable, a heavy band (1.102 - 1.119 g/ml) that is homogenous for pre-emergent spores and a light band (1.018 – 1.035 g/ml) that is is a mixture of early stages, immature spores and empty spore cases (Bacchi et al., 2001a). These techniques have allowed the study of polyamine synthesis and interconversion in intact pre-emergent spores and in cell-free extracts from mature spores.

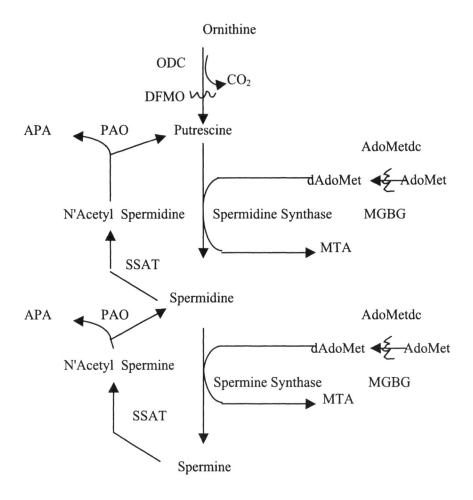

**Figure 2. Pathway of Polyamine Metabolism in Enc. cuniculi.**
*ODC, ornithine decarboxylase; AdoMet, S-adenosylmethionine; dAdoMet, decarboxylated AdoMet; MTA, methylthioadenosine; SSAT, spermidine/spermine N'acetyltransferase; PAO, polyamine oxidase; APA, acetamidoproprionaldehyde. Inhibitors are: DFMO, difluoromethylornithine; MGBG, methylglyoxal(bis)guanylhydrazone.*

Isolated pre-emergent spore preparations from the gradient procedure were analyzed for polyamine content using a reverse phase HPLC and fluorescence detection method able to quantitate 50 pmoles/sample (Yarlett and Bacchi, 1988). These preparations contained spermine 38.15 $\pm$ 27.8 and spermidine 23 $\pm$15.8 nmoles/mg protein (n = 7), but neither putrescine nor cadaverine were detected (Bacchi et al., 2001a; C. Bacchi, N. Yarlett, D. Rattendi, unpubl.), possibly due to metabolism during the harvesting and gradient procedures.

**Table 2. Activities of polyamine metabolizing enzymes in *Enc. cuniculi* mature spores[a].**

| Enzyme | Substrate | Activity (Range)[b] (nmoles/mg protein/2 h) | |
|--------|-----------|------------|------------|
| ODC | 1-[$^{14}$C]ornithine | 25.35 | (20-3 - 30.4) |
| AdoMetdc | S-[carboxy-$^{14}$C]AdoMet | 123.6 | (28.6 - 242.3) |
| SSAT | Spermine | 10.2 | (1.9 - 17.2) |
| | Spermidine | 4.4 | (1.0 - 9.5) |
| PAO | N'Acetylspermine | 1913 | (1240 - 2808) |

[a]Spores were obtained aseptically from supernatant culture media of infected RK-13 cells. They were washed in 1% SDS to remove any traces of adherent host cell protein. They were then disrupted in appropriate buffer using glass beads (Weiss et al., 1994).
[b]Activities represent the averages of four preparations. Assays were as follows: ODC, AdoMetdc, Bacchi et. al., 1993; SSAT: Libby et. al., 1989; PAO, Childs et. al., 1975).

Purified mature spores, washed in the presence of SDS, were homogenized using a bead-beater or Vortex-shaker with 425 - 600 μm glass beads, and the preparation clarified by centrifuging. These cell-free preparations of mature spores had ODC, AdoMetdc, SSAT and PAO activity (Table 2). DFMO, but not difluoromethylarginine (a specific inhibitor of arginine decarboxylase) inhibited ODC activity (Bacchi et. al., 2001a; C. Bacchi, D. Rattendi, unpubl.). MGBG, a competitive inhibitor of AdoMetdc blocked decarboxylation of AdoMet by 63% at 500 μM. SSAT activity was demonstrable with spermine the preferred substrate. PAO activity was elevated with respect to other enzymes of polyamine metabolism, and this activity differed from the mammalian enzyme with respect to substrate preference and susceptibility to inhibitors (C. Bacchi, D. Rattendi, in prepration): e.g., spermine was not a substrate for the *Enc. cuniculi* PAO, and

MDL-72527, an inhibitor of the mammalian enzyme, stimulated the *Enc.
cuniculi* PAO (C. Bacchi and D. Rattendi, unpubl.). Polyamine metabolism
was also examined in isolated, intact, pre-emergent spore preparations
harvested from RK-13 cells, and incubated for up to 2 h with radioactive
polyamines or precursors (Bacchi et. al., 2001a). These preparations
synthesized putrescine and spermidine from [$^3$H]ornithine + methionine.
DFMO at 500 μM inhibited putrescine synthesis 55%. [$^{14}$C]spermine was
assimilated at > 8 times the rate of ornithine and converted to spermidine and
putrescine over a 2h incubation period (Bacchi et al., 2001a). Isolated pre-
emergent spores were examined for ability to assimilate polyamines over a
short (15 min.) time period using rapid sampling techniques (Goldberg et al.,
1997). Results indicated that spermine was taken up with a $K_m$ value > 5
times lower than that of spermidine, putrescine or ornithine. Overall these
results indicate that polyamine uptake and interconversion have a more
important role than synthesis in *Enc. cuniculi* polyamine metabolism (C.
Bacchi, D. Rattendi, N. Yarlett, unpubl.).

## POLYAMINE METABOLISM AS A DRUG TARGET
Given the many functions of polyamines in cells and differences
found in their roles in procaryotic vs. eukaryotic cells, free-living vs.
parasitic organisms, normal vs. cancer cells, there has been an intense
interest in developing chemotherapy based on interference with polyamine
metabolism. Early studies focused on inhibition of polyamine synthesis as a
mechanism for reduction of polyamine levels. The first effective agent was
the enzyme-activated ODC inhibitor DL-α-difluoromethylornithine (DFMO,
eflornithine) which was developed by then Merrell-Dow (Metcalf et al.,
1978). Although DFMO was a potent, time-dependent irreversible inhibitor
of ODC, and was growth inhibitory to tumor cell lines, it found little clinical
use as an anti-tumor agent because of the rapid turnover of ODC in
mammalian cells, the resulting up-regulation of polyamine transport, and the
tendency of tumor cells exposed to DFMO to up-regulate ODC production
(Marton and Pegg, 1995).
One area in which DFMO has found application is the treatment of
laboratory models and clinical cases of African sleeping sickness (Bacchi
and McCann, 1987; Schechter and Sjoerdsma, 1989). The clinical success
of DFMO against *Trypanosoma gambiense* infections can be explained by
the presence of ODC with a long half-life (Phillips et al., 1987; Ghoda et.
al., 1992), the relatively low transport affinity for polyamines by African
trypanosomes (Bacchi and Yarlett, 1993), the reduction in trypanothione
content (Fairlamb et al., 1987), and compensatory over-production of
AdoMet and decarboxylated AdoMet by non-regulated trypanosome AdoMet

synthetase and AdoMetdc (Bacchi et al., 1995). These factors combine to block growth, block antigenic variation and present an antigenically stable population which is eliminated by the immunocompetent host (Bacchi and McCann, 1987).

A second, more potent inhibitor of polyamine synthesis was developed by Marion-Merrell-Dow in the late 1980's: 5'-{[(Z)-4-amino-z-butenyl]methylamino} 5'-deoxyadenosine (MDL 73811, AbeAdo: Danzin et. al., 1990). This agent is an enzyme-activated inhibitor of AdoMetdc, which rapidly inactivates the coenzyme by causing the transamination of the pyruvate residue at the active site (Shantz et. al., 1992). AbeAdo cured mouse model infections of *Trypanosoma brucei* and *T. rhodesiense* (Bacchi et al., 1992; Bitonti et al., 1990), although it had less activity in mouse tumor models (Pegg and McCann, 1992; Seiler et al., 1991). Several inhibitors of polyamine oxidase (PAO) were also synthesized by Marion Merrell-Dow in the mid-1980's (Bey et al., 1985). These included MDL-72527 [N',N$^4$-bis-(2,3-butadienyl)-1,4-butane diamine] and MDL-72521 [N'-methyl-N$^4$[2,3 butadienyl]-1,4 butane-diamine] which had $K_i$ values of 0.1 μM and 0.3 μM, respectively, for mammalian PAO (Seiler, 1995). However, inhibition of PAO by itself failed to inhibit mammalian cell growth (Bey et al., 1985; Seiler, 1995).

Today, the leading edge of research involving polyamine metabolism as drug-target concerns the development of polyamine analogues as opposed to inhibitors of specific enzymes of metabolism. This has arisen because of a greater understanding of the ability of cells to maintain homeostasis in polyamine levels. Analogues of polyamines are now available which enter the cell through polyamine transport, down-regulate ODC and AdoMetdc, and super-induce SSAT by up to several hundred fold (Porter et al., 1992; Frydman and Valasinas, 1999; Casero and Woster, 2001). Since the analogues may mimic spermine, the cell appears to be in a condition of polyamine excess. As a result, polyamine synthesis and uptake are shut down, remaining pools of spermine are acetylated and excretion is enhanced (Porter et al., 1991; Marton and Pegg, 1995). Because acetylation decreases the net positive charge of spermine, it also alters its binding capabilities and hence, function (Porter et al., 1992). Since these analogues do not substitute physiologically for the natural polyamines, the net result on the cell is blockade of division and apoptosis (Hu and Pegg, 1997).

The first series of polyamine analogues, developed in the 1980's, differed from the natural polyamines only in the length of the carbon chains between the amine groups (e.g., 3-3-3, 3-4-3). Other modifications included the addition of alkyl groups such as methyl and ethyl at the terminal nitrogens (Figure 3). These compounds were effective growth inhibitors for

L1210 and other cancer cell lines. Di- and tri-amine analogues based on spermidine, with or without terminal alkyl ($-CH_3$ or $-CH_2$ $CH_3$) substitutions proved less effective in tumor suppression (Edwards et. al., 1990; 1991), however, tetramines with the structure $R-NH-(CH_2)_3-NH-(CH_2)_7-NH-(CH_2)_3-NH-R$ (3-7-3) were active. Related compounds based on symmetrical addition of benzene and other ring structures to the 3-7-3 backbone resulted in increased activity. In particular, the bis(benzyl) 3-7-3 analogue MDL-27695 (Figure 3) was highly active (Edwards et al., 1990, 1991). It appeared that MDL 27695 and related compounds were substrates for PAO, since co-administration of a PAO inhibitor potentiated their effects (Bitonti et al., 1990). Spermine analogs such as bis(ethyl)norspermine (BE 3-3-3: Figure 3) were found to be highly active in inducing SSAT, reducing spermidine and spermine levels, and increasing excretion of acetylated polyamines (Porter et al., 1992). Polyamine uptake was also greatly inhibited and mitochondria were swollen and had reduced ATP (Bergeron et al., 1995). Although these agents were active as anti-tumor agents, the terminal bis (ethyl) aminopropyl groups are metabolized initially by PAO and converted to 3-aminoproprionaldehyde or additionally to acetamidoproprionaldehyde if both SSAT and PAO are operating (Bergeron et al., 1995). This metabolism served to shorten their effective time *in vivo* and decrease their efficacy. Analogues having lateral aminobutyl groups however, are not subject to metabolism and remain active *in vivo* for a longer period (Bergeron et al., 1996; Frydman and Valasinas, 1999). Present thinking on polyamine analogues indicates that those having superior anti-proliferative activity *in vitro* and *in vivo* share the following structural characteristics: carbon skeleton based on repeating N-butyl groups or a 3-7-3 configuration; terminal ethyl groups; symmetrical (bis) or unsymmetrical N-terminal alkylsubstitutions; conformational restriction about the central carbons, using double bonds and yielding cis- or trans-configurations (see Figure 3: Casero and Woster, 2001; Frydman and Valasinas, 1999; Valasinas et al., 2001).

## POLYAMINE ANALOGUES AS ANTI-MICROSPORIDIAL AGENTS

Developments in the area of parasite biochemistry have led to a study of antipolyamine agents as potential chemotherapeutic agents. As noted previously, the only clinically approved use of an inhibitor of polyamine synthesis is the ODC inhibitor DFMO for early and late stage gambian (West African) sleeping sickness. Heightened interest in opportunistic infections refractory to chemotherapy in immunocompromised persons has now led to examination of polyamine metabolism as potential drug target. Studies with *Cryptosporidium parvum, Enc. cuniculi* and

*Pneumocystis carinii* have indicated that this approach may be feasible (Bacchi et al., 2002; Waters et al.,2000; Merali et al., 2000).

With respect to the Microsporidia, two groups of polyamine analogues have been studied for activity vs. *Enc. cuniculi* in *in vitro* and *in vivo* models. One group consists of tetramine, pentamine, and oligoamine analogues based on repeating aminobutyl groups which are terminally bis-ethylated (Frydman and Valasinas, 1999; Valasinas et al., 2001). Some are sterically hindered by the introduction of a double or triple bonds, or cycloalkyl groups (Figure 3). The other group consists of bis- or mono-N'cyclobutyl, N'cyclopentenyl, cyclohexylmethyl and dibenzyl groups using a 3-7-3 backbone (Woster and Casero, 2001; Zou et al., 2001). We were drawn to these analogs as potentially interesting because of the significant uptake and interconversion of spermine observed in pre-emergent spores

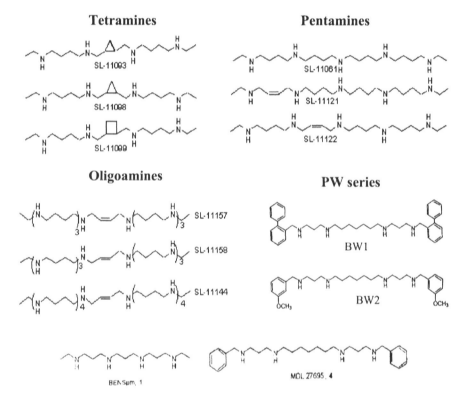

**Figure 3. Structure of some polyamine analogues.** Tetramines, Pentamines, Oligoamines synthesized by Dr. B. Frydman. Compounds #BW1, #BW2 synthesized by Dr. P. Woster. BENSpm is bisethyl norspermine or BE 3-3-3; MDL 27695 is bis-benzyl (BB) 3-7-3; SL11061 is BE-4-4-4-4.

(Bacchi et al., 2001a), and because DFMO failed to reduce *Enc. cuniculi* parasitemia in the RK-13 monolayer culture screen while bis-ethylnorspermine (3-3-3) eliminated the parasitemia (Coyle et al., 1996). More than 40 of these polyamine analogues were examined for ability to inhibit growth of *Enc. cuniculi* in the monolayer screen (Bacchi et al., 2002; Zou et al., 2001). $IC_{50}$ values were obtained for representatives of each group of compounds. Compounds emerging as most active had $IC_{50}$ values $\leq$ 10 μM, with some having activity at $\leq$ 1 μM. Activities for each class of compound in the *in vitro* screen are listed in Table 3

Of the three classes of (N-butyl)-based compounds studied, SL11093 (tetramine), SL 11061 (pentamine) SL-11144, and SL-11158 (oligoamines) were among the most active, and considering *in vivo* tolerance (toxicity) tests, were selected for *in vivo* testing (Bacchi et. al., 2002). Of the aryl-substituted 3-7-3 analogs, a bis(phenoxy) compound was not an effective growth inhibitor, but a bis(phenylbenzyl) derivative (BW-1) was effective at < 1 μM ($IC_{50}$ 0.47 μM: Table 3) and was also well tolerated in mice (Zou et. al., 2001). Two well-validated *Enc. cuniculi in vivo* screens were used to examine activity of the most active compounds, using C57Bl/6J ΔCD8 mice in one laboratory and BALB/c nude (nu/nu) mice in the other (Didier et al., 1994; Kahn et al., 1999). Curative doses for an *in vivo screen* in which mice were given a total of 10 daily i.p. doses, 5 days on, 2 days off, 5 days on, are given in Table 3. Although the curative doses approach $1/10^{th}$ of the maximum tolerated dose obtained on a parallel toxicity study, it is important to note that the minimum doses and times of dosing are still under investigation. For example, 100% cures were recently obtained with SL-11158 at a dose of 1 mg/kg for only 6 days duration, a reduction of 40% of the total dose (L.M. Weiss. B. Frydman, unpubl.).

In light of the significant uptake of spermine, as opposed to spermidine or putrescine, as shown in purified pre-emergent spore populations, we examined uptake of [$^{14}$C]spermine in the presence of analogues using a sensitive HPLC assay to detect radiolabeled products in intact cells and supernatant incubation medium (Yarlett and Bacchi, 1988; Bacchi et al., 2001b). At 10 μM, SL-11158 and SL-11144 reduced total [$^{14}$C]spermine incorporation by 63% and 66% respectively, after a 2 h incubation. Spermine is metabolized by the pre-emergent spores to spermidine and an unknown lower molecular weight compound, possibly acetaminoproprionaldehyde. Although significant amounts of this substance are found intracellularly, about 60% of the unknown is excreted into the incubation medium. In the presence of SL-11158, an overall reduction of 52% of the total unknown produced (13 nmoles/mg protein/2 h) was found. The amount of reduction of product was about the same in the extra-cellular as well as intracellular material. Both of the amine analogues at 10 µM significantly inhibited spermidine production as well as production of the unknown metabolite. In addition, SL-11158 was found to exhibit a mixed type of inhibition of SSAT activity from mature spore preparations, with a $K_m$ of 0.24 mM in the presence of spermine as substrate (C.J. Bacchi, N. Yarlett, & D. Rattendi, unpubl.).

**Table 3. Polyamine analogs active as inhibitors of growth *in vitro* and curative *in vivo* vs. *Enc. cuniculi*.[a]**

| Compound | Class | $IC_{50}$[b] (µM) | Maximum Tolerated[c] Dose (mg/kg) | Minimum Curative[d] Dose (mg/kg) |
|----------|-------|-------------------|-----------------------------------|----------------------------------|
| SL-11061 | Pentamine | 42.5 (2) | > 10 | 1.25 |
| SL-11093 | Tetramine | 270  (2) | > 50 | 1.0 |
| SL-11144 | Oligoamine | 0.62 (4) | 10 | 1.25 |
| SL-11158 | Oligoamine | 8.2  (3) | 10 | 1.0 |
| BW-1 | Aryl 3-7-3 | 0.47 (3) | 7.5 | 1.0 |

[a]Data from Zou et. al., 2001; Bacchi et.al., 2002.
[b]Growth inhibition in RK-13 cells grown to confluency with a 50 - 80% infection and treated for 8 days with the analogue; (n) number of trials.
[c]Groups of 3 mice were dosed once daily i.p. for 5 days, 2 days no dosing, 5 days dosing.
[d]Minimum dose, given by the above schedule in which mice infected with $10^6$ - $10^7$ *Enc. cuniculi* spores, mice survived > 30 days beyond death of infected untreated controls with no evidence of putrescine according to standard histologic and PCR study of tissues (Moreto et al., 2000).

These findings indicate that the amine analogues have multiple effects on polyamine metabolism in *Enc. cuniculi*. Other effects are possible, such as inhibition of parasite PAO, or the analogues themselves may be substrates for parasite PAO, yielding toxic metabolites, as was the case for MDl-27695 (Bitonti et al., 1990). Although it appears that the polyamine analogues interfere with spermine uptake and metabolism, there is no indication as to whether polyamine transport or SSAT activity are upregulated in *Enc. cuniculi* by exposure to these agents, as they are in mammalian cell lines (Casero and Woster, 2001). Moreover, since uptake studies were done only with pre-emergent spores, it is not known whether immature meront-like stages of the parasite accumulate the analogues more or less avidly. Since the amount of agent available to the parasite depends on accumulation by the host cell, any gauge of efficacy will also partly depend on the degree to which each amine is transported into host cells. The above considerations should help to explain the enigma of variability of *in vitro* $IC_{50}$ values and the seemingly equivalent *in vivo* efficacy in the mouse model infections (Table 3).

## FUTURE PROSPECTS

Since a number of polyamine analogues with different structural properties were effective against *Enc. cuniculi in vitro* and *in vivo* it is possible that the polyamine transporter(s) in microsporidia may be nonspecific. The activity of these compounds also increases the possibility that more effective analogues with higher therapeutic indices can be synthesized. Whether analogues that are active against *Enc. cuniculi in vivo* and *in vitro* will also be effective against other genera of Microsporidia remains to be determined. Recent studies with tetramine analogs (e.g., SL-11061) have indicated increased cytotoxicity was obtained by *cis*-type conformational restriction at one end of the linear molecule, possibly due to enhanced binding to DNA and displacement of natural polyamines (Valasinas et al., 2001). This fact further increases the probability that polyamine analogues can be successfully developed as therapeutic agents for microsporidiosis.

## REFERENCES

Aarons, E. J., D. Woodrow, W. S. Hollister, E. U. Canning, .Francis, and B. G. Gazzard. 1994. Reversible renal failure caused by a microsporidial infection. AIDS 8: 1119-1121.

Albrecht, H., I. Sobottka, H.J. Stellbrink, and H. Greten. 1995. Does the choice of Pneumocystis carinii prophylaxis influence the prevalence of *Enterocytozoon bieneusi* microsporidiosis in AIDS patients? [letter]. AIDS. 9: 302-3.

Anwar-Bruni, D.M., S.E. Hogan, D.A. Schwartz, C.M. Wilcox, R.T. Bryan, and J.L. Lennox. 1996. Atovaquone is effective treatment for the symptoms of gastrointestinal microsporidiosis in HIV-1-infected patients. AIDS 10: 619-23.

Armstrong, E. 1975. Fumidil B and benomyl: chemical control of *Nosema king* in *Drosophila willistoni*. Journal of Invertebrate Pathology 27: 363-366.

Asmuth, D. M., P. C. DeGirolami, M. Federman, C.R. Ezratty, D.K. Pleskow, G. Desai and CA Wanke. 1994. Clinical features of microsporidiosis in AIDS. Clinical and Infectious Disease 18: 819-825.

Azizul Haque, M., W. S. Hollister, A. Willcox, Canning E U. 1993. The antmicrosporidial activity of albendazole. Journal of Invertebrate Pathology 62: 171-177.

Bacchi, C.J, H. C. Nathan, N. Yarlert, B. Goldberg, P. P. McCann, Am J. Bitonti, and A. Sjoerdsma. 1992. Cure of murine *Trypanosma brucei rhodesiense* infections with an S-adenyosylmethione decarboxylase inhibitor. Antimicrobial Agents Chemother 36:2736-2740.

_____, J. Garofalo, M.A. Ciminelli, D. Rattendi, B. Goldberg, P.P. McCann, and N. Yarlett. 1993. Resistance to DL-α-difloromethylorntihine by clinical isolates of *Typansoma brucei rhodesiense*: role of S-adenosylmethionine. Biochemical Pharmacology 46: 471–481.

_____, B. Goldberg, J. Garofalo-Hannan, D. Rattendi, P. Lyte, and N. Yarlett. 1995. Fate of soluble methionine in African trypanosomes: effects of metabolic inhibitors. Biochemical Journal **309**: 737–743.

_____, _____, L.M. Weiss, N. Yarlett, P. Takvorian, A. Cali, and M. Wittner. 2001a. Polyamine synthesis and interconversion by the Microsporidian *Encephalitozoon cuniculi*. Journal of Eukaryotic Microbiology **48**: 374–381.

_____, and P.P. McCann. 1987. Parasitic protozoa and polyamines. In: Inhibition of polyamine metabolism: biological significance and basis for new therapies. P.P. McCann, A.E. Pegg, and A. Sjoerdsma (eds.) pp. 317-344. Academic Press, Orlando, FL.

_____, D. Orozco, L.M. Weiss, B. Frydman, V. Valasinas, N. Yarlett, L.J. Marton, M. Wittner. 2001b. SL-11158, a synthetic oligoamine, inhibits polyamine metabolism of *Encephalitozoon cuniculi*. Journal of Eukaryotic Microbiology **48**: 92S-94S.

_____, L.M. Weiss, S. Lane, B. Frydman, A. Valasinas, V. Reddy, J.S. Sun, L.J. Marton, I. Khan, M. Moretto, N. Yarlett, and M. Wittner. 2002. Novel Synthetic Polyamines are effective in the treatment of experimental microsporidiosis: an opportunistic AIDS-associated infection. Antimicrobial Agents and Chemotherapy **46**: 55-61.

_____, and N. Yarlett. 1993. Effects of antagonists of polyamine metabolism on African trypanosomes. Acta Tropica **54**: 225–236.

Beauvais, B., C. Sarfati, S. Challier, and F. Derouin. 1994. *In vitro* model to assess effect of antimicrobial agents on *Encephalitozoon cuniculi*. Antimicrobial Agents and Chemotherapy **38**: 2440-2448.

Bergeron, R.J., W.R. Wiemar, G. Luchetta, C.A. Sninsky, and J. Weigand. 1995. Metabolism and pharmacokinetics of $N^1,N^{11}$-diethylhomospermine. Drug Metabolism and Disposition **23**: 1117-1125.

_____, _____, _____, _____, _____, 1996. Metabolism and pharmacokinetics of $N^1,N^{14}$-diethylhomospermine. Drug Metabolism and Disposition **24**: 334-343.

Bey, P., F.N. Bolkenius, N. Seiler, and P. Casara. 1985. N-2,3 butadienyl-1,4-butanediamine derivatives: potent irreversible inactivators of mammalian polyamine oxidase. Journal of Medicinal Chemistry **28**: 1–2.

Blanshard, C, D. S. Ellis, S. P. Dowell, G. Towey, and B.G. Gazzard. 1993. Electron microscopic changes in *Enterocytozoon bieneusi* following treatment with albendazole. Journal of Clinical Pathology **46**: 898-902.

_____, _____, D.G. Tovey, S. Dowell, and B.G. Gazzard. 1992. Treatment of intestinal microsporidiosis with albendazole in patients with AIDS. AIDS **6**: 311-313.

Bitonti, A.J., J.A. Dumont, T.L. Bush, D.M. Stemerick, M.L. Edwards, and P.P. McCann. 1990. Bis(benzyl) polyamine analogs as novel substrates for polyamine oxidase. Journal of Biological Chemistry **265**: 382-388.

Cali, A, D. P. Kotler, and J. M. Orenstein. 1993. *Septata intestinalis* N.G., N. Sp., an intestinal Microsporidian associated with chronic diarrhea and dissemination in AIDS patients. Journal of Eukaryotic Microbiology **40**: 101-112.

_____, P. M. Takvorian, S. Lewin, M. Rendel, C Sian, M. Wittner, and L. M. Weiss. 1996. Identification of a new nosema-like microsporidian associated with myositis in an AIDS patient. Journal of Eukaryotic Microbiology **43**: 108S.

_____, _____, _____, _____, 1998. *Brachiola vesicularum*, N.G., N.Sp., a new microsporidium associated with AIDS and myositis. Journal of Eukaryotic Microbiology **45**:

240-251.

Casero, R.A., and A.E. Pegg. 1993. Spermidine/spermine $N^1$-acetyltransferase: the turning point in polyamine metabolism. FASEB Journal **7**: 653–661.

_____, and P.M. Woster. 2001. Terminally alkylated polyamine analogues as chemotherapeutic agents. Journal of Medicinal Chemistry **44**: 1–26.

Chopp, G. L., J. Alroy, L. S. Adelman, J. C. Breen, and P. R. Skolnik. 1993. Myositis due to *Pleistophora* (Microsporidia) in a patient with AIDS. Clinical Infectious Disease **16**: 15-21.

Childs, R.E., and W.G. Bardsley. 1975. The steady state kinetics of peroxidase with 2,2' azino-di-(3-ethyl-benzthiazoline-6-sulphonic acid) as chromogen. Biochemistry Journal **145**: 93–103.

Cohen, S. 1998. A guide to the polyamines. Oxford University Press, New York.

Colbourn, N. I., W. S Hollister, A.Curry, and E. U. Canning. 1994. Activity of albendazole against *Encephalitozoon cunicul in vitro*. European Journal of Protistology **30**: 211-220.

Corcoran, G. D., J. R. Isaacson, C. Daniels, and P. L. Chiodini. 1996. Urethritis associated with disseminated microsporidiosis: clinical response to albendazole. Clinical Infectious Disease **22**: 592-593.

Costa, S., and L. Weiss. 2000. Drug treatment of microsporidiosis. Drug Resistance Updates **3**: 1-16.

Coyle. C., C. Bacchi, N. Yarlett, H.B. Tanowitz, M. Wittner, and L.M. Weiss. 1996. Polyamine Metabolism as a Therapeutic Target for Microsporidia. Journal of Eukaryotic Microbiology **43**: 96S.

_____, M. Kent, H. B. Tanowitz, M. Wittner, and L. M. Weiss. 1998. TNP-470 is an effective antimicrosporidial agent. Journal of Infectious Disease **177**: 515-518.

Danzin, C., P. Marchal, and P. Casara. 1990. Irreversible inhibition of rat S-adenosylmethionine decarboxylase by 5'{[(Z)-4-amino-2-butenyl]methylamino}-5'-deoxyadenosine. Biochemical Pharmacology **40**: 1499–1503.

De Groote, M.A., G. Visvesvara, M.L. Wilson, N.J. Pieniazek, S.B. Slemenda, A.J. daSilva, G.J. Leitch, R.T. Bryan, and R. Reves. 1995. Polymerase chain reaction and culture confirmation of disseminated *Encephalitozoon cuniculi* in a patient with AIDS: successful therapy with albendazole. Journal of Infectious Disease **171**: 1375-1378.

Desportes, I., Y. Le Charpentier, A. Galian, F. Bernard, B. Cochand-Priollet B, A. Lavergne, P. Ravisse and R. Modigliani. 1985. Occurrence of a new microsporidian: *Enterocytozoon bieneusi* n.g., n.sp., in the enterocytes of a human patient with AIDS. Journal of Protozoology **32**: 250-254.

Didier. E. S. 1997. Effects of albendazole, fumagillin, and TNP-470 on microsoridial replication in vitro. Antimicrobial Agents and Chemotherapy **41**: 1541-1546.

_____, J. A. Maddry, C. D. Kwong, L. C. Greeen, K. F. Snowden, and J. A. Shadduck. 1998. Screening of compounds for antimicrosporidial activity *in vitro*. Folia Parasitologica **45**: 129-139.

_____, P.W. Varner, P.J. Didier, A.M. Aldras, N.J. Millichamp, and M. Murphey-Corb. 1994. Experimental microsporidiosis in immuno-compotent and immuno-deficient mice and monkeys. Folia Parasitologica **41**: 1–11.

Diesenhouse, M.C., L.A. Wilson, G.F. Corrent, G.S. Visvesvara, H.E. Grossniklaus, and R.T. Bryan. 1993. Treatment of microsporidial keratoconjunctivitis with topical fumagillin [see comments]. American Journal of Ophthalmology **115**: 293-298.

Dieterich, D. T., E. A. Lew, D. P. Kotler, M. A. Poles, and J. M. Orenstien. 1994. Treatment with albendazole for intestinal disease due to *Enterocytozoon bieneusi* in patients with AIDS. Journal of Infectious Disease **169**: 178-183.

Dionisio, D., L. I. Manneschi, S. Di Lollo, A. Orsi, G. Sterrantino, M. Meli, M. Gabbrielli, A. Tani, A. Papucii, and F. Leoncini. 1997. *Enterocytozoon bieneusi* in AIDS: symptomatic relief and parasite changes after furazolidone. Journal of Clinical Pathology **50**: 472-476.

Ditrich, O., Z. Kucerova, and B. Koudela, 1994. *In vitro* sensitivity of *Encephalitozoon cuniculi* and *E. hellem* to albendazole. Journal of Eukaryotic Microbiology **41**: 37S.

Dore, G. J., D. J. Marriott, M. C. Hing, J. L. Harkness, and A. S. Field. 1995. Disseminated microsporidiosis due to *Septata intestinalis* in nine patients infected with the Human Immunodeficiency Virus: response to therapy with albendazole. Clinical Infectious Disease **21**: 70-76.

Edlind T, S. Katiyar, G. Visvesvara, and J. Li. 1996. Evolutionary origins of microsporidia and basis for benzimidazole sensitivity: an update. Journal of Eukaryotic Microbiology **43**: 109S.

Edwards, M.L., N.J. Prakash, K.M. Stemerick, S.P. Sunkara, A.J. Bitonti, J.A. Dumont, P.P. McCann, P. Bey, and A. Sjoerdsma. 1990. Polyamine analogues with antitumor activity. Journal of Medicinal Chemistry **33**: 1369–1375.

_____, K.M. Stemetick, A.J. Bitonti, J.A. Dumont, P.P. McCann, P. Bey, and A. Sjoerdsma. 1991. Antimalarial polyamine analogs. Journal of Medicinal Chemistry **34**: 567–574.

Eeftinck Schattenkerk, J.K., T. van Gool, R.J. van Ketel, J.F. Bartelsman, C.L. Kuiken, W.J. Terpstra, and P. Reiss. 1991. Clinical significance of small-intestinal microsporidiosis in HIV-1-infected individuals [see comments]. Lancet **337**: 895-898.

Fairlamb, A.H., and A. Cerami. 1992. Metabolism and functions of trypanothionine in the kinetoplastida. Annual Reviews in Microbiology **46**: 695–729.

_____, G.B. Henderson, C.J. Bacchi, and A. Cerami. 1987. *In vivo* effects of difluoromethylornithine on trypanothione and polyamine levels in bloodstream forms of *Trypanosoma brucei*. Molecular and Biochemical Parasitology **24**: 185–191.

Fayer R, and R. Fettere. 1995. Activity of benzimidazoles against cryptosporidiosis in neonatal BALB/c mice. Journal of Parasitology **81**: 794-795.

Font, R L, A.N. Samaha , M. J. Keener, P. Chevez-Berrios, and J. D. Goosey. 2000. Corneal microsporidiosis: report of case, including electron microscopic observations. Ophthalmology **107**: 1769-1775.

Franssen, F.F., J.T. Lumeij, and F. van Knapen. 1995. Susceptibility of *Encephalitozoon cuniculi* to several drugs *in vitro*. Antimicrobial Agents and Chemotherapy **39**: 1265-8.

Franzen, C., D. A Schwartz, G. S. Visvesvara, A. Muller, A. Schwenk, B. Salzberger, G. Fatkenheuer, P. Hartmann, G. Mahrle, and V. Diel. 1995. Immunologically confirmed disseminated, asymptomatic *Encephalitozoon cuniculi* infection of the gastrointestinal tract in a patient with AIDS. Clinical of Infectious Disease **21**: 1480-1484.

Friedberg, D. N., and D. C. Ritterband. 1999. Ocular microsporidiosis. The microsporidia and microsporidiosis (Murray Wittner, editor, and Louis M. Weiss, contributing editor) American Society for Microbiology, Washington D.C.

Field, A S, J. Marriot, S. T. Milliken, B. J. Brew, E. U. Canning, J. G. Kench, P. Darveniza, and J. L. Harkness. 1996. Myositis associated with a newly described microsporidian, *Trachipleistophora hominis* in a patient with AIDS. Journal of Clinical Microbiology **34**: 2803-2811.

Frydman, N., and A. Valasinas. 1999. Polyamine based chemotherapy of cancer. Expert

Opinion and Therapeutic Patents. **9**: 1055–1068.

Ghoda, L., D. Sidney, M. Macrae, and P. Coffino. 1992. Structural elements of ornithine decarboxylase required for intracellular degradation and polyamine-dependent regulation. Molecular and Cellular Biology **12**: 2178–2185.

_____, H.S. Basu, C.W. Porter, L.J. Marton, and P. Coffino. 1992. Role of ornithine decarboxylase suppression and polyamine depletion in the antiproliferative activity of polyamine analogs. Molecular Pharmacology **42**: 302–306.

Goldberg, B., N. Yarlett, J. Sufrin, D. Lloyd, and C.J. Bacchi. 1997. A unique transporter of S-adenosylmethionine in African trypanosomes. FASEB Journal **11**: 256–260.

Green, L.C., P.J. Didier, and E.S. Didier. 1999. Fractionation of sporogonial stages of the microsporidian *Encephalitozoon cuniculi* by Percoll gradients. Journal of Eukaryotic Microbiology. **46**: 434–438.

Griffith E.C, Z. Su, B.E. Turk, S.P. Chen, Y.H. Chang, Z.C. Wu, K. Biemann, and J.O. Liu. 1997. Methionine aminopeptidase (type 2) is the common target of angiogenesis inhibitors AGM-1470 and ovalicin. Chemistry and Biology **4**: 61-471.

_____, Z. Su, S. Niwayama, A. Ramsay, Y. Chang, and J. O. Liu. 1998. Molecular recognition of angiogenesis inhibitors fumagillin and ovalicin by methionine aminopeptidase 2. Proceedings of the National Academy of Sciences (USA) **95**: 15183-15188.

Gritz, D. C., D. S. Holsclaw, , R. E. Neger, J. P. Whitcher Jr., and T. P. Margolis. 1997. Ocular and sinus microsporidial infection cured with systemic albendazole. American Journal of Ophthalmology **124**: 241-243.

Guerrard, A., M. Rabodonirina, L. Cotte, O Liguory, M. A. Piens, S. Dauod, S. Picot, and J. L. Touranine. 1999. Intestinal microsporidiosis occurring in two renal transplant recipients treated with mycophenolate mofetil. Transplantation **68**: 699-701.

Gunnarsson, G., D. Hurlbut, P. C. DeGirolami, M. Federman, C. Wanke. 1995. Multiorgan microsporidiosis: report of five cases and review. Clinical Infectious Disease **21**: 37-44.

Han, C.K., S.K Ahn, N.S. Choi, R.K. Hong, S.K. Moon, H.S.Chun, S.J. Lee, J.W. Kim, C.I. Hong, D. Kim, J.H. Yoon, and K.T. No. 2000. Design and synthesis of highly potent fumagillin analogues from homology modeling for a human MetAP2. Bioorganic and Medicinal Chemistry Letters **10**: 39-43.

Haque, A., W.S. Hollister, A. Willcox, and E.U. Canning. 1993. The antimicrosporidial activity of albendazole. Journal Invertebrate Pathology **62**: 171-177.

He, Y., T. Suzuki, and K. Kashiwagi. 1994. Antizyme delays the restoration by spermine of growth of polyamine-deficient cells through its negative regulation of polyamine transport. Biochemical and Biophysical Research Communications **203**: 608–614.

Higgins, M. J., M. L. Kent, J. D. Moran, L.M. Weiss, S. C. Dawe. 1998. Efficacy of the fumagillin analog TNP-470 for *Nucleospora salmonis* and *Loma salmonae* infection in chinook salmon *Oncorhynchus tsawytscha*. Diseases of Aquatic Organisms **11**: 45-49.

Hirt, R. P., J. M. Logsdon, Jr., B. Healy, M.W. Dorey, W.F. Doolittle, and T.M. Embley, 1999. Microsporidia are related to Fungi: evidence from the largest subunit of RNA polymerase II and other proteins. Proceedings of the National Academy of Sciences (USA) **96**: 580-585.

Hu, R.H., and A.E. Pegg. 1997. Rapid induction of apoptosis by down regulated uptake of polyamine analogues. Biochemical Journal **328**: 307–316.

Joste, N. E., J.D. Rich, K.J. Busam, and D.A. Schwartz. 1996. Autopsy verification of *Encephalitozoon intestinalis* (Microsporidiosis) eradication following albendazole therapy. Archives of Pathology and Laboratory Medicine **120**: 199-203.

Kahn, I.A. J.D. Schwartzman, L.H. Kasper, and M. Moretto. 1999. CD8[+] CTL's are essential for protective immunity against *Encepahlitozoon* infection. Journal of Immunology **162**: 6086-6091.

Kano, T., and H.Fukui. 1982. Studies on Pleistophora infection in eel, *Anguilla japonica* – I. experimental induction of microsporidiosis and fumagillin efficacy. Fish Pathology **16**: 193-200.

_____, T. Okauchi, and H. Fukui. 1982. Studies on Pleistophora infection in eel, *Anguilla japonica* – II. Preliminary tests for application of fumagillin. Fish Pathology **17**: 107-114.

Katinka, M.D., S. Duprat, E. Cornillot, G. Metenier, F. Thomarat, G. Prensier, V. Barbe, E. Peyretaillade, P. Brottier, P. Wincker, F. Delbac, H. El Alaoui, P. Peyret, W. Saurin, M. Gouy, J. Weissenbach, and C.P. Vivares. 2001. Genome sequence and gene compaction of the eukaryote parasite *Encephalitozoon cuniculi*. Nature **414**: 450-453.

Katiyar, S.K., V.R. Gordon, G.L. McLaughlin, and T.D. Edlind. 1994. Antiprotozoal activities of benzimidazoles and correlations with *B*-tubulin sequence. Antimicrobial Agents and Chemotherapy **38**: 2086-2090.

_____, and T.D. Edlind. 1997. *In vitro* susceptibilities of the AIDS-associated microsporidian *Encephalitozoon intestinalis* to albendazole, its sulfoxide metabolite, and 12 additional benzimidazole derivatives. Antibicrobial Agents and Chemotherapy **41**: 2729-2732.

Katznelson, H., and C.A. Jamieson. 1953. Control of nosema disease of honeybees with fumagillin. Science **115**: 70-71.

Keeling, P.J., M.A. Luker, and J.D. Palmer. 2000. Evidence from beta-tubulin phylogeny that microsporidia evolved from within the fungi. Molecular Biology and Evolution **17**: 23-31.

Keithly, J.S., G. Zhu, S.J. Upton, K.M. Woods, M.P. Martinez, and N. Yarlett. 1997. Polyamine biosynthesis in *Cryptosporidian parvum* and its implications for chemotherapy. Molecular Biochemical Parasitology **88**: 35–42.

Kent, M.L., and S.C. Dawe. 1994. Efficacy of fumagillin DCH against experimentally induced *Loma salmonae* (Microsporea) infections in chinnok salmon *Oncorhynchus tsawytscha*. Diseases of Aquatic Organisms **20**: 231-233.

Kotler, D.P., and J.M. Orenstein. 1998. Clinical syndromes associated with microsporidiosis. Advances in Parasitology **40**: 322-343.

Lacey, E. 1990. Mode of action of benzimidazoles. Parasitology Today **6**: 112-115.

Lauren, D.L., A. Wishkovsky, J.M. Groff, R.P. Hedrick, and D.E. Hinton. 1989. Toxicity and pharmacokinetics of the antibiotic fumagillin in yearling rainbow trout (*Salmo gairdneri*). Toxicology and Applied Pharmacology **98**: 444-453.

Lecuit, M., E. Oksenhendler, and C. Sarfati. 1994. Use of albendazole for disseminated microsporidian infection in a patient with AIDS. Journal of Infectious Diseases **19**: 332-333.

Leder, K., N. Ryan, D. Spelman, and S.M. Crowe. 1998. Microsporidial disease in HIV-infected patients: A report of 42 patients and review of the literature. Scandinavian Journal of Infectious Diseases **30**: 331-338.

Libby, P.R., M. Henderson, R.J. Bergeron, and C.W. Porter. 1989. Major increases in spermidine/spermine N[1]-acetyltransferase activity by spermine analogues and their relationship to polyamine depletion and growth inhibition in L-1210 cells. Cancer Research **49**: 6226–6231.

Liu, S., Widom, J., C.W. Kemp, C.M. Crews, and J. Clardy. 1998. Structure of human methionine aminopeptidase-2 complexed with fumagillin. Science **282**: 1324-1327.

Lowther,W.T., D.A. McMillen, A.M. Orville, and B.W. Matthews. 1998. The anti-angiogenic agent fumagillin covalently modifies a conserved active-site histidine in the *Escherichia coli* methionine aminopeptidase. Proceedings of the National Academy Sciences (USA) **95**: 12153-12157.

Marton, L.J., and A.E. Pegg. 1995. Polyamines as targets for therapeutic intervention. Annual Review of Pharmacology and Toxicology **35**: 55–91.

Merali, S., S. Muhamed, K. Chin, and A.B. Clarkson, Jr. 2000. Effect of a bis-benzyl polyamine analogue on *Pneumocystis carinii*. Antimicrobial Agents and Chemotherapy **44**: 337–343.

Metcalf, B.W., P. Bey, C. Danzin, M.J. Jung, P. Casara, and J.P. Vevert. 1978. Catalytic irreversible inhibition of mammalian ornithine decarboxylase (E.C. 4.1.1.17) by substrate and product analogues. Journal of American Chemical Society. **100**: 2551–2553.

Metcalf, T.W., Doran, R.M.L., P..L Rowlands, A. Curry, and C.J.N. Lacey. 1992. Microsporidial keratoconjunctivitis in a patient with AIDS. British Journal of Ophthalmology **76**: 177-178.

Modigliani R, C. Bories, Y. Le Charpantier, M. Salmeron, B. Messing, A. Galian, J. C. Rambaud, A. Lavergne, B. Cochand-Priollet, and I. Desportes. 1985. Diarrhoea and malabsorption in Acquired Immune Deficiency Syndrome: a study of four cases with special emphasis on opportunisitic protozoan infestations. Gut **26**: 179-187.

Molina, J.M., C. Chastang, J. Goguel, J.F. Michiels, C. Sarfati, I. Desportes-Livage, J. Horton, F. Derouin, and J. Modai. 1998. Albendazole for treatment and prophylaxis of microsporidiosis due to *Encephalitozoon intestinalis* in patients with AIDS: a randomized double-blind controlled trial. Journal of Infectious Diseases **177**: 1373-1377.

_____, J. Goguel, C. Sarfati, C. Chastang, I. Desportes-Livage, J.F. Michiels, C. Maslo, C. Katlama, L. Cotte, C. Leport, F. Raffi, F. Derouin, and J. Modai. 1997. Potential efficacy of fumagillin in intestinal microsporidiosis due to *Enterocytozoon bieneusi* in patients with HIV infection: results of a drug screening study. The French Microsporidiosis Study Group. AIDS **11**: 1603-10.

_____, _____, _____, J.F. Michiels, I. Desportes-Livage, S. Balkan, C. Chastang, L. Cotte, C. Maslo, A. Struxiano, F. Derouin, and J.M. Decazes. 2000. Trial of oral fumagillin for the treatment of intestinal microsporidiosis in patients with HIV infection. ANRS 054 Study Group. Agence Nationale de Recherche sur le SIDA. AIDS **14**: 1341-1348.

_____, _____, _____, C. Chastang, I. Desportes-Livage, J. F. Michiels, C. Maslo, C. Katlama, L. Cotte, C. Leport, F. Raffi, F. Derouin, and J. Modai. 1997. Potential efficacy of fumagillin in intestinal microsporidiosis due to *Enterocytozoon bieneusi* in patients with HIV infection: results of a drug screening study. AIDS **11**: 1603-1610.

_____, E. Oksenhendler, B. Beauvais, C. Sarfati, A. Jaccard, F. Derouin, and J. Modai. 1995. Disseminated microsporidiosis due to *Septata intestinalis* in patients with AIDS: clinical features and response to albendazole therapy. Journal of Infectious Diseases **171**: 245-249.

_____, Tourneur, M., C. Sarfari, S. Chevret, A. de Gouvello, J.G. Gobert, S. Balkan, and F. Deroun. 2002. Fumagillin treatment of intestinal microsporidiosis. New England Journal of Medicine **346**: 1963-1969.

Molnar, K., F. Baska, and C.Szekely. 1987. Fumagillin, an efficacious drug against renal sphaerosporosis of the common carp *Cyprinus carpio*. Diseases of Aquatic Organisms **2**: 187-190.

Moreto, M., I. Casciotti, B. Borell, and I.A. Khan. 2000. Lack of CD4+ T cells does not affect induction of CD8$^+$ T-cell immunity against *Encephalitozoon cuniculi* infection. Infection and Immunity **68**: 6223–6232.

Murakami, Y., K. Tanakam, S. Matsufuji, Y. Miyazaki, and S. Hayashi. 1992. Antizyme, a protein induced by polyamines, accelerates the degradation of ornithine decarboxylase in Chinese-hamster ovary-cell extracts. Biochemical Journal **283**: 661–664.

Orenstein, J.M., J. Chang, W. Steinberg, P.D. Smith, H. Rotterdam, and D.P. Kotler. 1990. Intestinal microsporidiosis as a cause of diarrhea in Human Immunodeficiency Virus-infected patients. Human Pathology **21**: 475-481.

_____, M. Tenner, A. Cali, and D.P. Kotler. 1992. A microsporidian previously undescribed in humans, infecting enterocytes and macrophages, and associated with diarrhea in an Acquired Immunodeficiency Syndrome patient. Human Pathology **23**: 722-728.

Phillips, M.A., P. Coffino, and C.C. Wang. 1987. Cloning and sequencing of the ornithine decarboxylase gene from *Trypanosoma brucei*. Journal of Biological Chemistry **262**: 8721–8727.

Pol, S., C.A. Romana, S. Richard, P. Amouyal, I. Desportes-Livage, F. Carnot, J. F. Pays, and P. Berthelot. 1993. Microsporidia infection in patients with the Human Immunodeficiency Virus and unexplained cholangitis. New England Journal of Medicine **328**: 95-99.

Porter, C.W., B. Ganis, P.R. Libby, and R.J. Bergeron. 1991. Correlations between polyamine analog-induced increases in spermidine/spermine N$^1$-acetyltransferase activity and growth inhibition in human melanoma cell lines. Cancer Research **51**: 3715–3720.

_____, U. Regenass, and R.J. Bergeron. 1992. Polyamine inhibitors and analogues as potential anticancer agents. In: Polyamines in the gastrointestinal tract, Falk Symposium 62. R.H. Dowling, U.R. Folsch, and Chr. Loser (eds.) Kluwer Academic Publishers, Dordrecht pp. 301–322.

Rabeneck, L., F. Gyorkey, R.M. Genta, P. Gyorkey, L.W. Foote, J.M.H. Risser. 1993. The role of *Microsporidia* in the pathogenesis of HIV-related chronic diarrhea. Annals of Internal Medicine **119**: 895-899.

Rastrelli, P.D., E. Didier, and R.W. Yee. 1994. Microsporidial keratitis. Ophthalmology Clinics of North America **7**: 617-633.

Rosberger, D.F., O.N. Serdarevic, R.A. Erlandson, R.T. Bryan, D.A. Schwartz, G.S. Visvesvara, and P.C. Keenan. 1993. Successful treatment of microsporidial keratoconjunctivitis with topical fumagillin in a patient with AIDS. Cornea **12**: 261-265.

Rossi, R.M., C. Wanke, and M. Federman. 1996. Microsporidian sinusitis in patients with the Acquired Immunodeficiency Syndrome. Laryngoscope **106**: 966-971.

Rossi, P., Urbani, C., Gianfranco, D., and Pozio, E. 1999. Resolution of microsporidial sinusitis and keratoconjunctivitis by itraconazole treatment. American Journal of Ophthalmology **127**: 210-212.

Sbottka, I., H. Albrecht, H. Schafer, J. Schottelius, G. S. Visvesvara, R. Laufs, and D. A. Schwartz. 1995. Disseminated *Encephalitozoon (Septata) intestinalis* infection in a patient with AIDS: novel diagnostic approaches and autopsy-confirmed parasitological cure following treatment with albendazole. Journal of Clinical Microbiology **33**: 2948-2952.

Schechter, P.J., and A. Sjoerdsma. 1989. Therapeutic utility of selected enzyme-activated irreversible inhibitors. In: Enzymes as targets for drug design. M.G. Palfreyman, P.P. McCann, W. Lovenberg, J.G. Temple, and A. Sjoerdsma (eds.) Academic Press, San Diego. pp. 201-210.

Seiler, N. 1995. Polyamine oxidase, properties and functions. In: Progress in brain research. P.M. Yu, K.F. Tipton, and A.A. Boulton (eds.) Progress in Brain Research106: 333–344.

_____, S. Sarhan, P. Mamont, P. Vasara, and C. Danzin. 1991. Some biological consequences of S-adenosylmethionine decarboxylase inhibition by MDL-73811. Life Chemistry Reports. 9: 151–162.

Shadduck, J.A. 1980. Effect of fumagillin on *in vitro* multiplication of *Encephalitozoon cunuculi*. Journal Protozoology 27: 202-208

Shantz, L.M., B.A. Stanley, J.A. Secrist, and A.E. Pegg. 1992. Purification of human S-adenosylmethionine decarboxylase expressed in *Escherichia coli* and use of this protein to investigate the mechanism of inhibition by the irreversible inhibitors, 5'-deoxy-5'-[(3-hydrazinopropyl)methylamino]adenosine and 5'{[(Z)-4- amino-2-butenyl]methylamino-5' deoxyadenosine. Biochemistry 31: 6848–6855.

Sharpstone, D., A. Rowbottom, M. Nelson, and B. Gazzard. 1995. The treatment of microsporidial diarrhoea with thalidomide [letter]. AIDS 9: 658-659.

_____, _____, N. Francis, G. Tovey, D. Ellis, M. Barrett, and B. Gazzard. 1997. Thalidomide: a novel therapy for microsporidiosis [published erratum appears in Gastroenterology 1997 Sep;113(3):1054]. Gastroenterology 112: 1823-1829.

Silveira, H., E.U. Canning. 1995. *Vittaforma corneae* N. Comb. for the human microsporidium *Nosema corneum* Shadduck, Meccoli, Davis & Font, 1990, based on its ultrastructure in the liver of experimentally infected athymic mice. Journal of Eukaryotic Microbiology 42: 158-165.

Simon, D., L.M.Weiss, H.B. Tanowitz, A. Cali, J. Jones, and M. Wittner. 1991. Light microscopic diagnosis of human microsporidiosis and variable response to octreotide. Gastroenterology 100: 271-273.

Sin, N., L. Meng, , M.Q.W. Wang, J.J. Wen, W.G. Bornmann, C. M. Crews. 1997. The anti-angiogenic agent fumagillin covalently binds and inhibits the methionine aminopeptidase, MetAP-2. Proceedings of the National Academy of Sciences (USA) 94: 6099-6103.

Soule, J.B., A.L. Halverson, R.B. Becker, M.C. Pistole, and J.M. Orenstein. 1997. A patient with Acquired Immunodeficiency Syndrome and untreated *Encephalitozoon (Septata) intestinalis* microsporidiosis leading to small bowel perforation: response to albendazole. Archives of Pathology and Laboratory Medicine 121: 880-887.

Sprague, V., J.J. Becnel, and E.I Hazard. 1992. Taxonomy of phylum Microspora. Critical Reviews in Microbiology 18: 285-395.

Taylor, A. 1993. Aminopeptidases: structure and function. FASEB Journal 7: 290-298

Terada, S., K. Rajender Reddy, L.J. Jeffers, A. Cali, and E. Schiff. 1987. Microsporidian hepatitis in the Acquired Immunodeficiency Syndrome. Annals of Internal Medicine 107: 61-62.

Valasinas, A., A. Sarkar, V.K. Reddy, L.J. Marton, H.S. Basu, and B. Frydman. 2001. Conformationally restricted analogues of $N^1,N^{14}$ Bis(ethyl)homospermine (BE-4-4-4-4): synthesis and growth inhibitory effects on human prostate cancer cells. Journal of Medicinal Chemistry 44: 390–403.

Vavra, J., A., T.Yachnis, J. A. Shadduck, and J. M. Orenstein. 1998. Microsporidia of the genus Trachipleistophora--causative agents of human microsporidiosis: description of *Trachipleistophora anthropophthera* n. sp. (Protozoa: Microsporidia). Journal of Eukaryotic Microbiology 45: 273-83.

Visvesvara, G.S., G.J. Leitch, H. Moura, S. Wallace, R. Weber, and R.T. Bryan. 1991. Culture, electron microscopy and immunoblot studies on a microsporidian parasite isolated from the urine of a patient with AIDS. Journal of Protozoology **38**: 105S–111S.

_____, _____, A.J. da Silva, G.P. Croppo, H. Moura, S. Wallace, S.B. Slemenda, D.A. Schwartz, D. Moss, and R.T. Bryan1994. Polyclonal and monoclonal antibody and PCR-amplified small-subunit rRNA identification of a microsporidian, *Encephalitozoon hellem*, isolated from an AIDS patient with disseminated infection. Journal of Clinical Microbiology **32**: 2760-2768.

Vujcic, S., P. Diegelman, C.J. Bacchi, D.L. Kramer, and C.W. Porter. 2002. Identification and characterization of novel mammalian spermine oxidase.    Biochemical Journal **367**: 665-675.

Waters, R.W., B. Frydman, L.J. Marton, A. Valasinas, V.K. Reddy, J.A. Harp, M.J. Wannemuehler, and N. Yarlett. 2000. ['N,$^{12}$N]bis(ethyl)-Cis-6,7-dehydrospermine: a new drug for treatment and prevention of *Cryptosporidium parvum* infection of mice deficient in T-cell receptor Alpha. Antimicrobial Agents and Chemotherapy **44**: 2891–2894.

Weber, R., R.T. Bryan, R.L. Owen, C.M. Wilcox, I. Gorelkin, and G.S. Visvesvara. 1992. Improved light-microscopical detection of microsporidia spores in stool and duodenal aspirates. New England Journal of Medicine **326**: 161-166.

_____, P. Deplazes, M. Flepp, **et al.** 1997. Cerebral microsporidiosis due to *Encephalitozoon cuniculi* in a patient with Human Immunodeficiency Virus infection. New England Journal of Medicine **336**: 474-478.

_____, M. Flepp, and W. Wichmann. 1997. Cerebral microsporidiosis due to *Encephalitozoon cunuculi* – reply to letter. New England Journal of Medicine **337**: 640-641.

_____, B. Sauer, M.A. Spycher, P. Deplazes, R. Keller, R. Ammann, J. Briner, and R. Luthy. 1994. Detection of *Septata intestinalis* in stool specimens and coprodiagnostic monitoring of successful treatment with albendazole. Clinical and Infectious Disease **19**: 342-345.

Weiss, L.M., S.F. Costa, and H. Shang. 2001. Microsporidian methionine aminopeptidase type 2. Journal of Eukaryotic Microbiology Suppl:88-90S.

_____, E. Michalakakis, C.M. Coyle, H.B. Tanowitz, and M. Wittner. 1994. The activity of albendazole against  *Encephalitozoon cuniculi*. Journal of Eukaryotic Microbiology **41**: S65.

_____, T.D. Edlind, C.R. Vossbrinck, and T. Hashimoto. 1999. Microsporidian molecular phylogeny: the fungal connection. Journal of Eukaryotic Microbiology **46**: 17S-18S.

_____, and C.R. Vossbrinck 1998. Microsporidiosis: molecular and diagnostic aspects. Advances in Parasitology **40**: 351-395.

Wittner, M., and L.M. Weiss (eds). 1999. The Microsporidia and Microsporidiosis. ASM Press. Washington DC.

Xiao, L. and R.A. Casero. 1995. Regulation of spermidine/spermine N'-acetyltransferase. In: Polyamines: regulation and molecular interaction. R.A. Casero (ed.) pp. 77–98. Springer-Verlag, New York.

Yarlett, N., and C.J. Bacchi. 1988. Effect of DL-α-difluoromethylornithine on polyamine synthesis and interconversion in *Trichomonas vaginalis* grown in a semi defined medium. Molecular and Biochemical Parasitology **31**: 1–10.

Yee, R.W., F.O. Tio , J. A. Martinez, K. S. Held, J. A. Shadduck, and E.S. Didier. 1991. Resolution of microsporidial epithelial keratopathy in a patient with AIDS. Ophthalmology **98**: 196-201.

Yokoyama, H., K. Ogawa, and H. Wakabayashi. 1990. Chemotherapy with fumagillin and toltrazuril against kidney enlargement disease of goldfish caused by the myxosporean *Hoferellus carassii*. Fish Pathology **25**: 157-163.

Zender, H.O, E. Arrigoni, J. Echert, and Y. Kapanici. 1989. A Case of *Encephalitozoon cuniculi* peritonitis in a patient with AIDS. American Journal of Clinical Pathology **92**: 352-356.

Zou, Y., Z. Wu, N. Sirisoma, P.M. Woster, R.A. Casero, L.M. Weiss, D. Rattendi, S. Lane, and C.J. Bacchi. 2001. Novel alkylpolyamine analogues that possess both antiparasitic and antimicrosporidial activity. Bioorganic and Medicinal Chemistry Letters **11**: 1613–1617.

# PHYLOGENETICS: TAXONOMY AND THE MICROSPORIDIA AS DERIVED FUNGI

Charles R. Vossbrinck[1], Theodore G. Andreadis[1] and Louis M. Weiss[2]

[1]The Connecticut Agricultural Experiment Station, New Haven, CT, USA. The Departments of [2]Pathology and [3]Medicine, Albert Einstein College of Medicine, Bronx, NY, USA.

## ABSTRACT

Microsporidia are a group of obligate eukaryotic intracellular parasites first recognized over 100 years ago with the description of *Nosema bombycis* the parasite from silkworms that caused the disease pebrine in these economically important insects. Microsporidia infect almost all animal phyla. Among the more than 144 described genera, several have been demonstrated in human disease: *Nosema, Vittaforma, Brachiola, Pleistophora, Encephalitozoon, Enterocytozoon, Septata* (reclassified to *Encephalitozoon*) and *Trachipleistophora*. In addition, the genus *Microsporidium* has been used to designate microsporidia of uncertain taxonomic status. The recognition of microsporidia as opportunistic pathogens in humans has led to increased interest in the molecular biology of these pathogens. Recent work has focused on the determination of the nucleotide sequences for ribosomal RNA (rRNA) genes, which have been used as diagnostic tools for species identification as well as for the development of a molecular phylogeny of these organisms. Microsporidia have historically been considered to be "primitive" protozoa, however, molecular phylogenetic analysis has led to the recognition that these organisms are not "primitive" but degenerate and that they are related to the fungi and not to other protozoa. Such molecular phylogeny has also led to the recognition that the traditional phylogeny of these organisms based on structural observations may not reflect the "true" relationships among the various microsporidia species and genera. This chapter reviews the data on the taxonomy of the microsporidia and the relationship of these organisms to other eukaryotes.

**Key words:** Microsporidia, rRNA genes, molecular phylogeny, fungi, taxonomy, genome analysis, evolution

## INTRODUCTION

Microsporidia evolved a most remarkable adaptation to intracellular parasitism, the polar filament. This spore organelle is wrapped around the periphery of the inside of the spore and upon germination extends, by eversion, many times the length of the spore, forming a hollow tube through which the organism injects itself into the host cell. This invasion strategy is highly successful and there are about 1200 species of Microsporidia currently recognized. This single apomorphic feature is unique and unites all of the Microsporidia although a morphologically similar structure of different origin evolved separately in the Myxosporidia. Another remarkable feature of the Microsporidia is their reduced cellular complexity. A degenerate or reduced state, is often found in parasites in general, and in the case of Microsporidia may be in part due to their need to rapidly reproduce. Microsporidia of the genus *Encephalitozoon* have the smallest known eukaryotic genomes, with that of *Encephalitozoon cuniculi* being 2.9 Mb. This genome has been sequenced (Katinka et al., 2001) and provides some insight into how this eukaryotic pathogen functions. One striking finding has been that the average gene size in this organism is 15% less than that of similar genes in yeast and that its genes are often missing domains involved in complex protein interactions. This suggests that study of this organism may provide insights into the minimal protein motifs required for metabolic and other cellular processes. These organisms are clearly an excellent model for the study of degeneration due to parasitism.

## TAXONOMIC PLACEMENT OF THE MICROSPORIDIA

Electron microscopic studies on a microsporidia were first reported by Krieg (1955) who reported on a *Pleistophora sp.* from white grubs, by Weiser (1959) who presented electron microscopic images of *Nosema laphygmae* from *Laphygma (Spodoptera) frugiperda* and by Huger (1960) who demonstrated clearly the ultrastructural characteristics of the microsporidian spore. Images of the laminar structure of the "polaroplast" (a term proposed by Huger (1965)), the inner and outer spore coat, the nuclei, polar granules and clear cross sections of the polar filament were demonstrated. Huger also presented an accurate schematic diagram of the polar filament coiled around the inside of the periphery of the spore. These ultrastructural characters were used for comparative phylogenetic purposes both to study the relationship of the Microsporidia among eukaryotes and to generate phylogenies within the Microsporidia. Studies by Lom and Vavra (1961), Kudo (1963) and others soon followed.

The presence of nematocyst-like structures in the Myxosporidia and Actinomyxida resulted in the placement of these two groups with the

Microsporidia early in the systematics of the Protozoa. Doflein (1901) grouped the Suborder Microsporidia with the Suborders Myxosporidia and Actinomyxida in the Order Cnidospora based on the presence of a spore containing polar filament. The Order Cnidosporidia was placed in the Subclass Neospora, in the Class Sporozoa. Lom and Vavra (1962) concurred with Dolflein's classification, but noted that the only character common to the Subphylum Cnidospora was the polar filament. They further indicated that the Myxosporidia and Actinomyxida are much more closely related to each other than to the microsporidia because for both the Myxosporidia and Actinomyxida (Class Heteronucleida) " During development the nuclei are twice differentiated: 1) into vegetative nuclei of the 'plasmodium' and into 'sporogenous' nuclei, which 2) give rise to nuclei of shell valves, polar capsules, and sporoplasm. The spores thus originating are polycellular ones."

While Microsporidia were originally grouped with the Myxosporidia and Actinomyxida based on the presence of a polar filament- like structure inside the spore, eventually both the Myxosporidia and Actinomyxida were separated from the Microsporidia and placed into the classes Myxosporea (fish parasites) and Actinosporea (parasites of annelids) in the Phylum Myxozoa. Relatively recently, two interesting discoveries were made about the Myxozoa. First it was discovered that the Actinosporea were actually alternate hosts of the Myxosporea (Markiw and Wolf, 1983) and it was suggested that the term Actinosporea be suppressed (Kent et al., 1994). Secondly, based on rDNA analysis the Myxozoa have been demonstrated to be degenerate Cnidarians (Siddal et al., 1995). This relationship was suggested much earlier and it was stated at that time that the polar filament-like organelle in the Myxozoa was a nematocyst (Weill, 1938). The function of this structure is to aid a Myxozoan in penetrating a host cell. Myxozoa do not use this filament to inject themselves into host cells as it is not a hollow structure.

The Class or Order Microsporidia depending on the spelling, were elevated to the Phylum Microspora, by Sprague and Vávra (1977). Sprague and Becnel (1998) have suggested the term Microsporidia for the phylum as the original name, Microsporidies, (Balbiani, 1882) for the group. In this article we also use the name Microsporidia for the phylum and as such the term "Microsporidia" is a proper name and should be capitalized. For common reference, the term microsporidian should be used (we avoid the term microsporidium because it is also use as a term at the generic level for Microsporidia of unknown phylogenetic placement, e.g. *Microsporidium lamproglenae*) and as an adjective the term microsporidial is reasonable. "The microsporidial infection was diagnosed as a microsporidian based on the presence of the coiled polar filament inside the spores. While the spores are definitely Microsporidia, it's placement within the phylum was uncertain and it was designated as *Microsporidium lamproglenae*." As we will discuss below the consensus based on available evidence is that Microsporidia should be considered a Phylum in the kingdom of Fungi.

## FROM WHAT DID THE MICROSPORIDIA EVOLVE?

At present there is a wealth of evidence supporting the hypothesis that Microsporidia evolved from fungi and that they have secondarily lost mitochondria. We examine the molecular evidence for this statement from a historical perspective below.

Prior to the advancements in comparative molecular analysis, Microsporidia were thought to be in their own phylum unrelated to any known protist group. The advent of protein sequencing (Sanger and Tuppy, 1951) meant that comparative molecular techniques could be developed Dayhoff et al., 1972. Comparative protein analysis was based on proteins, such as cytochrome c, which could easily be isolated and sequenced. Microsporidia were not included in any of those analyses.

The earliest ribosomal analysis of a microsporidian was the observation that after nucleic acid isolation and polyacrylamide gel electrophoresis not only were the ribosomal RNAs smaller but the 5.8s rRNA was missing (Vossbrinck and Woese, 1986). It had become established that all prokaryotes were missing the 5.8s rRNA (the analogous piece was actually attached to the large subunit rRNA) and that all eukaryotes had a 5.8s rRNA (the analogous piece assembles in the proper position as the ribosome is formed). Direct sequencing of the 5' end of the large subunit rRNA demonstrated that in the case of *Vairimorpha necatrix*, as with the prokaryotes the region analogous to the 5.8s was covalently linked to the large subunit rRNA.

Because of its universal presence in all life forms, including both prokaryotes and eukaryotes and its high copy number Woese et al. (1977) catalogued ribosomal RNAs from 13 organisms including prokaryotes and eukaryotes. The cells were grown in the presence of $^{32}$P, then were harvested and the small subunit ribosomal RNA was cut with T1 RNase and subjected to two-dimensional electrophoresis to produce an oligonucleotide fingerprint. The individual oligonucleotides on each fingerprint were then sequenced by RNase digestion to produce an oligonucleotide catalogue of each organism. Comparison of small subunit rRNA catalogues led to the discovery that Methanobacteria were as different from Bacteria such as *Escherichia coli* as they were from Eukaryotes such as *Saccharomyces cerevisiae*.

With the advent of cloning and DNA sequencing (Sanger and Coulson, 1977; Maxim and Gilbert, 1977), interest developed in obtaining complete sequences of small subunit rDNA. By 1987 at least 7 eukaryotic small subunit rRNAs, representing a diversity of species including plants, animals, fungi, flagellates and ciliates, had been sequenced. The small subunit rDNA was sequenced for *Vairimorpha necatrix* (Vossbrinck et al., 1987) and comparative phylogenetic analysis based on both a distance method and maximum parsimony indicated that the Microsporidia were different eukaryotes indeed. The authors (Vossbrinck et al., 1987) speculated that if the analysis were correct then the Microsporidia may have separated from the eukaryotic line before the advent of the mitochondria, before the split of the 5.8s rRNA and before the Earth's atmosphere contained oxygen. Additional ssrDNA studies by Leipe et al. (1993) adding *Giardia lamblia* and *Hexamita inflata* (diplomonads lacking mitochondria) agreed with the early divergence of Microsporidia and focused on whether Microsporidia or the Diplomonada diverged from the tree of life first and whether the G+C content and the nature of the parasitism affected the phylogenies generated by the analysis. Artifically long branch length due to rapid evolutionary factors was also discussed (Leipe et al., 1993). Elongation factor 1α (EF-1α) and the eubacterial homologue (EF-Tu) and elongation factor 2 (EF-2) and its eubacterial homologue (EF-G) (Hashimoto and Hasegawa, 1996; Kamaishi et al., 1996a, 1996b) were the next genes sequenced for a microsporidian. This microsporidia, *Glugea plecoglossi*, a parasite of the ayu fish, has a G+C ratio of about 50%, so there should be no errors based on nucleotide bias. These researchers focused on a rigorous maximum likelihood analysis testing various models of amino acid substitution. They addressed the problem of dealing with an organism which may have a long branch length due to more rapid changes in protein or nucleic acid structure. Kamaishi et al., 1996b, concluded that concluded that the Microsporidia branched first from the eukaryotic tree followed by the Diplomonads followed by various mitochondrial containing protist groups.

Based upon the *Vairimorpha necatrix* analysis, Cavalier-Smith (1987) proposed that other amitochondriate protozoa may also have diverged early and may also be primitively lacking mitochondria. Consistent with Cavalier-Smith's (1987) proposal, *Giardia lamblia*, an amitochondriate parasitic diplomonad, is also a highly divergent protistan based upon rDNA analysis (Sogin et al., 1989). The identification of prokaryotic homologues of cpn60 and hsp70, i.e. mitochondrial heat shock protein genes, in all three mitochondria lacking groups (Diplomonada: *G. lamblia* cpn 60 (Soltys and Gupta 1994; Roger et al., 1998); Trichomonada: *Trichomonas vaginalis* cpn 60 (Bui et al., 1996; Horner et al., 1996; Roger et al., 1996) and hsp70 (Bui et al., 1996; Germot et al., 1996); and Microspora: *N. locustae* hsp70(Germot et al., 1997), *V. necatrix* hsp70 (Hirt et al., 1997), *Enc. cuniculi* (Peyretaillade et al., 1998) and *Enc. hellem* hsp 70 (Arisue et al., 2002)) suggests that these organisms did possess mitochondria at some point in their evolution. In addition, mitochondrial type cpn60 has also been identified in *E. histolytica* (Clark and Roger, 1995). The lack of mitochondria is thus probably an apomorphic rather than a pleisomorphic character. This is not in itself proof that these organisms once had mitochondria because all life forms appear to have this gene. However, the fact that phylogenetic analysis places the eukaryotic hsp70 gene with the α-Proterobacteria (Germot et al., 1997; Hirt et al., 1997) as does mitochondrial rRNA analysis (Yang et al., 1985) provides convincing evidence for the idea that these organisms once had mitochondria and that the common ancestor of know eukaryotes already had a mitochondria. Further support for this idea is the recognition that the Trichomonada hydrogenosomes were derived from mitochondria (Embly et al., 1997).

During this period other evidence surfaced for Microsporidia having some odd characteristics. It was shown for example that *Vairimorpha necatrix* lacked 7 methyl-G, 2,2,7 trimethyl-G, or gama monomethyl phosphate caps on either it's small nuclear RNA (snRNA) or its messenger RNA (mRNA) (DiMaria et al., 1996). The study demonstrated that Microsporidia did have caps on their snRNA but the identity of the cap was not determined. Hausmann et al. (2002) have recently identified the capping apparatus of *Encephalitozoon* cuniculi. This consists of a triphosphatase guanlylytransferase (EcCet1) and methyltransferase. EcCet1 belongs to a family of metal-dependent phosphohydralases found in fungi, DNA virus and *Plasmodium falciparum*, that are distinct from the capping enzymes found in metazoans and plants. The U2snRNA did show significant homology to other eukaryotic sequences and contained the highly conserved GUAGUA branch point binding sequence but did not demonstrate a sm-Binding site. A secondary model of the U2snRNA was highly divergent from other eukaryotes. Since the snRNAs are part of the intron splicing machinery, the implication of the study was that Microsporidia have or had introns in their genes. Recent completion of the *Encephalitozoon cuniculi* has confirmed that introns are rare but do occur in this microsporidian's genes (Katinka et al., 2001), although they are small and atypical. Fast et al. (1998) published similar results, for *Nosema locustae*, implying again that the machinery for intron splicing was present in Microsporidia. Fast et al. (1998) also determined a secondary structure for U2snRNA, demonstrated a sm-Binding site and presented the U6snRNA sequence of *N. locustae*.

## FUNGAL ORIGINS OF MICROSPORIDIA

As early as 1994 Edlind had hypothesized that based on β-tubulin sequence analysis (Edlind et al., 1996), Microsporidia were not ancient eukaryotes but were related to the fungi. While the data in the original paper were not conclusive the hypothesis has been supported by additional studies by other laboratories. Keeling and Doolittle (1996) reported similar results for α-tubulin from three species of Microsporidia, *Encephalitozoon hellem*, *Nosema locustae* and *Spraguea lophii* and for β-tubulin from *Encephalitozoon hellem*. In their manuscript Keeling and Doolittle point out that the alpha and beta tubulins do not, however, represent independent confirmations of the nature of microsporidian phylogeny (as alpha and beta subunits have a close association in the cell). They build a case for this phylogenic associate with possible synapomorphic characters between fungi and Microsporidia. These include chitin in the spore wall, an insertion in the EF-1 alpha gene that is seen in fungi but not protozoa and similarities in meiosis and life cycles between fungi and Microsporidia. These studies, thus, initiated the hypothesis that Microsporidia are highly derived fungi.

The heat shock protein gene (hsp70) gene yielded the first confirmatory evidence of the β-tubulin based phylogeny for the evolution of the Microsporidia. Two groups (Germot et al., 1997; Hirt et al., 1997) demonstrated that the gene for hsp70 existed in Microsporidia and that through comparative phylogenetic analysis of this gene base4d on Maximum Liklehood analysis the Microsporidia were related to the fungi. The hsp70 protein is one of three chaperone proteins found in all life forms and it is well conserved across all groups. The analysis of hsp70 included sequences from various eubacterial groups as well as those for mitochondria (Germot et al., 1997; Hirt et al., 1997). The hsp70 sequence analysis for *Nosema locustae* and *Vairimorpha necatrix* demonstrated that both branched next to the yeasts (*Saccharomyces cereviscae* and *Schizosaccharomyces pombe*). Maximum likelihood analysis gave much clearer evidence for this relationship (99% bootstrap values versus 58% for maximum parsimony, Germot et al., 1997). Both studies pointed out that only the Maximum Likelihood analysis gave a close relationship between Microsporidia and the fungi, but based on the Kishino-Hasegawa test alternate positions for the Microsporidia within the eukaryotes were possible. These papers also pointed out that based on Maximum Parsimony and Least Squares analysis there were no clear relationships between Microsporidia and the fungi. Subsequently (Peyretaillade et al., 1998) the hsp70 of *Encephalitozoon cuniculi* (Peyretaillade et al., 1998), *Nosema locustae* and *Encephalitozoon hellem* (Arisue et al., 2002) were sequenced and analysis of these hsp70 genes confirmed a relationship of Microsporidia to fungi. This analysis of hsp70 genes gave conclusive support to the idea that Microsporidia once had mitochondria.

In addition, to the mitochondrial hsp70, valyl-tRNA synthetase (ValRS) genes consistent with the secondary absence of mitochondria have been found in *T. vaginalis* and *G. lamblia* (Hashimoto et al., 1998), as well as in the Microsporidia. The ValRS genes were cloned from *Enc. hellem* and *Enc. cuniculi* using homology PCR employing degenerate primers followed by screening the corresponding genomic libraries with these PCR products. The microsporidial ValRS genes contain a 37-residue insert that is only present in

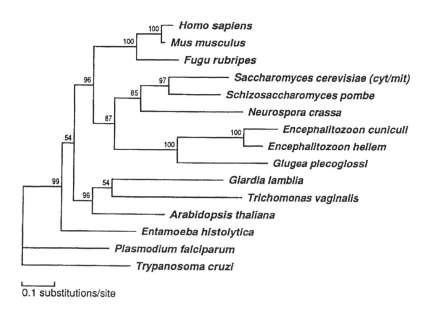

Fig. 1: Phylogenetic analysis of Valyl-tRNA synthetase genes (eukaryotic tree with no outgroup) ML method of protein phylogeny (JTT-F model) (Hashimoto, 1998).

eukaryotic and β/γ protobacterial ValRS. These data also imply that the microsporidia did not diverge from other eukaryotes before the advent of mitochondrial symbiosis. The ValRS phylogeny also supports the placement of the microsporidia with the fungi (Figure 1).

Evidence continues to mount for both the relationship of Microsporidia to the fungi and for the former presence of mitochondria in the

Microsporidia.    The *Enc. cuniculi* genes for thymidylate synthase and dihydrofolate reductase are separate genes, although these genes are on the same chromosome (Vivares et al., 1996). The small subunit rRNA gene of Microsporidia lacks a paromomycin binding site (Edlind, 1998). The EF-1α sequence of the microsporidian *Glugea plecoglossi* has an insertion that is found only in fungi and animals but not in protozoa (Edlind, 1998; Hirt et al., 1999; Kamiashi et al., 1996b). Microsporidia also display similarities to the fungi in mitosis, e.g. closed mitosis and spindle pole bodies, (Desportes, 1976) and meiosis (Flegel and Pasharawipas, 1995). Microsporidia have chitin in their spore wall and store trehalose, as do fungi. Comparative analysis of the largest subunit of the RNA polymerase II (RPB1) gene (Hirt et al., 1999) indicates the Microsporidia are related to fungi and produced a phylogeny similar to the result obtained from an analysis of hsp70 or β-tubulin genes. In these cases *Vairimorpha necatrix* and *Nosema locustae* branched, with high bootstrap values, as the sister group to *Saccharomyces cerevisiae* and *Schizosaccharomyces pombe* (yeast) groups.    Hirt et al. (1999) also reanalyzed the data from EF-1α (Hashimoto et al., 1996; Kamaishi et al., 1996a, 1996b) and concluded that the original analysis could be attributed to artifacts and Hirt's reanalysis suggested that EF-1α also supported a relationship of the Microsporidia and the fungi. For example, Hirt (Hirt et al., 1999) pointed out that microsporidian EF-1α genes have an 11 amino acid insertion that is found in fungi but not protozoa. Upon reanalysis, taking into consideration long branch artifacts, the rRNA data did not support a deep placement of the Microsporidia and provided some (although weak) support for a relationship with fungi (Hirt et al., 1999; Van de Peer et al., 2000). In addition, analysis of glutamyl-tRNA synthetase, seryl-tRNA synthetase, vacuolar ATPase and transcription factor IIB (Fast et al., 1999; Hirt et al., 1999; Katinka et al., 2001) support a relationship of the Microsporidia and fungi.

Analysis of the TATA box binding protein (TBP) (Fast et al., 1999) adds another protein for comparative analysis of Microsporidia (*Nosema locustae*) in the tree of eukaryotic life. According to these authors the TBP sequence for *Nosema locustae* is a much more conserved sequence than some of the other molecules that have been sequenced and analyzed. The implication being that analysis of TBP would not suffer from analytical anomalies due to long branch length. Their results demonstrated that by both Distance and Maximum Parsimony methods *N. locustae* branched with the fungi. It should be noted, however, that on testing the microsporidial-fungal relationship using the Kihino-Hasegawa tests using PUZZLE, version 4.0, this analysis demonstrated that the basal position of the Microsporidia was no worse than the position of Microsporidia as fungi.

Keeling (2003) presented a recent analysis of α-tubulin data that included additional species of Microsporidia and more fungal phyla employing Maximum Likelihood analysis using Quartet Puzzling as a heuristic method for finding the shortest tree for a large number of taxa. This analysis demonstrated varying outcomes depending on which outgroups were included. When both Plants and animals were included as outgroups the Microsporidia were the sister group to the Zygomycota. When just the animals were included in the analysis as the outgroup clade the Microsporidia were the sister group to the Ascomycota. While they did not report good bootstrap values for the specific fungal relationships, the study supports a clear relationship between the Microsporidia and the fungi. This study also suggests that the Microsporidia are most closely related to the Zygomycycetes in which they form a strongly supported monophyletic group. In addition, this analysis indicated that Microsporidia are not closely related to the Harpellales (Tichomycetes). A relationship with the Harpellales had been suggested based on superficial similarities of the polar tube with structures seen in the harpellalean apical spore body.

Fast and Keeling (2001) have demonstrated the presence of a pyruvate dehydrogenase complex E1 in the Microsporidia providing more evidence for the secondary loss of mitochondria in these organisms. Using α- Proteobacteria as the outgroup, they demonstrated conclusively that the α- and β- subunits of pyruvate dehydrogenase from *N. locustae* fall within the mitochondrial clade. The authors state that Microsporidia are derived fungi, but their data linking Microsporidia to the fungi for this gene are weak. For the α-subunit *Trypanosoma cruzi* is the sister group to a plant fungus clade with *Nosema locustae* the outgroup followed by the animal clade. For the β-subunit *T, cruzi* is not included and *N. locustae* is the sister group to an animal fungus clade with the plants as the sister group. As a result the position of the Microsporidia for this gene can be explained as an artifact due to divergence. According to Fast and Keeling (2001) most amitochondriate eukaryotes do not have the PDH complex and use instead the pyruvate:ferredoxin oxidoreductase complex (PFOR). Both systems are capable of catalyzing the decarboxylation of pyruvate, in one case the oxidant is ferredoxin and in the other it is $NAD^+$. Fast and Keeling make a number of interesting speculations on the metabolic function of pyruvate dehydrogenase in Microsporidia. Williams et al. (2002) report a mitochondrial "remnant" in the microsporidian *Trachipleistophora hominis*. Using an antibody produced to a recombinant *T. hominis* mitochondrial hsp70 gene expressed in *E. coli* they were able to localize this hsp70 protein to double walled organelles seen using immunogold labeling, by immuno-electron microscopy. They present an additional phylogenetic analysis of hsp70 gene from six microsporidia, including *T. hominis,* that demonstrates that the Microsporidia to be the sister group to the fungi.

In their analysis of the *Enc. cuniculi* genome Katinka et al. (2001) point out that many of the proteins are most similar to fungal homologues. In addition, they state that the most robust phylogenies on many of these genes (e.g. seryl-tRNA synthetase, transcription initiation factor IIB, subunit A of vacuolar ATPase, and a GTP bidning protein) support a clear relationship of the Microsporidia and fungi. The presence in *Enc. cuniculi* of the principal enzymes for the synthesis and degradation of trehalose confirm that this disaccharide could be the major sugar reserve in Microsporidia, as is seen in many fungi. Analysis of glycosylation pathways suggest that O-mannosylation (e.g. O-linked glycosylation with mannose) as seen in fungi, may occur in the Microsporidia. Recent evidence suggests that such O-mannosylation does indeed occur on the major polar tube protein PTP1 (Weiss LM, unpublished data). An interesting problem in the study of Microsporidia is the question of from what structure the polar filament evolved from. It is possible that genome wide analysis of additional organisms will identify a relationship perhaps between a germ tube and plug over the germ tube of a fungus and the polar filament and anchoring disk of a Microsporidia. There are clearly superficial similarities between the germination of some fungi, including spore coat proteins and preparation for germination, that clearly support the argument for a connection between the Microsporidia and fungi.

Bürglin (2002) pointed out that based on his analysis of homeobox genes that Microsporidia display TALE and normal homeobox sequences very near each other on the same chromosome. Bürglin makes the argument that this arrangement is similar to that found in yeast and other fungi and is indicative of mating types found in fungi. He suggests that Microsporidia like fungi, have mating types. This paper, however, does not discuss the relationships of other homeobox sequences which are found in almost all life forms including plants and animals. While Microsporidia are most likely derived fungi, the mating type connection at this point has not been definitely established.

Perhaps the best evidence for a relationship between Microsporidia and fungi is not the mathematical analysis which often weakly links these groups, but the fact that a deep branching position of Microsporidia is most likely demonstrating a lack of connection. That is to say if a molecule such as the small subunit rDNA (Vossbrinck et al., 1986) indicates a deep branch it may just be a default due to a lack of evidence for a specific relationship whereas if one or more proteins indicate a weak link to fungi the positive link to a group is somewhat convincing. There are a few considerations that those interested in the base of the eukaryotic tree may think about: The explanation of the discrepancy between the initial description of Microsporidia as very divergent eukaryotes and analyses showing the Microsporidia to be derived fungi is generally considered to be one of rapid evolution in at least some of the genes from the Microsporidia. The idea being that it is difficult to tell if organisms have numerous nucleotide or amino acid differences in a given molecule when compared to the same molecule of other organisms. The difference could be due to the fact that the organism has diverged long before the other organisms or because it is rapidly changing from its nearest relatives and therefore from all organisms. Simply put the difference between deep branching and long branch length is difficult to resolve. Logically one would think that Distance Methods of analysis would be least sensitive than Maximum Parsimony methods to detecting long branch length because placing an organism with it's closest relative would result in fewer steps than placing it lower on the phylogenetic tree. However, Felsenstein (1978) reported "Cases in which parsimony or compatibility methods will be positively misleading". That is to say that sequence data can give an incorrect phylogeny, gathering more data results in a more positive view of the incorrect phylogenetic relationship. Maximum Likelihood methods were implemented to avoid long branch attraction (Felsentstein, 1988) and became the method of choice when long branch attraction was a concern. Swofford et al. (2001) pointed out that Felsenstein's assumptions were correct only if the underlying model of sequence evolution fit the data well which is not thought to be true in most cases.

It should be appreciated; however, that if a new apomporphic feature results in an evolutionary enhancement followed by a rapid radiation it is entirely possible that no molecular analysis, even of the entire genome, will allow the resolution to determine the true phylogeny. This is simply due to the fact that the only characters that change rapidly enough to distinguish among closely related taxa have changed. If the lines leading to the ciliates, flagellates, fungi, plants and animals separated over a relatively short period of time due to a radiation because of the advent of the nucleus, and if those nucleotides changed again in the next 2 or 3 billion years, then the relationship between these groups will be impossible to resolve and as more and more date are obtained an incorrect will be confirmed again and again. Whether the phylogeny can be resolved then will depend on which lineages died out. Phillipe (2000) has referred to this a "big bang" in evolution that may prevent clear phylogenic relationships from emerging among many of the protists.

Nonetheless, at present the evidence is convincing that Microsporidia once had mitochondria and have secondarily lost them. The best conclusion from all of the available data is that the Microsporidia should be considered derived fungi. The analysis by Vossbrinck et al. (1987) included the small subunit rDNA from the microsporidian *Vairimorpha necatrix* and the seven other eukaryotes for which this gene had been sequenced at the time. In addition one bacterium and one Archaean were included as outgroups. Today there is an entire genome sequenced for Microsporidia and another one, for *Nosema locustae,* nearly completed. In addition at least 10 other eukaryotic genomes are completed including two from fungi. The only group represented in the Vossbrinck et al. (1987) study that does not have a genome sequence is the ciliates. The time for cloning and sequencing specific gene genes from Microsporidia for comparative purposes is over. The next, exciting, phase in resolving the relationship of Microsporidia in the eukaryotic world will be determined by computer analysis of the complete genomes. With such complete genome sequence analysis perhaps the fungi to which the Microsporidia are most closely related will be determined.

## EVOLUTION WITHIN THE MICROSPORIDIA

Currently, taxonomy and species classification is based on ultrastructural and ecological features including size and morphology of the spores, the number of coils of the polar tube, the developmental life cycle, the host-parasite interface and the developmental cycle in the host. A brief review of the modern classification schemes based on these characters is presented below. Tuzet et al. (1971), Sprague (1977), Issi (1986), Larsson (1986, 1988), and Sprague et al. (1992) are key references for classification based on the biology and ultrastructure of the microsporidia. Molecular analysis of rDNA and other genes is changing our view of the taxonomic significance of these structural and ecological characters and it is clear that the classification scheme for the microsporidia will need to be altered to incorporate insights from this data.

The Microsporidia are usually divided into three basic groups: (1) The "primative" (Metchnikovellidae), hyperparasites of gregarines in annelids, separated from the other microsporidia by the presence of a rudimentary polar filament (a short, thick, manubrium-shaped tube) and the absence of a polaroplast; (2) The Chytridopsidae, Hesseidsae and Burkeidae may be seen as "intermediate" Microsporidia described as having a short polar filament and minimal development of the polaroplast and endospore; and (3) the "higher" Microsporidia which have a well developed polar filament, polaroplast and posterior vacuole.

Tuzet et al. (1971) separated the Microsporidia suborders into two based on the presence (Pansporoblastina) or absence (Apansporoblastina) of a membrane surrounding the sporoblast. The next divisions were based on whether a sporogonial plasmodium was present (in the Apansporoblastina) and the number of spores produced in the pansporoblast (in the Pansporoblastina). This resulted in the following classification scheme:

Class Microsporidea Corliss and Levine, 1963
   Order Microsporida Balbiani, 1882
      Suborder Apansporoblastina
         Family Caudosporidae Weiser, 1958
            Genus *Caudospora*
         Family Nosematidae Labbe, 1899
      Suborder Pansporoblastina
         Family Monosporidae
            Genus *Tuzetia*
         Family Telomyxidae Leger and Hess, 1922
            Genus *Telomyxa*
         Family Polysporidae
            Genera *Glugea, Gurleya, Thelohania,*
                *Heterosporis, Duboscqia,*
                *Trichoduboscqia, Plistophora,*
                *Weiseria, Pyrotheca.*

Sprague (1977) also separated the higher microsporidia based on the presence of a pansporoblastic membrane. The suborders were broken down into families based on the details of sporogony and the nuclear condition. Sprague's work is the standard in the field that defines the Microsporidia taxa including the genera. Information including host and site, vegetative stages, sporulation stages, spore, and locality are given for each species. In 1992, Sprague et al. (1992) further revised the taxonomy of the microsporidia and separated out the metchnikovellids (including *Hessia*) as *incertae setis taxa*. The Class Dihaplophasea, has diplokaryotic stages and undergoes a pairing of gametes which then proliferate and end by undergoing haplosis to produce gametes again. Haplosis can occur either by meiosis (Order Meiodihaplophasida) or by nuclear dissociation (Order Dissociodihaplophasida). The Class Haplophasea, is entirely haplophasic. Further taxonomic divisions and definitions are based upon the presence of a pansporoblastic membrane, the number of nuclei, the number of spores in a sporophorous vesicle and other details of the life cycle.

Classification of Microsporidia according to Sprague et al., 1992.
    Phylum Microspora
            Class 1. Diphaplophasea
                Order 1.  Meiodihaplophasida
                Order 2.  Dissociodihaplophasida
            Class 2. Haplophasea.

Classification of the Microsporidia according to Sprague, 1977.
    Phylum Microspora
        Class Rudimicrosporea
            Order Metchnikovellidae
                Family Metchnikovellidae
                    Genera *Metchnikovella, Amphiacantha,*
                        *Ambliamblys.*
        Class Microsporea
            Order Chytridopsida
                Family Chytridiopsidae
                    Genera *Chytridiopsis, Steinhausia*
                Family Hessidae
                    Genus *Hessea*
                Family Burkeidae
                    Genus *Burkea*
            Order Microsporida
              Suborder Pansporoblastina
                Family Pleistophoridae
                Family Pseudopleistophoridae
                Family Duboscquiidae
                Family Thelohaniidae
                Family Gurleyidae
                Family Telomyxidae
                Family Tuzetiidae
              Suborder Apansporoblastina
                Family Glugeidae
                Family Unikaryonidae
                Family Caudosporidae
                Family Nosematidae
                Family Mrazekiidae

Weiser (1977) placed Chytridiopsis and Hessea with the Metchnikovellids based on the presence of a rudimentary polar filament with spherical spores closed in persistent thick walled pansporoblasts. He divided the rest of the Class Microsporididea into the Order Pleistophoridida (sporogony and spores uninuclear) and the Order Nosematidida (sporogony and spores diplokaryotic). Levine et al. (1980) used Sprague's (1977) divisions and listed Apansporoblastina and Pansporoblastina as the primary divisions of the higher microsporidia. Issi (1986) defined eleven characters visible at the light and electron microscopic level and defined these character states for 68 genera of microsporida. She separated the microsporidia into four subclasses: Metchinikovellidea, Chytridiopsidea, Cylindrosporids, and Nosematidea. Larson (1986, 1988) also defined eleven characters describing the character states for 66 species from 51 genera. The important morphological characters used for taxonomic purposes include: spore shape; number of sporoblasts per sporont; shape number and location of the nuclei; structure of the polaroplast; structure of the parasitophorous vacuole (host origin); structure of the sporophorous vesicle (parasite origin); shape and number of coils in the polar filament; and details of the exospore.

Present methods for phylogenetic analysis involve a list of characters (i.e. spore coat and polar filament) which have two or more character states (i.e. present or absent and isofillar or anisofillar). Missing characters are acceptable for various taxa as long as there are enough characters common to all taxa examined. In order to determine the correct phylogeny the only characters that are useful are those which have changed when organisms diverged and must not have changed subsequently. All other characters will either result in no information or give misleading results. The inclusion of fossil data is often the key in determining the minimum age of a particular characteristic and for concluding whether character states are primitive (pleisomorphic) or derived (apomorphic). Unfortunately there is no fossil data for the Microsporidia and the morphological, life cycle and ecological characters are too few to allow definitive determination of phylogenetic relationships at higher taxonomic levels (above the family level) and thus to determine the significance of each character as well as which characters are derived.

It is possible that with the development of the polar filament and accessory apparatus for injecting the parasite into the host cell, the Microsporidia underwent a rapid evolutionary process, radiating into a wealth of hosts. The "long branch attraction" discussed previously (Germot et al., 1997) would agree with this hypothesis. The mutational event that led to the direct injection of cells into their host cytoplasm would have opened up the entire non-cell walled world as hosts for these parasites. If, as is likely, the Microsporidia underwent rapid radiation, then the basic relationships of the Microsporidian phylogeny may be even more difficult to resolve than the question of the origin of the Microsporidia.

Cloning of many microsporidian rRNA genes has been accomplished. From the microsporidian rRNA phylogenies presented in figure 2 it is seen that with a few exceptions the microsporidia can be divided into three groups: (1) a group of marine fish parasites with a sister group infecting marine crustaceans; (2) a group of terrestrial parasites made up of mostly insect parasites but also including a group of vertebrate parasites; and (3) a group of freshwater aquatic parasites infecting mostly insects and Crustacea that also includes a group of bryozoan parasites. It is not certain at this point whether these groups, based on host and environment, will endure under further analysis. These groupings may be a result of sampling bias. For instance there have been a lot of *Nosema/Vairimorpha* parasites from lepidopteran pests sequenced and only one *Antonospora* sequenced. The recent sequencing of parasites of the Bryozoa may indicate that there are Microsporidia yet to be discovered and sequenced from many aquatic animals. Sprague (1977) indicates that there are many taxa with many species of Microsporidia yet to be sequenced. However, based on present taxonomic categories we would not have expected that the Microsporidia would separate as cleanly as they do along host and environmental lines.

In discussing these groups it should be kept in mind that these groupings may be biased by sampling. We do not have representative samples of many of the known Microsporidian groups. For example, representatives from the Chytridiopsidae and the Metchnikovellidae have not been sequenced and there has been speculation that these groups contain species with more pleisomorphic ("primitive") polar filaments. Microsporidia are reported from numerous phyla of animals (Sprague, 1977), but some such as the insects, may have been sampled more heavily than others such as the Bryozoa, making it appear that the insect Microsporidia is a larger group than it truly is. The newly reported microsporidia from the bryozoans possibly indicates the presence of additional species of these parasites from a multitude of aquatic organisms. A broader group of hosts needs to be sampled and their Microsporidia sequenced before we can determine how this phylogeny represents the basic taxa which exist.

Figure 2 (next page) is a neighbor joining tree (PAUP) of the small subunit rRNA gene of 82 Microsporidia and two outgroup taxa. The alignment was created by Clustal X and the last 50 or so 3' terminal nucleotides of the rDNA genes were eliminated from the analysis.

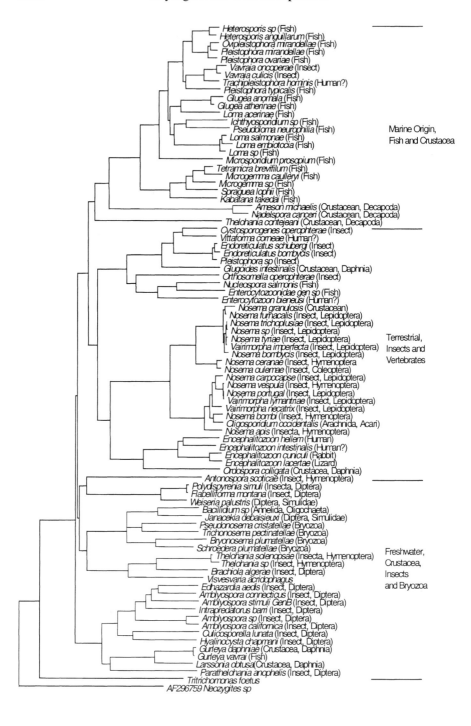

One of the main observations that is becoming more and more apparent form the comparative molecular phylogenetic work is that the ultrastructural and life cycle characters being used to create higher taxonomic categories are not those which separate taxa along evolutionary lines. This is probably because those characters which have been deemed as important for delineating taxonomic lines change character states too rapidly to be useful. Comparative rDNA analysis has shown that some species which were thought to be closely related are instead distant relatives. The definition of *Nosema* as a microsporidian that is diplokaryotic throughout its life cycle is probably no longer valid, because there are divergent species which share this character trait. The type species for *Nosema* is *Nosema bombycis*, from *Bombyx mori*. It is clear that many distinct genera can be separated out from the genus *Nosema*. It has been shown, for example, that other completely separate lines such as *Nosema corneum* (now *Vitiforma corneae*) and *Nosmea algerae* (now *Bracheola algerae*) share the character of being diplokaryotic throughout most if not all of their lifecycle. A separate group related to *N. bombycis* contains *Vairimorpha necatrix*, the type species for the genus *Vairimorpha*. The genus *Vairimorpha* is distinguished from *Nosema* by having a part of its life cycle that contains octospores, at least under some temperatures. Octospores, which are uninucleate, may represent the sexual stage for the species which contains this characteristic. This is not, however, clear as relatives of both *Nosema bombycis* and *Vairimorpha necatrix* show the presence or absence of octospores. Figure 2 demonstrates that *Vairimorpha inperfecta* (which has an octosporus sequence) is found in the *Nosema bombycis* group while the *Vairimorpha nexcatrix* group contains many species which are considered *Nosema* species by virtue of lacking uninucleate octospores. The environmental conditions which lead an organism to discontinue sexual reproduction are not clear and neither the union of haploid octospores nor the infection of a host by octospores has been demonstrated for the *Vairimorpha*. In this case defining a higher level taxon based on the characteristic of octospores appears not to be useful. Unfortunately the alternative is to define these taxa based on DNA sequence information and a good method for grouping Microsporidia into categories based on molecular data has yet to be established.

Within the freshwater parasites are the *Amblyospora*, a group of mosquito parasites which generally have alternate hosts. Species which are derived from members within this group are *Edhazardia aedis* and *Intrapredatorus barri*. *Edhazardia aedis*, a parasite of *Aedes egypti* has been placed in this separate genus by the fact that it does not have an intermediate host a character that it could have easily lost due to some ecologic factors.

*Aedes egypti* breeds rapidly in containers, under these conditions there may not be an intermediate host available. *Intrapredatorus barri*, (Chen et al., 1998) a parasite of *Culex fuscanus* has a several spore types almost identical to that of *Amblyospora*, but because two spore types were found in a single larva a separate genus designation was given. Because this organism was morphologically similar to *Amblyospora trinus*, a parasite of *Culex halifaxi*, and because both species are from predaceous mosquitoes it was suggested the *Amblyospora trinus* be changed to *Intrapredatorus trinus*. It is clear that *Amblyospora*-like spores infecting *Culex* and *Aedes* mosquitoes are clearly *Amblyospora* species. While it is acceptable to make paraphyletic groupings, the rate at which monotypic genera are being described for the aquatic Microsporidia is resulting in genera, such as *Ambylospora*, being paraphyletic.

The most prevalent human microsporidial parasite is *Enerocytozoon bieneusi*. As seen in Figure 2, comparative rDNA analysis demonstrates a close relationship between *Ent. bieneusi* and *Nucleospora salmonis* (syn *Enterocytozoon salmonis*), a parasite of salmonid fish. Ultrastructural similarities include precocious development of the polar tube before the division of the sporogonial plasmodium into sporoblasts, and the lack of a pansporoblastic membrane (the growth of all stages of the parasite in direct contact with the host). The primary distinguishing feature is the growth of *Nuc. salmonis* in the nucleus of the host cell rather than in the cytoplasm as seen in *Ent. bieneusi*. Nucleospora salmonis has proven useful as an animal model for screening drugs for activity against the Enterocytozoonidae (Coyle et al., 1998). Molecular data also demonstrates that *Vittaforma corneae* is related to the family Enterocytozoonidae. As *Ent. bieneusi* cannot be cultivated continuously in vitro, *Vit. cornea* has been used for in vitro screening of drugs for their activity against *Ent. bieneusi* (Didier, 1997). The rDNA analysis (see Figure 2) demonstrates that *Vit. corneae* is related to *Endoreticulatus schubergi* and other insect Microsporidia. *Trachipleistophora hominis*, another human pathogen of unknown origin, is related to *Vavraia oncoperae* and *Pleistophora angtillarum*. Such molecular relationships may be useful in suggesting the environmental reservoirs for the microsporidia found in humans.

## SUMMARY

We are starting to elucidate complex and evolving relationships between the Microsporidia and their hosts. It is clear from several lines of evidence that these organisms are highly evolved and specialized relatives of the fungi rather than "primitive" eukaryotes. With the advent of molecular sequencing techniques the definition of what should comprise a species description may be called into question. Molecular data provides an excellent means for the identification of a species with a set of characters that is usually unique to that organism. In addition, sequence information provides a superb data set for proposing evolutionary relatedness through phylogenetic analysis. Therefore, if possible, a species description should contain a complete rRNA sequence including the ITS region to provide the means to unequivocally confirm whether the same species is isolated subsequently. Such molecular information is particularly useful in identifying intermediate hosts of the same species. Comparison of this sequence data should have greater reliability than comparison of electron micrographs. Ultrastructural analysis (e.g. electron microscopy) also provides information that is often unique to the species, although some closely related species may be indistinguishable. Ultrastructural characters, however, do not provide information for producing accurate phylogenies, particularly at the higher taxonomic levels. It appears that characters such as the number of nuclei, the number of spores/sporonts, the length and structure of the polar filament, the presence of a pansporoblast and details of the life cycle are attributes which can change relatively rapidly as Microsporidia adapt to different hosts and tissue types. This may be the reason for the large number of monotypic genera described in the Microsporidia. Detailed life cycle and ecological studies as well as ultrastructural changes observed in different hosts provide additional characters for comparison of the Microsporidia. In the future, considerations of species descriptions should be based upon what information is most important for furthering our understanding of the Microsporidia and their phylogeny.

## ACKNOWLEDGEMENT

This work was supported by NIH AI31788 (LMW).

## REFERENCES

Arisue, N., L.B. Sanchez, L.M. Weiss, M. Muller, and T. Hashimoto. 2002. Mitochondrial-type 70 genes of the amitochondriate protists, *Giardia intestinalis*, *Entamoeba histolytica* and two microsporidians. Parasitology International **51:** 9-16.

Balbiani, G., 1882. Sur les microsporidies ou psorospermies des articules. Comptes rendus de l'Acadedmie des sciences **95 :**1168-1171.

Bui, E.T., P.J. Bradley, and P.J. Johnson. 1996. A common evolutionary origin for mitochondira and hydrogenosomes. Proceedings of the National Academy of Sciences USA **93:**9651-9656.

Bürglin, T.R.. 2002. The homeobox genes of *Encephalitozoon cuniculi* (Microsporidia) reveal a putative mating-type locus. Developmental Genes and Evolution. **213:** 50–52.

Cavalier-Smith T. 1987. Eukaryotes with no mitochondria. Nature **326:** 332-333.

Chen, W, T. Kuo, and S. Wu. 1998. Development of a new microsporidian parasite, *Intrapredatorus barri* n.g., n.sp. (Microsporidia: Amblyosporidae) from the predacious mosquito *Culex fruscanus* Wiedemann (Diptera: Culicidae) Parasitology International **47:** 183-193.

Clark, C.G., and A.J. Roger. 1995. Direct evidence for secondary loss of mitochondria in *Entamoeba histolytica*. Proceedings of the National Academy of Sciences USA **92**: 6518-6521.

Coyle, C.M., M. Kent, H.B. Tanowtiz, M. Wittner, and L.M. Weiss. 1998. TNP-470 is an effective anti-microsporidial agent. Journal of Infectious Disease **177**: 515-518.

Dayhoff, M.O., R.V. Eck, and C.M. Park. 1972. A model of evolutionary change in proteins, P89-99. *In* M.O. Dayhoff (ed.) Atlas of Protein Sequence and Structure, vol. 5. National Biomedical Research Foundation, Washington D.C.

Desportes, I. 1976. Ulatrastructure de *Stempellia mutabilis* leger et Hess, microsporidie parasite de l'ephemere *Ephemera vulgatta*. L. Protistologica **12**: 121-150.

Didier, E.S. 1997. Effects of albendazole, fumagillin and TNP-470 on microsporidial replication in vitro. Antimicrobial Agents and Chemotherapy **41**:1541-1546.

DiMaria, P., L. Palic,. B.A. Debrunner-Vossbrinck, J. Lapp, and C.R. Vossbrinck. 1996. Characterization of the highly divergent U2 RNA homolog in the microsporidian *Vairimorpha necatrix*. Nucleic Acids Research **24**: 515-522.

Dolflein, F. 1901. Die Protozoen als Parasiten und Krankheitserreger nach biologischen Gesichtspunkten dargestellt. Verlag von Gustav Fisher.

Edlind, T.D., J. Li, G.S. Visvesvara, M.H. Vodkin, G.L. McLaughlin, and S.K. Katiyar. 1996. Phylogenetic analysis of beta-tubulin sequences from amitochondrial protozoa. Molecular Phylogenetics and Evolution **5**: 357-367.

_____. 1998. Phylogenetics of protozoan tubulin with reference to the amitochondriate eukaryotes. In: Coombs, G.H., K. Vickerman, M.A. Sleigh and A. Warren. Eds. Evolutionary Relationships Among Protozoa. Chapman & Hall, London, pp. (91-108)

Embley, T.M., L.J. Horner, and R.P. Hirt. 1997. Anaerobic eukaryote evolution: hydrogenosomes as biochemically modified mitochondria? TREE **12**: 437-441.

Fast, N.M., A.J. Roger, C.A. Richardson and W. F. Doolittle. 1998. U2 and U6 snRNA genes in the microsporidian. Nucleic Acids Research **26**: 3202-3207.

_____, J.M. Logsdon, and W.F. Doolittle. 1999. Phylogenetic analysis of the TATA box binding protein (TBP) gene from *Nosema locustae*: Evidence for a Microsporidia-Fungi relationship and splicosomal intron loss. Molecular Biolology Evolution **6**: 1415-1419.

Fast, N. M., and P. J. Keeling. 2001. Alpha and beta subunits of pyruvate dehydrogenase E1 from the microsporidian *Nosema locustae*: mitochondrion-derived carbon metabolism in microsporidia. Molecular and Biochemical Parasitology. **117**: 201-209.

Felsenstein, J. 1978. Cases in which parsimony or compatibility methods will be positively misleading. Systematic .Zoology **27**: 401-410.

_____. 1988. Phylogenies from molecular sequences: inference and reliability. Annual Review of Genetics **22**: 521-65.

Flegel, TW and T.A. Paharawipas. 1995. A proposal for typical eukaryotic meiosis in microsporidians. Canadian Journal of Microbiology **41**: 1-11.

Germot, A., H. Phillppe, and H. LeGuyader. 1996. Presence of a mitochondrial-type 70-kDa heat shock protein in *Trichomonas vaginalis* suggest a very early mitochondrial endosymbiosis in eukaryotes. Proceedings of the National Academy of Sciences USA **93**: 14614-14617.

_____, _____, and _____. 1997. Evidence for loss of mitochondria in Microsporidia from a mitochondrial-type HSP70 in *Nosema locustae*. Molecular and Biochemical Parasitology **87**: 159-168.

Hashimoto, T., and M. Hasegawa. 1996. Origin and early evolution of eukaryotes inferred from the amino acid sequences of translation elongation factors 1α/Tu and 2/G. Advances in Biophysics **32**: 73-120.

_____, L.B. Sanches T. Shirakura, M. Muller, and M. Hasegawa. 1998. Secondary absence of mitochondria in *Giardia lamblia* and *Trichomonas vaginalis* revealed by valyl-tRNA synthetase phylogeny. Proceedings of the National Academy of Sciences USA **95**: 6860-6865.

Hausemann S, C.P. Vivares, and S. Shuman. 2002. Characterization of the mRNA capping apparatus of the microsporidian parasite *Encephalitozoon cuniculi*. Journal of Biological Sciences **277**: 96-102.

Hirt, R.P., B. Healy, C.R. Vossbrinck, E.U. Canning, and T.M. Embley. 1997. A mitochondrial HSP70 orthologue in *Vairimorpha necatrix*: Molecular evidence that microsporidia once contained mitochondria. Current Biology **7**: 995-998.

_____, J.M. Logsdon, B. Healy, M.W. Dorey, W.F. Doolittle, and T.M. Embley. 1999. Microsporidida are related to Fungi: Evidence from the largest subunit of RNA polymerase II and other proteins. Proceedings of the National Academy of Sciences USA **96**: 580-585.

Horner, D.S., R.P. Hirt, S. Kilvington, D. Lloyd, and T.M. Embley. 1996. Molecular data suggest an early acquisition of the mitochondrion endosymbiont. Proceedings Royal Society London B Biolologic Science **263**: 1053-1059.

Huger, A. 1960. Electron microscope study on the cytology of a microsporidian spore by means of ultrathin sectioning. Journal of Insect Pathology **2**: 84-105.

Issi, I. V. 1986. Microsporidia as a phylum of parasitic protozoa. Academy of Science U.S.S.R. (Leningrad), **10**: 6-136.

Kamiashi, T., T. Hashimoto, Y. Nakamura, F. Nakamura, S. Murata, N. Okada, D. Okamoto, M. Shimizu and M. Hasegawa. 1996a. Protein phylogeny of translation elongation factor EF-1 alpha suggests microsporidians are extremely ancient eukaryotes. Journal of Molecular Evolution **42**: 257-263.

_____, _____, _____, Y. Masuda, F. Nakamura, K. Okamoto, M. Shimizu, and M. Hasegawa. 1996b. Complete nucleotide sequence of the genes encoding translation elongation factors 1α and 2 from a microsporidian parasite, *Glugea plecoglossi*: Implications for the deepest branching of eukaryotes. Journal of Biochemistry **120**: 1095-1103.

Katinka, M. D., S. Duprat, E. Cornillot, G. Metenier, F. Thomarat, G. Prensier, V. Barbe, E. Peyretaillade, P. Brottier, P. Wincker, F. Delbac, H. El Alaoui, P. Peyret, W. Saurin, M. Gouy, J. Weissenback, and C.P. Vivares, 2001. Genome sequence and gene compaction of the eukaryote parasite *Encephalitozoon cuniculi*. Nature **414**: 450-453.

Keeling P.J., and W. F. Doolittle, 1996. Alpha-Tubulin from early-diverging eukaryotic lineages and the evolution of the tubulin family. Molecular Biology and Evolution **13**: 1297-1305.

_____ and G.I. McFadden. 1998. Origins of microsporidia. Trends in Microbiology **6**: 19-23.

_____, M. A. Luker, and J. D. Palmer. 2000. Evidence from beta-tubulin phylogeny that microsporidia evolved from within the fungi. Molecular Biology and Evolution **17**: 23-31.

_____, and N. M. Fast. 2002. Microsporidia: biology and evolution of highly reduced intracellular parasites. Annual Review of Microbiology **56**: 93-116.

_____. 2003. Congruent evidence from α-tubulin and β- tubulin gene phylogenies for a sygomycete origin of microsporidia. Fungal Genetics and Biology **38**: 298-309.

Kent, M.L., L. Margolis, and J.O. Corliss. 1994. The demise of a class of protists: taxonomic and nomenclatural revisions proposed for the protist phylum Myxozoa Grassè, 1970. Canadian Journal of Zoology **72**: 932-937.

Krieg, A. 1955. Ueber Infektionskrankheiten bei Engerlingen von Melolontha sp. unter besonderer Berucksichtigung einer Mikrosporidien-Erkrankung. Zentralblatt fur Bakteriologie Parasitenkunde, Infektionskrankheiten und Hygiene. II Abt **108**: 533-538.

Kudo, R.R., and E.W. Daniels. 1963. An electron microscope study of the spore of a Microsporidian, *Thelohania californica*. Journal of Protozoology **10**: 112-120.

Larsson, J. I. R. 1986. Ultrastructure, function, and classification of microsporidia. Progress in Protistology **1**: 325-390.

Larsson, J.I.R. 1988. Identification of microsporidian genera (Protozoa, Microspora)- a guide with comments on taxonomy. Archiv fur Protistenkunde **136**: 1-37.

Leipe, D.D., J. H. Gunderson, T.A. Nerad, and M.L. Sogin. 1993. Small subunit ribosomal RNA of *Hexamita inflata* and the quest for the first branch in the eukaryotic tree. Molecular and Biochemical Parasitology **59**: 41-48

Levine, N.D., J.O. Corliss, F.E. Cox, G. Deroux, J. Grain, B.M. Honigberg, G.F. Leedale, A.R. Loeblich 3d, J. Lom, D. Lynn, EG Merinfeld, F.C. Page, G. Poljansky, V. Sprague, J. Vavra, and F.G. Wallace. 1980. A newly revised classification of the protozoa. Journal of Protozoology **27**: 37-58.

Lom, J., and J. Vavra. 1961. Niektore Wyniki Baden Nad Ultrastruktura Spor Posozyta Ryb Plistophora hyphessobrycornis (Microsporidia). Wiadomosci parazytologiczne **7**: 828-832.

_____, and _____. 1962. A Proposal to the Classification within the Subphylum Cnidospora. Systematic Zoology **11**: 172-175.

Markiw, M.E., and K. Wolf. 1983. Mysoxoma cerebralis (Myxozoa: Myxosporea) etiologic agent of salmonid whirling disease requires tubificid worms (Annelida: Oligochaeta) in its life cycle. Journal of Protozoology **30**: 561-564.

Maxim, A.M., and W. Gilbert. 1977. A new method for sequencing DNA. Proceedings of the National Academy of Sciences **74**: 560-564.

Peyretaillade, E., V. Broussolle, P. Peyret, G. Metenier, M. Gouy, and C.P. Vivares. 1998. Microsporidia, amitochondrial protists, possess a 70-kDa heat shock nprotein gene of mitochondrial evolutionary origin. Molecular Biology and Evolution **15**: 683-689.

Philippe, H. A. Germot, and D. Moreira. 2000. The new phylogeny of eukaryotes. Current Opinions in Genetic Development **10**: 596-601.

Roger, A.J., C.G. Clark, and W.F. Doolittle. 1996. A possible mitochondrial gene in the early-branching amitochondriate protist *Trichomonas vaginalis*. Proceedings of the National Academy of Sciences USA **93**: 14618-14622.

_____, S.G. Svard, J. Tovar, C.G. Clark, M.W. Smith, F.D. Gillin, and M.L. Sogin. 1998. A mitochondrial-like chaperonin 60 gene in *Giardia lamblia*: evidence that diplomonads once harbored an endosymbiont related to the progenitor of mitochondria. Proceedings of the National Academy of Sciences USA **95**: 229-234.

Sanger, F., and H. Tuppy. 1951. The amino-acid sequence in the phenylalanyl chain of insulin. 1. The identification of lower peptides from partial hydrolysates. Biochemistry **49**: 463-481.

_____ and A. R. Coulson. 1977. DNA sequencing with chain-terminating inhibitors. Proceedings of the National Academy of Sciences **74**: 5463-5467.

Siddall, M.E., D.S. Martin, D. Bridge, S.S. Desser, and D.K. Cone. 1995. The demise of a phylum of protists: Phylogeny of myxozoa and other parasitic cnidaria. Journal of Parasitology **81**: 961-967.

Sogin, M.L., J H. Cunderson, H.J. Elwood, R. A. Alonso, and D. A. Peattie. 1989. Phylogenetic meaning of the kingdom concept: an unusual RNA from *Giardia lamblia*. Science **243**: 75-77.

Soltys, B.J., and R.S. Gupta. 1994. Presence and cellular distribution of a 60-Kda protein related to mitochondrial HSP 60 in *Giardia lamblia*. Journal of Parasitology **80**: 580-590.

Sprague, V., and Vavra. 1977. "Systematics of the Microsporidia." In: 'Comparative Pathobiology Vol. 2'. (L.A. Bulla and T.C. Cheng. eds.) pp. 1-30. Plenum Press, New York.

_____, J.J. Becnel and E.I. Hazard. 1992. Taxonomy of the phylum microspora. Critical Review of Microbiology **18**: 285-395.

_____, and _____. 1998. Note on the Name-Author-Date combination of the taxon Microsporidies. Balbiani, 1882, When Ranked as a Phylum. Journal of Invertebrate Pathology **71**: 91-4.

Swofford, D.L, P. J. Waddell, J. P. Huelsenbeck, P. G. Foster, P.O.Lewis, and J. S. Rogers. 2001. Bias in phylogenetic estimation and its relevance to the choice between parsimony and likelihood methods. Journal of Systematic Biology **50**: 525-39.

Tuzet, O., J. Maurand, A. Fize, R. Michel, and B. Fenwich. 1971. Proposition d'un nouveau cadre systematique por les genres de Microsporidies. Comptes rendus de l'Acadedmie des sciences (Paris) **272**: 1268-1271.

Williams B.A.P., R.P. Hirt, J.M. Lucocq, and T.M. Embley. 2002. A mitochondrial remnant in the microsporidian *Trachipleistophora hominis*. Nature **418**: 865-869.

Weill, R. 1938 L'interpretation des Cnidosporides et la valeur taxonomique de leur cnidome. Leur cycle comparé à la phase larvaire des Narcomeduses Cuninides. Travaux de la Station Zoologique de Wimereaux. **13**: 727-744.

Weiser, J. 1959. *Nosema laphygmae* n. sp. and the internal structure of the microsporidian spore. Journal of Insect Pathology **1**: 52-59.

_____. 1977. Contribution to the classification of microsporidia. Vestnick Ceskoslovenske Spolecnost Zoologica. **41**: 308-320.

Woese, C.R., and Fox, G.E. 1977. Phylogenetic structure of the prokaryotic domain: The primary kingdoms. Proceedings of the National Academy of Sciences USA **74:** 5088-5090.

Van de Peer Y, A. Ben Ali, and A. Meyer. 2000. Microsporidia: accumulating molecular evidence that a group of amitochondriate and suspectedly primitive eukaryoties are just curious fungi. Gene **246**: 1-8.

Vossbrinck, C.R., and C.R. Woese. 1986. Eukaryotic ribosomes that lack a 5.8s RNA. Nature **320**: 287-288.

_____, J.V. Maddox, S. Friedman, B.A. Debrunner-Vossbrinck, and C.R. Woese. 1987. Ribosomal RNA sequence suggests microsporidia are extremely ancient eukaryotes. Nature **326**: 411-414

Vivares, C., C. Biderre, F. Duffieux, E. Peyretaillade, P. Peyret, G. Metenier, and M. Pages. 1996. Chromosomal localization of five genes in *Encephalitozoon cuniculi* (microsporidia). Journal of Eukaryotic Microbiology. **43**: 97S.

_____, M. Gouy, F. Thomarat, and G. Metenier. 2002. Functional and evolutionary analysis of a eukaryotic parasitic genome. Current Opinions in Microbiology **5**: 499-505.

Yang D., Y. Oyaizu, H. Oyaizu, G.J. Olsen, and C.R. Woese. 1985. Mitochondrial origins. Proceedings of the National Academy of Sciences U.S.A. **82**: 4443-4447.

# THE MICROSPORIDIA GENOME: LIVING WITH MINIMAL GENES AS AN INTRACELLULAR EUKARYOTE

Christian P. Vivarès and Guy Méténier

*Laboratoire de Parasitologie moléculaire et cellulaire, LBP, UMR CNRS 6023, Université Blaise Pascal - Clermont-Ferrand, 63177 Aubière Cedex, France.*

## ABSTRACT

*Encephalitozoon cuniculi* is a mammal-infecting microsporidian harbouring an extremely reduced nuclear genome (2.9 Mbp). Following its complete sequencing, gene annotation has revealed possible transport and metabolic pathways in the parasite. A clathrin-independent route of endocytosis is predicted. Anaerobic glycolysis and pentose-phosphate pathway may account for generation of ATP and reducing power, respectively, but the fates of pyruvate and NADH remain to be explored. A mevalonate pathway may produce geranylgeranyl and dolichyl groups, not sterols. Although the microsporidian Golgi apparatus is unstacked, several encoded proteins are specific of *cis* and *trans* Golgi compartments. *Encephalitozoon cuniculi* should be unable to perform N-linked glycosylation, a unique case among eukaryotes.

**Key words:** Microsporidia, *Encephalitozoon cuniculi*, genome, metabolism,

## INTRODUCTION

Comprising several species now recognized as emerging pathogens of humans (Weiss, 2000), the microsporidia are characterized by important reductive adaptations to their life inside other eukaryotic cells (Keeling and Fast, 2002). The study of their metabolism and physiology is hindered by the obligate character of their development inside other eukaryotic cells and by the difficulties of isolating the non-sporal forms of these parasites. Thus, when the first analysis of the molecular karyotype of *Encephalitozoon cuniculi* revealed an haploid genome size of 2.9 Mbp (Biderre et al., 1995), we thought that sequencing this miniaturized genome should be a good means to expedite the molecular characterization of a microsporidian pathogen and to define a minimal gene repertoire for eukaryotic parasites. The microsporidian genome project was initiated in 1998 with a random sequencing strategy used at

Genoscope (Evry) and the completion of the annotated genome sequence was reported three years later (Katinka et al., 2001). This was the first sequenced genome of a eukaryote parasite. It is worth noting that the year 2001 was marked by fully sequenced eukaryotic genomes having rather extreme lengths: ~3000 Mbp for the human nuclear genome (Lander et al., 2001; Venter et al., 2001) and 551 kbp for the genome of the vestigial nucleus (nucleomorph) of an algal endosymbiont (Douglas et al., 2001). Hence, the *E. cuniculi* genome represents in size less than one thousandth of the human genome and only 5.2 times the nucleomorph genome. About 2000 predicted protein-coding genes were found to be densely packed over 11 chromosomes.

Here, we review some major activities that may be attributed to *E. cuniculi* on the basis of sequence similarity criteria. Of primary importance for the knowledge of host-microsporidia interactions, the potentialities for solute transport and endocytosis are first discussed. The inferred metabolic conversions of carbohydrates and lipids, essential for energy production and membrane organization, are then presented. Genes encoding proteins assigned to the endomembrane system are finally considered to propose a minimal conception of its functioning in microsporidian secretion processes.

## THE HOST-PARASITE INTERFACE AND MEMBRANE TRANSPORTERS

Upon entering an host cell, the *E. cuniculi* sporoplasm is surrounded by a plasma membrane closely adherent to the membrane of a parasitophorous vacuole (PV) that is commonly assumed to be of host origin, perhaps as a result from partial endocytosis of the extruded polar tube (Bigliardi and Sacchi, 2001). However, meronts are always seen to be applied to the PV membrane, each of these cells establishing a contact resembling a "mobile junction", whereas sporonts detach from this membrane and subsequent sporogonial stages are free within the lumen of the vacuole (Vavra and Larsson, 1999). The genesis of the PV seems therefore dependent on continuous interactions between host cytoplasm and meronts but very little is known about these interactions (relations with host cell vesicular trafficking, recruitment of host mitochondria, insertion of lipid and protein components, membrane permeability, parasite cell coat changes, etc.). In *Plasmodium falciparum*-infected erythrocytes, both a transmembrane receptor and glycosylphosphatidylinositol (GPI)-anchored proteins characteristic of microdomains in host cell membranes can be recruited into the PV membrane, first revealing a vacuolar uptake of host membrane proteins possibly mediated by lipids such as cholesterol and sphingomyelin (Lauer et al., 2000). Conversely, a rhoptry protein secreted by *Toxoplasma gondii* can be inserted into the PV membrane (Beckers et al., 2002). Thus, although still fragmentary, the data derived from other intracellular parasites suggest that the PV membrane of *Encephalitozoon* should be a "host-parasite mosaic".

In *E. hellem*-infected cells, the concentrations of $H^+$ and $Ca^{2+}$ ions in the PV do not differ from those in the host cytoplasm and the large anion calcein (623 Da) enters the PV, indicating that the PV membrane acts as a molecular sieve (Leitch et al., 1995). Like in apicomplexan parasites (Saliba and Kirk, 2001), *Encephalitozoon* therefore seems to render the PV membrane permeable to a wide range of low molecular weight solutes, via the insertion of a "porin-like" channel. The annotation of the *E. cuniculi* genome sequence has revealed several homologues of known transporters but any assignment to the PV membrane remains elusive. Harboring a six-transmembrane topography and belonging to the MIP (*major intrinsic protein*) family, a putative *E. cuniculi* channel similar to aquaporins could allow the diffusion of water in the direction of an osmotic gradient and possibly also the passage of some small uncharged solutes. The presence of such a channel in microsporidia was previously inferred from observations on aquaporin-like intramembrane particles at a very high density in the plasma membrane of *Nosema* spores (Undeen and Frixione, 1991) and from the inhibition of spore germination by $Hg^{2+}$ ions (Frixione et al., 1997). In lens fiber cells, the MIP (or AQP0) protein is characterized by adhesive properties in addition to its transport function, as shown by atomic force and cryo-electron microscopy data revealing the double-layered nature of MIP lattices with a tight "tongue-and-groove" fit between the apposing MIP molecules of adjacent membranes (Fotiadis et al., 2000). Whether aquaporins contribute to the adherence of meronts to the PV membrane awaits investigations.

Very few ion transporters have been identified through significant homologies with sequences from other organisms. A multi-subunit vacuolar $H^+$-ATPase (V-ATPase) is predicted but a parsimonious hypothesis is to consider a location to only internal membranous structures such as endosomes or vacuole. Active transport of some inorganic cations may occur via two large-sized P-ATPases (or E1-E2 ATPases) that do contain the invariant phosphorylation site DKTGTLT and at least ten transmembrane spans. Falling in the subfamily of unknown cation specificity that is called "type V" (= "type 5", not "vacuolar"), these ATPases are possibly of recent emergence (Axelsen and Palmgren, 1998). They are likely essential for maintaining ionic homeostasis and generating electrochemical gradients, as proposed for some "type-V" homologs in *Plasmodium* (Rozmajzl et al., 2001) and *Cryptosporidium* (LaGier et al., 2002). A homolog of fungal $Na^+/H^+$ antiporters should contribute to pH regulation and salt tolerance, like in some yeast (Kinclova et al., 2002). The uptake of inorganic phosphate may be mediated by a transporter similar to yeast PHO88. The coding sequence ECU08_0530 represents a possible sulfate permease. With only 536 amino acid (aa) residues, this protein is much shorter than *Schizosacharomyces pombe* SulP (958 aa) but within the same size range as various bacterial sulfate transporters.

Most members of the ATP-binding cassette (ABC) superfamily play a major role in export-import processes. Their diversity is however reduced in *E. cuniculi*. Twelve different ABC proteins encoded by 13 genes (two identical copies on chromosome I) are indeed representative of only four subfamilies and none of these is organized as a "full-size" ABC transporter, *i.e.* comprising two membrane spanning domains (IM) and two cytoplasmic ATPase domains (ABC) (Cornillot et al., 2002). Five "half-size" ABC transporters sharing an ABC-IM topology were assigned to the WHI subfamily, suggesting a participation to sterol and phospholipid transport and/or to drug resistance if referring to the functions allocated to some human homologs (review: Dean et al., 2001). The five other "half-size" transporters (IM-ABC type) were assigned to the HMT subfamily, characterized by yeast and human mitochondrial transporters involved in the export of Fe-S clusters from mitochondria to cytosol. Because several *E. cuniculi* genes are related to some major steps of Fe-S cluster assembly, one of these ABC transporters was assumed to be associated with the inner membrane of a mitochondrion-derived organelle or "mitosome" (Katinka et al., 2001; Vivarès et al., 2002). The mitosome hypothesis is supported by the recent finding of small double-membrane structures reactive to anti-mtHSP70 antibodies in the cytoplasm of *Trachipleistophora hominis* meronts (Williams et al., 2002). However, it seems unlikely that the five HMT paralogues have the same function and the same localization. Note that the HMT subfamily belongs to a large family (DPL) comprising various prokaryotic transporters of IM-ABC type that are involved in drug, peptide/protein and lipid export (Dassa et al., 2001). As antibodies raised against the mammalian P-glycoprotein recognize epitopes in merogonial stages, it has been suggested that *Encephalitozoon* species possess a multi-drug resistance pump (Leitch et al., 2001). Two distinct sequences for ABC transporters (EiABC1 and EiABC2) have been reported in *E. intestinalis* (Bonafonte et al., 2001). EiABC1 appears as the ortholog of one *E. cuniculi* HMT protein (ECU03_0240) (Cornillot et al., 2002). By contrast, EiABC2 that was assumed to represent a multi-drug resistance pump (Bonafonte et al., 2001) has no clear counterpart in *E. cuniculi* (Cornillot et al., 2002). Its domain organization is of IM-ABC type, however. This case of high intrageneric divergence as well as the lack of potential mitochondrial targeting signal in both EiABC2 and two *E. cuniculi* HMT proteins strongly argue against a mitosomal location and, therefore, offer a large choice of putative transport functions to be explored. The two remaining ABC systems of *E. cuniculi* lack transmembrane domain (ABC-ABC type) and correspond to two well-conserved proteins that are respectively homologous to human RNase L inhibitor (RLI) and yeast translational regulator GCN20 (Cornillot et al., 2002). The function of non-mammalian RLI proteins is unknown. The conservation of a [4Fe-4S]-binding motif in all prokaryotic and eukaryotic RLI homologs led us to postulate a role in an ubiquitous electron transfer that would have been retained by microsporidia (Méténier and Vivarès, in press). The GCN20 homolog in *Plasmodium falciparum* has been shown to be exported to the lumen of the parasitophorous vacuole and associated with both the tubulovesicular network in the erythrocyte and membranous structures of the parasite, indicating that GCN20 may also function either as an ATP-binding subunit of a unknown multimeric transporter or as a chaperone for protein translocation across multiple membranes (Bozdech et al., 1998; Bozdech and Schurr, 1999). This is a good illustration of the complexity of host-parasite interactions and should stimulate further searches for the intracellular localization, subunit organization and function of the different ABC proteins encoded by the *E. cuniculi* genome.

The entry of monosaccharides and nucleosides across the *E. cuniculi* plasma membrane may involve four permeases of the MF (major facilitator) superfamily. One of these (ECU07_1100) is similar to the mammalian high-affinity glucose transporter GLUT1, but we must be cautious for inferring strict substrate specificities because of the dependence on subtle amino acid residues substitutions (Walmsley et al., 1998). The coding sequence ECU08_0640, having some similarity to the C-terminal region of yeast nicotinic acid permease and now appearing closer to the putative *Plasmodium* permease PF07_0070, is assumed to represent a purine transporter. Another example of weak similarity is offered by ECU11_1600 in which the N-terminal region fits to a domain characteristic of the carriers of reduced folate, especially of murine RFC-1 (Tolner et al., 1997). Six genes are candidate amino acid permeases, each harboring 10-11 transmembrane helices, but only one oligopeptide transporter is specified by the *E. cuniculi* genome. The latter is representative of a ubiquitous proton-coupled transporter of di-and tripeptides that can be ranged within the PTR family (Steiner et al., 1995). This contrasts with the four genes required to the functioning of two distinct peptide transport systems in *Saccharomyces cerevisiae*, known as the PTR system for di-/tripeptides and the OPT system for tetra-/pentapeptides (Hauser et al., 2001). As yet commented in previous papers (Katinka et al., 2001; Vivarès et al., 2002; Méténier and Vivares, in press), the import of host ATP by the microsporidian parasite is conceivable. Recall that among prokaryotes, the intracellular parasites of the *Rickettsia* and *Buchnera* genera are unique through the possession of ADP/ATP translocases permitting the uptake of ATP from host cell (Andersson, 1998). This import process resembles that mediated by chloroplast ADP/ATP translocases for gaining ATP from the cytosol of a plant cell, leading to hypothesize horizontal transfer events (Wolf et al.,

1999). Four paralogous ADP/ATP exchanger genes having the highest homologies with the bacterial and plastidial types are found in *E. cuniculi*. The evolutionary relationship between all these translocases is not clear, but there is no doubt that the expression and localization of *Encephalitozoon* ADP/ATP translocase(s) deserve studies for a better understanding of energy requirements during microsporidian development.

## A POTENTIAL PROTEIN MACHINERY FOR ENDOCYTOSIS

The occurrence of endocytosis in *Toxoplasma gondii* tachyzoïtes has been documented by the internalization of surface lipids as monitored by styryl dyes (Robibaro et al., 2001) and the accumulation of cholesterol derived from low-density lipoprotein (LDL) particles having transited through host lysosomes (Coppens et al., 2000). Are microsporidia also able to internalize some nutrients via an endocytic pathway? Discharged sporoplasms of *Spaguea lophii* can incorporate several tracers (ferritin particles, dextran and albumin) during incubation in an extracellular medium (Weidner et al., 1999) and comparative analyses of lipid compositions are suggestive of an uptake of host-derived fatty acids and cholesterol (Vivarès et al., 1980; Biderre et al., 2000; El Alaoui et al., 2001). The existence of an endocytic process in *E. cuniculi* is especially supported by genes encoding two dynamin-like GTPase proteins and a Rab5 homolog that are commonly considered as markers of early endosomes in mammalian cells. However, the functions currently assigned to members of the dynamin family are rather multiple (Schmid et al., 1998; Sever, 2002). The dynamin-like protein ECU10_1700 is strongly similar to yeast VPS1 that participates to the sorting of vacuolar proteins as well as to the regulation of peroxisome abundance (Sever, 2002) but ECU01_1210 remains a good candidate to the association with pits at the plasma membrane. The small GTPase Rab5 is known to be involved in the control of the fusion of early endosomes. It is interesting to also note that a recent study in *T. gondii* has demonstrated that Rab5 homolog is localized in tubulovesicular structures distinct from the Golgi apparatus and enhances exogenous cholesterol uptake (Robibaro et al., 2002).

The clathrin-mediated pathway is the most commonly used mechanism of endocytosis but, in contrast to that of *P. falciparum* (Gardner et al 2002), the genome of *E. cuniculi* does not encode clathrin heavy chain homolog. Three so-called "clathrin-associated proteins" are however representative of the three kinds of subunits (or "adaptins") able to form a heterotetrameric adaptator protein (AP) complex. In mammals, the AP-1 and AP-2 complexes participate to the recruitment of clathrin to membranes whilst the outer shells of the coats containing two other complexes (AP-3 and AP-4) are of unknown nature (review: Boehm & Bonifacino 2002). In *S. cerevisiae*, AP-4 is absent but AP-3 is specifically involved in the clathrin-independent "ALP (alkaline phosphatase) pathway" characteristic of vesicles coated with the protein Vps41 that could fuse directly with the vacuole (Rehling et al., 1999). Curiously, the potential *E. cuniculi* adaptins are more similar to those of the AP-1 complex but, in the absence of clathrin, it is more logical to consider that the microsporidian adaptator is in fact located to the vesicles of a clathrin-independent pathway. Much remains to be learned about endocytic routes in eukaryotes. As recently shown by Sabharanjak et al. (Sabharanjak et al., 2002), GPI-anchored proteins can be transported to the recycling endosomal compartment via a non-clathrin and non-caveolin pathway regulated by a GTPase of the Rho family (CDC42), the GPI anchor appearing as the signal for the internalization into endosomes responsible for a major part of fluid-phase uptake in higher eukaryotes. Moreover, evidence of a GPI-specific endosomal pathway depending on a Rab5 isoform has been reported in *Trypanosoma brucei*, in which the GPI anchor is well known to play an essential role in membrane attachment of surface antigens (Pal et al., 2002). Thus, considering that *Encephalitozoon* likely expresses Rab5 as well as GPI-anchored proteins and Rho GTPases, we assume that a GPI-specific pathway is preserved in microsporidia.

## CARBOHYDRATE METABOLISM

Most reactions for the predicted carbohydrate metabolism of *E. cuniculi* are indicated in Figure 1. All the genes encoding the enzymes for anaerobic glycolysis, from glucose to pyruvate, have been identified. The putative microsporidial phosphofructokinase is pyrophosphate-dependent, excludes regulations associated with an ATP-dependent enzyme. Citrate inhibition is evidently dispensable inasmuch as the citric acid cycle does not occur. Thus, the rate of glycolysis should be regulated mainly at the level of hexokinase and pyruvate kinase activities. Gluconeogenesis starting from either pyruvate, fructose 1,6-*bis*P or glucose 6-P cannot be inferred. The capacity to produce NAPH during the two dehydrogenation steps characteristic of the oxidative branch of the pentose phosphate pathway is retained, as expected for the coupling with reductive biosyntheses. In the non-oxidative branch of this pathway, a problem arises from the apparent lack of transaldolase gene. If transaldolase is really absent, a three-carbon unit cannot be transferred

**Figure 1** (next page). Predicted carbohydrate metabolism in *Encephalitozoon cuniculi*. Dashed arrows indicate uncertainties relevant to the fate of the pyruvate, electron transport from NADH and utilization of UDP-GlcNAc for direct O-linked glycosylation. Homologs of both cytoplasmic (C) and mitochondrial (M) glycerol 3-P dehydrogenases are present.

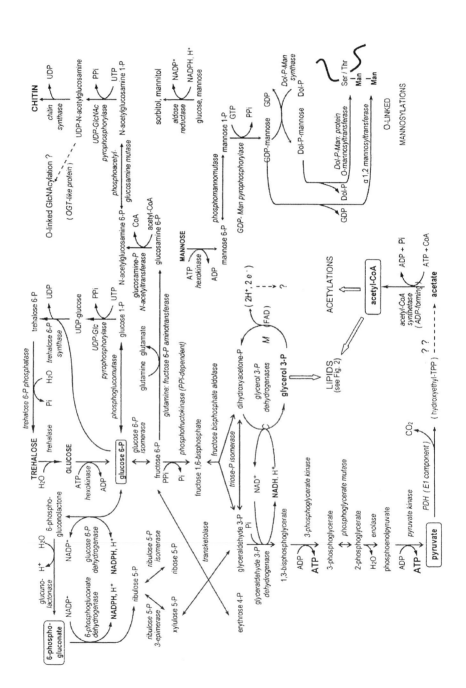

from sedoheptulose 7-P to glyceraldehyde 3-P, the products of the first reversible reaction catalyzed by transketolase. In *P. falciparum*, no transaldolase homolog is found but the requirement of erythrose 4-P for chorismate pathway indicates that tranketolase would mediate the second reaction (Gardner et al., 2002). We assume that *E. cuniculi* transketolase uses xylulose 5-P only for the reaction involving erythrose 4-P as two-carbon unit acceptor and the two phosphorylated sugars rejoining the glycolytic pathway (Figure 1). The need for ribose 5-P is certainly less crucial than for NADPH, considering the incapacity of *E. cuniculi* to perform ribose-5-P-dependent syntheses of ribonucleosides or aromatic amino acids. Mannose may be phosphorylated to produce GDP-mannose and dolichol phosphate mannose that will serve as substrates for two types of mannosyltransferases involved in O-linked glycosylations. No gene for the utilization of other sugars such as galactose, fucose, sialic and uronic acids have been identified.

$NAD^+$ regeneration cannot be justified by aerobic respiration or conversion of pyruvate to either lactate or ethanol. A possible way is however offered by the reduction of dihydroxyacetone-P via cytosolic NAD-dependent glycerol-3-P dehydrogenase (GPDH-C) producing glycerol 3-P, a major precursor for lipid biosyntheses. The fate of the pyruvate is obscure. Genes encoding the two subunits of the E1-component of the pyruvate dehydrogenase (PDH) complex has been identified in *Nosema locustae* (Fast and Keeling, 2001), which is surprising because the PDH complex is normally located inside the matrix of mitochondria whereas these organelles are not recognized in microsporidia. If the E1 component is really functional, the problem of the fate of the pyruvate is displaced toward that of its decarboxylation product (hydroxyethyl-thiamine pyrophosphate), owing to the absence of candidates for the two other components. The mitosome hypothesis was however attractive because several putative gene products of *E. cuniculi* should fulfill an ancestral mitochondrial function represented by iron-sulfur cluster biosynthesis (Lill and Kispal, 2000). This biosynthesis requires ferredoxin-dependent reduction steps, which encouraged us to propose that the ferredoxin-ferredoxin reductase system might also play a key role in a novel pyruvate-to-acetate pathway, as the counterpart of the unique reaction catalyzed by pyruvate-ferredoxin oxidoreductase in trichomonads (Katinka et al., 2001, Vivarès and Méténier, 2002). Then, formation of acetyl-CoA from acetate may be mediated by a specific ADP-forming synthetase (Figure 1). The finding of a homolog of mitochondrial FAD-dependent GPDH (GPDH-M) was another surprise. In mitochondriate eukaryotes, GPDH-M indeed belongs to the mitochondrial inner membrane (catalytic site facing the intermembrane space) and participates to the "glycerol-3-P shuttle" conveying reducing equivalents from cytosolic NADH toward ubiquinone in the respiratory chain. In the absence of such chain and oxidative phosphorylation, the significance of the preservation of GPDH-M in microsporidia is problematic. Likewise, a manganese superoxidase dismutase (SOD) gene is retained by *E. cuniculi* whilst mitochondrial respiration cannot be invoked as being the major source of superoxide anion. The unique microsporidian SOD is likely required to minimize oxidative damages. As molecular oxygen oxidizes easily the reduced forms of flavins, the enzyme could be especially useful for the optimal functioning of an electron transfer from GPDH-M toward an undefined acceptor.

Biosynthesis and degradation of glycogen should not occur, in agreement with the lack of cytochemical or biochemical evidence of this store polysaccharide in microsporidia. This is also the case in *P. falciparum* (Gardner et al., 2002). Through a recent comparison of fully sequenced prokaryotic genomes, the lack of glycogen metabolism has been considered to be characteristic of bacteria with a parasitic or symbiotic lifestyle and the extension of this view to eukaryotes has been suggested (Henrissat et al., 2002). Likewise, *E. cuniculi* should be unable to synthesize 1,3-β-glucan, unlike model yeasts and several other fungi in which this glucan is mainly responsible of the elasticity of a permanent cell wall. By contrast, the hexosamine pathway leading to the biosynthesis of chitin can account for the important accumulation of this polymer within the electron-lucid inner region (endospore) of the rigid microsporidian spore wall (Vavra and Larsson, 1999). The formation of 1,4-β-linkages between N-acetylglucosamine (GlcNAc) residues is mediated by a single putative plasma membrane-bound chitin synthase (class IV) that should be activated during sporogony. Control of the activation of chitin deposition likely exists at different levels. In *Saccharomyces* mutants with cell-wall defects, an increased accumulation of chitin mainly results from the overexpression of glutamine-fructose-6P amidotransferase (GFA), the first enzyme of the hexosamine pathway (Lagorce et al., 2002). In the aquatic fungus *Blastocladellia emersonii*, chitin production occurs during zoospore differentiation and is controlled at the post-translational level, depending on the availability of UDP-GlcNAc that inhibits the active posphorylated form of GFA and prevents its dephosphorylation by protein phosphatases (Maia, 1994). The C-terminal part of a long coding sequence (ECU09_1320, 597 aa) matches clearly with some plant endochitinases (family 19) that participate to the defense of plants against pathogenic fungi and insects. We hypothesize that the putative *E. cuniculi* chitinase is involved in a discrete disruption of the wall at the anterior spore prior to polar tube extrusion. One can have some doubt about the function of a sequence related to the 110-kDa subunit of O-GlcNAc transferases (OGT) that add GlcNAc from UDP-GlcNAc to nuclear and cytosolic proteins through O-linkage with hydroxy-aminoacid residues. The homology mainly corresponds to ubiquitous tetratricopeptide repeats in the N-terminal domain of OGT, not to the catalytic C-terminal domain.

Except for chitin synthesis, no alternate utilization of GlcNAc would be possible.

Trehalose is the best candidate as the major store carbohydrate in microsporidia and trehalose hydrolysis is assumed to generate an osmotic potential required for the induction of germination in aquatic species (Undeen and Vander Meer, 1999). Sorbitol has been also detected. Both sugars may be also viewed as stress protective agents. In *E. cuniculi*, sorbitol could derive from the activity of an aldose reductase and the two-step synthesis of trehalose is supported by two genes representative of trehalose 6-P synthase and trehalose 6-P phosphatase. These genes are similar to those encoding the two catalytic subunits of the yeast trehalose synthase complex regulated by two additional subunits playing an interchangeable stabilizing role (Bell et al., 1998). As the four subunits share 33% identity over a large common region, it seems likely that the regulatory subunits are absent in the microsporidian. Another example of simplified situation is provided by the potential of trehalose degradation, compared to that of fungi. Indeed, whilst most fungal species have at least two types of trehalase (neutral and acidic) with distinct locations (cytosolic and vacuolar, respectively), *E. cuniculi* has only one trehalase gene. Its sequence does not present the large N-terminal extension with a regulatory phosphorylation domain characteristic of yeast neutral trehalase (Nwaka and Holzer, 1998). In fact, the best matches are observed with metazoan trehalases (32-35% identity) that are less characterized than in fungi but have generally an acidic optimum pH. The entry of *E. cuniculi* trehalase into the endoplasmic reticulum lumen is suggested by an N-terminal cleavable signal peptide. An unusual C-terminal extension is present that may indicate a need for post-translational processing for trehalase activation.

## LIPID METABOLISM

The *E. cuniculi* genome has no apparent potentialities for the initial carboxylation of acetyl-CoA to malonyl-CoA and the complete elongation cycle in fatty acid biosynthesis, as well as for the β-oxidation pathway. Long-chain fatty acids may be activated in acyl-CoA esters by a specific ligase. Conversely, an acyl-CoA thioesterase similar to bacterial TesB may release fatty acids from these esters. Although the latter shares sequence homologies with eukaryotic peroxisomal enzymes, its C-terminal tripeptide (SKK) is not predicted as being a peroxisomal targeting signal. Triacylglycerols have been detected in the spores of *E. cuniculi* and two other microsporidia (less than 10% of total lipids) (El Alaoui et al., 2001a). The unique committed step in triacylglycerol biosynthesis is indicated by ECU10_0300 coding for a homolog of various metazoal diacylglycerol O-acyltransferases. A potential triacylglycerol lipase (lipase of class 3) matches with Aut5/Cvt17, a yeast enzyme involved in intravacuolar lysis of membrane-enclosed autophagic bodies (Epple et al., 2001).

Possible biosynthetic routes for the four major glycerophospholipid classes are mainly depicted in Figure 2. Three distinct synthases can account for the *de novo* pathway based on the formation of CDP-diacylglycerol and the use of this activated phosphatidyl unit for the synthesis of acidic phospholipids (phosphatidylserines or PS and phosphatidylinositols or PI). The expression of *E. cuniculi* CDP-diacylglycerol synthase has been demonstrated by cDNA library screening and RT-PCR amplification, supporting the functionality of the *de novo* pathway (El Alaoui et al., 2001b). From a comparison of the incorporation rates of radiolabeled choline, ethanolamine, serine and methionine in the phospholipids extracted from uninfected and *E. cuniculi*-infected mammalian cells, it has been assumed that the microsporidian parasite can undergo PS decarboxylation producing phosphatidylethanolamines (PE) then PE methylation yielding phosphatidylcholines (PC) (El Alaoui et al., 2001b). However, the candidate genes remain to be identified. Alternate reactions characteristic of the salvage pathway are inferred from the genes encoding phosphatidate phosphatase, aminoalcohol-(choline and ethanolamine)-phosphotransferase and phosphocholine cytidyltransferase. We failed to predict formation of CDP-ethanolamine and initial phosphorylation of each aminoalcohol, but it is difficult to decide whether this reflects true metabolic gaps or only highly divergent genes. In the first hypothesis, choline and ethanolamine would not be directly used by the parasite, which may justify the absence of significant differences between uninfected and infected host cells as regard to the incorporation levels of exogenous [14C]choline and [14C]ethanolamine in PC and PE fractions, respectively (El Alaoui et al., 2001b ). The increased incorporation of [14C]ethanolamine in the PC fraction of parasitized cells would then be due to a stimulated PE-to-PC conversion in host cytoplasm. Two encoded phospholipid-transporting P-type ATPases are representative of aminophospholipid (PS and PE) "flippase" activities assisting in the maintenance of lipid asymmetry of membranes. One of these is possibly necessary for the preferential translocation of PS from the external to the cytofacial monolayer of the plasma membrane (Daleke and Lyles, 2000).

**Figure 2** (next page). Potential pathways for lipid biosynthesis in *Encephalitozoon cuniculi*. Continuous dashed lines represent the steps that are not clearly supported by gene data. Ceramide synthesis is inferred from the presence of a LAG1 homolog.

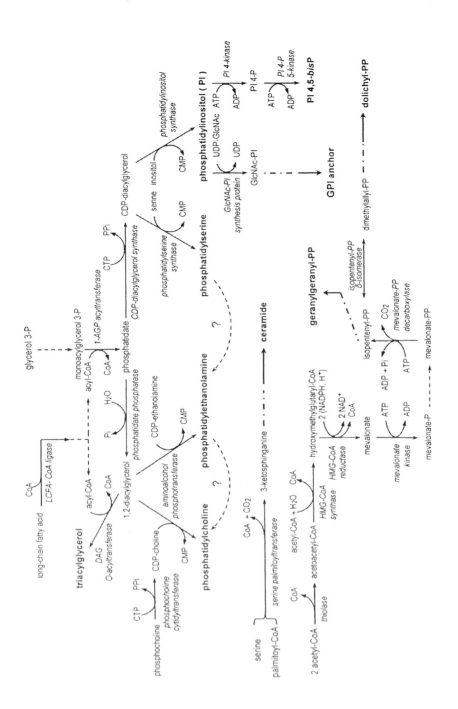

The committed step in sphingolipid synthesis (formation of 3-ketosphinganine from serine and palmitoyl-CoA) is predicted. Further steps are difficult to precise for this pathway that is not well documented in eukaryotic microbes. However, like various eukaryotes, *E. cuniculi* displays a gene related to the *Saccharomyces* longevity-assurance gene 1 (LAG1) encoding a protein first shown to facilitate the transport of GPI-anchored proteins from the ER to the Golgi (Barz and Walter, 1999). The Lag1 protein is now viewed as an essential component of the acyl-CoA-dependent ceramide synthase, functioning with very long chain fatty acids and justifying a role in ceramide signalling that affects various physiological processes (Jazwinski and Conzelmann, 2002). No complex glycosphingolipid should be formed. In fungi including *Cryptococcus neoformans*, phosphoinositol is transferred from PI to ceramide to produce inositol phosphoceramide (IPC) that is further modified by addition of mannose and of a second phosphoinositol group ( Luberto et al., 2001). We failed to find an *E. cuniculi* sequence homologous to the fungus-specific IPC synthase, suggesting that the model for a key role of this enzyme in the pathogenesis of *Cryptococcus* cannot be applied to micosporidia. The synthesis of sphingomyelin, the major phosphosphingolipid in mammals, seems to be also excluded. Sphingomyelin represents a significant proportion of the sporal lipids in *E. cuniculi* (5%) but is absent in the fish microsporidian *Glugea atherinae* (El Alaoui et al., 2001a). This argues for a process of translocation of sphingomyelin from host cells, occurring in a variable extent according to the considered parasitic species.

In the case of phosphoinositides, two specific kinases may catalyze the consecutive phosphorylations of PI to PI 4-phosphate and PI-4,5-*bis*phosphate, and a phospholipase D similar to *S. cerevisiae* Spo14 may release phosphatidic acid from PC (Figure 2). Spo14 is known to be activated by PI-4,5-*bis*phosphate and required for both meiosis and spore formation. In addition, a fully stimulated Spo14 is needed for secretion in the absence of PI/PC transfer protein (Rudge et al., 2002). The putative *E. cuniculi* phospholipase D is therefore a good candidate for linking secretion to cellular signalling during sporogony. Although several proteins of the PI3/PI4 kinases family are encoded, none of these is specific of the phosphorylation of the only 3-hydroxyl group in the inositol ring. The key enzyme responsible for the production of the early intermediate (GlcNac-PI) in GPI-anchor biosynthesis is represented. An isoprenoid pathway from acetyl-CoA to mevalonate seems to be disconnected from the formation of the interconvertible precursors (isopentenyl diphosphate or IPP and dimethylallyl diphosphate) for geranylgeranyl groups and dolichol, because no phosphomevalonate kinase homolog was detected (Figure 2, lower part). However, as the three enzymes for synthesis of IPP from mevalonate contain the same fold and mediate phosphorylation of similar substrates (Bonanno et al., 2001), it may be envisaged that the second step of this synthesis in the microsporidian is catalyzed by one of the two other enzymes. Cholesterol represents about 1% of the *E. cuniculi* sporal lipids (El Alaoui et al., 2001a) but, owing to the lack of candidate genes for its synthesis from IPP, this sterol is probably of mammalian host origin.

## ENDOMEMBRANE SYSTEM AND SECRETORY PATHWAYS

The most recognizable cytoplasmic membrane system in all microsporidian stages is the endoplasmic reticulum (ER) associated with ribosomes. A morphological entity assimilated to the Golgi apparatus is present but difficult to identify, especially in meronts. Discrete vesicles are observed but smooth ER elements, Golgi vesicles and endosomes cannot be easily discriminated. During sporogony, the Golgi apparatus appears as an vesiculo-tubular network that increases in size and may contribute to the biogenesis of the sporal polar tube, polaroplast membranes and posterior vacuole (Bigliardi & Sacchi, 2001). The progressive accumulation of an electron-dense material inside the vesicles that fuse to produce polar tube coils is the best illustration of the occurrence of a secretory traffick. The presence of *cis* and *trans* compartments is suggested by cytochemical data in *Glugea stephani* (Takvorian and Cali, 1994) and *Nosema grylli* (Sokolova et al., 2002). Recently, *Brachiola algerae* spores and discharged sporoplasms have been shown to contain a new structure referred as to the "multilayered interlaced network" that accounts for the peculiar "thickening" of the plasmalemma in the *Brachiola* genus and could be of Golgi origin (Cali et al., 2002). The *E. cuniculi* genome sequence predicts various proteins that are well representative of the functioning of an ER-Golgi-endosome system in protein translocation, folding and modification associated with vesicle-mediated trafficking. Our current conception is outlined in Figure 3 (next page).

Although 7SL RNA has not been identified, the signal recognition particle (SRP) as well as the SRP receptor must be present. The passage of proteins across the ER membrane via a transmembrane channel or translocon (Johnson and van Waes, 1999) should depend on three major proteins (Sec61, Sec62, Sec63) and a chaperone of the HSP70 family. An auxiliary transmembrane protein (Sec66/71) is preserved. The removal of signal peptides probably involves a signal processing peptidase of type I (ECU02_0760). The formation of disulfide bridges and rotation of proline residues can be catalyzed by specific isomerases. As expected from the presence of potential *E. cuniculi* ER-resident proteins harbouring a C-terminal sequence obeying to the KDEL consensus, a homolog of the so-called KDEL receptor (Erd2), a good marker of the ER, is evidently important for the retention of these proteins in the ER. A selective degradation of misfolded secretory proteins may be mediated by a protein similar to yeast Der1 (Knop et al., 1996). In addition, an Npl4 homolog should function in association with an AAA ATPase (CDC48) for a post-ubiquitination but pre-proteasome step in ER-associated protein degradation (Bays and Hampton, 2002).

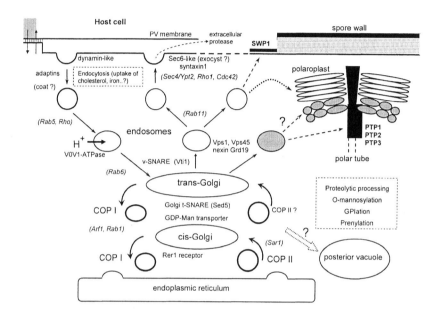

**Figure 3.** Possible trafficking pathways in the *E. cuniculi* endomembrane system. Clathrin-independent endocytosis is assumed to occur during merogony. GTPases regulating transport steps are indicated in italics and in brackets. A Bet3-like protein (similar to Plasmodium A41-2) should be associated with cis-Golgi. No N-linked glycosylation is predicted. Sporogony-specific vesicular transports (dashed arrows) can be associated with the biogenesis of the spore wall (for SWP1 protein secretion, see Bohne et al., 2000) and that of the invasion apparatus. Lamellar polaroplast is thought to result from the accumulation of membrane stacks as precursors for the new plasma membrane of discharging sporoplasm. The three polar tube proteins identified so far (PTPs) should be major cargo molecules of large electron-dense vesicules. A secretion route specifically related to vesicular polaroplast is still uncertain. The components and origin of the posterior vacuole are not defined.

The Golgi apparatus of *E. cuniculi* somewhat resembles that of *S. cerevisiae* inasmuch the Golgi compartments of the budding yeast are not stacked and are dispersed throughout the cytoplasm. An overexpression of the gene for the GDP-mannose transporter located to early and medial Golgi leads to the formation of mammalian-like stacked cisternae in yeast cells, without mislocalization of the transporter in the ER or vacuole (Hashimoto et al., 2002). This induced progression of Golgi stacks illustrates the possibility that spatial organization of the organelle may depend on the relative amount of some Golgi-resident proteins. We hypothesize that Golgi remodeling at the onset of sporogony in microsporidia is associated with an increased gene expression of such proteins. The occurrence of vesicular transport in *E. cuniculi* can be easily correlated with the conservation of genes for subunits of the COPI and COPII coats and for some GTPases of the Arf and Rab/Ypt families. Comprising two complexes (Sec13-Sec31 and Sec23-Sec24) and the GTPase Sar1, COPII-coated vesicles mediate the transport from the ER to the Golgi. A Sec24 homolog having escaped the initial annotation is represented by ECU02_1230. COPI-coated vesicles are involved in both retrograde transport from the Golgi to the ER and within the Golgi but their possible role in anterograde intra-Golgi transport is still debated (Nickel et al., 2002). Genes for the coatomer subunits, two small GTPases (Arf1 and Rab1/Ypt1) and the transmembrane protein p23 (or Tmp21) support a COPI pathway that should be critical for recycling Golgi enzymes in *E. cuniculi*. A Rer1 homolog corresponding to a sorting receptor in the *cis* Golgi is expected to recognize signals in transmembrane domains of some ER proteins for retrieving them to the ER (Sato et al., 2001).

Several components of the SNARE complexes relevant to vesicle docking and fusion, commonly divided into t-SNAREs (on target membranes) and v-SNAREs (on transport vesicles), are predicted. A target for ER-derived vesicles can be inferred from an *E. cuniculi* counterpart of animal syntaxin-5 or yeast Sed5, a major *cis* t-SNARE heavy chain. A Bos1-like protein (ECU07_1620) is a potential light chain. A Sec1-related protein could be a key regulatory factor, if considering the recent demonstration of the role of a yeast homolog in the assembly and specificity of fusion complexes (Peng & Gallwitz, 2002). Because Sed5 is the only syntaxin needed as a t-SNARE heavy chain for the transport throughout the yeast Golgi and has various partners to form distinct t-SNAREs (Tsui et al., 2001; Parlati et al., 2002), it is important to consider the significance of two other syntaxin-like sequences in *E. cuniculi*. One of these is distantly related to yeast Sft2 that facilitates the fusion of endosome-derived vesicles with the Golgi (Conchon et al., 1999). The other is a syntaxin-1 homolog, indicating a role in the fusion of secretory vesicles at the plasma membrane. It is interesting to note that *S. cerevisiae* contains two homologs (SSO1 and SSO2) that may differ functionally, SSO1 being required for sporulation but not SSO2 (Jantti et al., 2002). Thus, the microsporidian Golgi seems to have only one t-SNARE heavy chain. The most conserved v-SNARE in *E. cuniculi* is a Vti1 protein. In *S. cerevisiae*, this protein is required for transport from the Golgi to the vacuole but can interact with Sed5 in retrograde transport to the *cis* Golgi and with the t-SNARE protein Pep12 in transport from the *trans* Golgi network to the prevacuolar compartment (Fischer von Mollard and Stevens, 1999). The functioning of Vti1 in a late Golgi or endosomal compartment in *E. cuniculi* is likely. A predicted Sec1-related protein is indeed similar to yeast Vps45, a major component of the complex interacting with Vti1 at the endosomal level (Pelham, 1998). A Vps1-like dynamin may serve to the genesis of endosomes from the Golgi and a Rab11 homolog may regulate the Golgi-to-endosome transport as well as an export to the cell surface (Rodman and Wandinger-Ness, 2000). One sorting nexin (Grd19 homolog) contains the PX domain that interacts with PI 3-phosphate, a phospholipid now considered as a good marker for endosomal membranes (Pelham, 2002). The acidic pH of endosomes can be maintained by the V0-V1 ATPase complex. A single Rab6 homolog is the counterpart of yeast Ypt6, recently shown to be involved in both endosome-to-Golgi and intra-Golgi retrograde transports (Luo and Gallwitz, 2002). In *T. gondii*, Rab6 appears as a regulator of the retrograde transport from secretory "dense granules" to late Golgi (Stedman et al., 2003). We assume that Rab6/Ypt6 participates to the control of a major secretion route during microsporidian sporogony (Figure 3). It should be stressed that a vacuole similar to that in *S. cerevisiae* is never seen during merogony. A large "posterior vacuole" differentiates in late sporogony to become a permanent structure in the *E. cuniculi* spore, but there is no experimental proof of its equivalence to the yeast vacuole or mammalian lysosome. We have been unable to detect an homolog of any component of the *Saccharomyces* vacuolar protein transport system, *e.g.* syntaxin Vam3 and Rab GTPase Ypt7. Likewise, there is no coding sequence with significant homology to mammalian mannose 6-P receptors or to two distinct yeast vacuolar receptors (Vps10 and Mrl1) for the sorting of lysosomal/vacuolar hydrolases. Obviously, the possibility that *E. cuniculi* uses an unrelated receptor system cannot be ruled out.

The machinery for exocytosis is of peculiar interest to consider in *E. cuniculi*. A polarized growth of the meronts might indeed depend on their partial adherence to the PV membrane, and polar tube extrusion is characterized by a dramatic discharge of materials across a very restricted apical area of the spore. In both *S. cerevisiae* and mammalian cells, the Sec4 protein located at the surface of exocytic vesicles can bind specifically to Sec15 that is one of the eight proteins of the "exocyst" particle. The exocyst is peripherally associated with plasma membrane domains of active secretion and cell growth: yeast bud, tips of growing neuritis, regions close to tight junctions in epithelial cells (see review in Lipschutz and Mostov, 2002). A putative *E. cuniculi* Rab GTPase resembles *S. pombe* Ypt2, a Sec4-related protein recognized as an actor of the exocytic pathway (Craighead et al., 1993). In addition, two GTPases of the Rho family (Rho1 and Cdc42) are candidate to direct interactions with Sec3, the exocyst subunit providing the "spatial landmark for polarized secretion" in *S. cerevisiae* (Finger et al., 1998). No gene for exocyst subunits in *E. cuniculi* was defined but since the protein sequences are moderately conserved in fungi and mammals (20-24% identity) a re-examination was necessary. In a preliminary specific search, ECU10_0390 and ECU11_0390 were found to be related to Sec5 and Sec6, respectively. In fact, only ECU11_0390 (690 aa) fits with the Sec6 sequences from different organisms, the highest amino acid identity being 19%. This finding is obviously insufficient to conclude that a full exocyst complex is formed in *E. cuniculi*. During neuronal development, Sec6 expression correlates with discrete membrane domains representing nascent presynaptic specializations and is downregulated in mature synapses (Hsu et al., 1999; Chin et al., 2000). Studying the distribution of Sec6-like protein should be interesting to evaluate probable changes in exocytic sites in relation with the transitions between different developmental stages. Pathogenic fungal species are well known to secrete several types of endoproteases and peptidases that are clearly related to the virulence in the case of dermatophytes (review: Monod et al., 2002). A potential *E. cuniculi* serine protease of the subtilase family (ECU01_1130) is homologous to several fungal extracellular proteases of vacuolar origin. We hypothesize that the microsporidian protease may be released in the close environment of the parasites represented by the PV space, thus playing a role in the digestion of some host-derived proteins.

The last years have provided an accumulation of data supporting roles of the actin cytoskeleton in vesicular transport, especially in the scission of vesicles and the donor-to-acceptor translocation process, but much remains to be understood (review: Stamnes, 2002). In *E. cuniculi*, actin has been shown to be located to the cortical cytoplasm in various cell stages and also close to developing and mature polar tubes (Bigliardi et al., 1999). The translated sequence that was submitted to EMBL/GenBank databases as being representative of *E. cuniculi* actin is too long (ECU01_0460: 407 aa) because of the omission of a change in the initiator codon position. Indeed, the true start codon corresponds to the second methionine (position 33), giving the expected length of 375 aa for a conserved cytoplasmic actin. In contrast, a highly divergent actin-like sequence (ECU04_1090) is reduced to 341 aa. A possible activation of actin assembly during Golgi vesicle formation in *E. cuniculi* could be related to an ARF-mediated stimulation of the production of phosphoinositides (PI-4,5-*bis*phosphate) allowing the recruitment of actin and of a spectrin-like protein (ECU09_0290) to the Golgi apparatus. An ankyrin-like protein could serve as an adaptor for linking spectrin to other membrane proteins and a putative motor for short-range actin-based motility is offered by a unconventional myosin (class V). We assume that interactions with microtubules are especially required for the extension of membranous structures over long distances during sporogenesis. The set of potential microtubule-associated motors in *E. cuniculi* is represented by a single dynein and six members of the kinesin superfamily that are currently investigated in our laboratory.

A major metabolic activity of the eukaryotic endomembrane system is to insure the glycosylation of various proteins through two well-distinct pathways (O- and N-glycosylations). The glycosylation of two spore wall proteins (SWP1 and SWP2) destined to the exospore region of mature spores has been demonstrated through their reactivity with two lectins (ConA and WGA) in *E. intestinalis* and the presence of an asparagine-linked oligosaccharide has been hypothesized (Hayman et al., 2001). In *Saccharomyces*, two rather simple N-glycan structures are formed in the Golgi and the so-called "mannan" structure is characteristic of the outermost layer of the yeast cell wall (Munro, 2001). Mannose may be utilized by *E. cuniculi* but the capacity to perform N-glycosylation is quite improbable. A Dpm1 homolog for an ER-associated dolichol-phosphate mannose (Dol-P-Man) synthase might be thought to produce the key mannosyl donor for the synthesis of N-linked oligosaccharides. However, none of the subunits of the oligosaccharyltransferase complex catalyzing the transfer of a preassembled high-mannose oligosaccharide onto asparagine residues is predicted. In addition, there are no gene candidates for specific glycosidases acting in the ER for the removal of glucose and mannose residues to give the usual N-linked core structure. Dol-P-Man is known to also serve for both O-linked oligosaccharides and GPI anchor (Orlean, 1992). Two potential Dol-P-mannose protein mannosyltransferases of the PMT family can effectively mediate the transfer of mannose from Dol-P-Man to serine/threonine residues (see Figure 1). Thus, like in fungal species (review: Ernst and Prill., 2001), O-glycosylation of *E. cuniculi* proteins should begin from Dol-P-Man in the ER, rather than from a sugar nucleotide in the Golgi apparatus. Further addition of mannose residues in the Golgi is relatable to genes for a GDP-mannose transporter similar to *Candida* or *Saccharomyces* Vrg4 protein (formerly designated as Gog5) and only one mannosyltransferase (KTR family) that uses GDP-mannose as substrate and specifically forms α-1,2 linkages. This still argues against a processing of N-glycosylated proteins and contrasts with the diversity of *S. cerevisiae* mannosyltransferases represented by nine members of the KTR family and six members of the MNN family (Lussier et al., 1999). As recently demonstrated in *C.*

*albicans* (Nishikawa et al., 2002), Vrg4 is essential for fungal viability and the lack of mammalian homolog suggests a potential target for antifungal therapy. This view could extend to microsporidia. After mannose transfer, the hydrolysis of GDP by a putative GDPase (ECU07_1260) will give GMP that may exit the Golgi lumen in an equimolar exchange with cytosolic GDP-mannose. Clearly, the gene equipment indicates that simple O-linked oligosaccharides of mannose are elaborated in *E. cuniculi*. Mannoproteins are found at an high proportion in the cell wall of various fungi and it would not be surprising that the exospore proteins so far identified in *E. cuniculi* (Bohne et al., 2000) and *E. intestinalis* (Hayman et al., 2001) are in fact O-linked mannoproteins.

The attachment of some proteins to membranes, especially that of small GTP-binding proteins and nuclear lamins, depend on a post-translational process of prenylation involving the linkage of an isoprenyl pyrophosphate group to a cysteine residue near or at the carboxy terminus. Two *E. cuniculi* genes are characteristic of distinct subunits found in the two different types of geranylgeranyltransferases (GGTases): the type I, that reacts preferentially with the cysteine of the carboxy-terminal motif CaaL (a, aliphatic amino acid), and the type II, reacting with the cysteines of terminal -CC or -CXC motifs mainly present in Rab proteins. In the case of GTTase I substrates, the removal of the three C-terminal amino acids can be mediated specifically by a putative CaaX prenyl protease (ECU02_1380). Another important protein lipidation predicted to occur in *E. cuniculi* is the covalent linkage to a GPI anchor (review: Ikezawa, 2002). Biosynthesis of GPI precursors and post-translational modification take place in the ER. In yeasts, three proteins (Gpi7, Gpi13 and Mcd4) are involved in the transfer of phosphoethanolamine to the three mannoses of the GPI anchor (Toh-e and Oguchi, 2002). An *E. cuniculi* integral membrane protein (ECU07_0310) similar to *Candida* Gpi7 (24.7 % identity in 626 aa overlap) is candidate to a transferase acting at the level of at least one mannose residue. Its role in spore wall biogenesis may be crucial, considering that *Gpi7* gene deletions in *S. cerevisiae* and *C. albicans* have been shown to entail the accumulation of wall-targeted proteins in the growth medium (Richard et al., 2002). The GPI attachment step is a transamidation reaction that replaces the C-terminal hydrophobic signal peptide with GPI in the substrate protein. Differences in subunit composition of GPI transamidase complexes exist between protozoans and mammals, possibly because of subtle differences in C-terminal signal. The sequence ECU09_1070 is characterized by a peptidase domain of the C13 family and should be the counterpart of Gpi8, the catalytic component of GPI transamidase.

## CONCLUSIONS:

We expect that the present survey of *E. cuniculi* gene sequences, although far from exhaustive, has been informative for predicted transport and metabolic capacities in a microsporidian parasite and may be useful for encouraging experimental studies. A considerable simplification of the core metabolism and of some compartmentalized activities is evident. Obviously, the lack of Krebs cycle or of oxidative phosphorylation is not surprising in the case of an amitochondriate eukaryote. There is no doubt that the loss of genes involved in various mitochondrial processes has strongly contributed to reduce the diversity of the *E. cuniculi* nuclear genes encoding known functions. However, gene deficits are also noted for non-mitochondrial functions. A remarkable example is provided by the lack of putative *E. cuniculi* enzymes responsible for N-linked glycosylations, a ubiquitous pathway of protein modification throughout eukaryotes and involving the ER-Golgi system. Note that this dispensability renders rather difficult the acceptance of the idea that the "raison d'être" of the N-glycan core attached to internal proteins in other eukaryotes is related to only folding and quality control in the ER. Among a great number of problems to be resolved for a better understanding of microsporidian physiology, we can cite the relative importance of ADP/ATP carrier proteins in the uptake of host-derived ATP, the persistence of a mitochondrion-derived organelle and its possible role in some unrevealed terminal oxydoreduction processes, as well as the contribution of potential O-linked mannoproteins and GPI-anchored proteins to spore wall biogenesis.

Microsporidia are very diverse in terms of life cycle, host specificity, host-parasite interface, nuclear configuration, spore morphology. With its 1200 species so far described, the evolution of the so-called "microsporidian world" (Vivarès, 2001) has been characterized by various adaptive strategies that may have implied more or less important gene losses or acquisitions. Interestingly, from a first analysis of new *Vittaforma cornea* gene sequences, significant differences seem to occur at the metabolic level (Mittleider et al., 2002). Two *Vittaforma* gene sequences have been shown to match with reverse transcriptase and topoisomerase IV, which is not the case in *E. cuniculi*. Another possible inter-genus metabolic divergence is supported by recent biochemical evidence for catalase in *Spraguea lophii*, this enzyme being located to the posterior vacuole of the spore (Weidner and Findley, 2002). If further confirmed at the sequence level, this would signify that peroxisome-like organelles may exist in some microsporidia. This may be the "tip of the iceberg" of a wide metabolic diversification in these parasites.

### REFERENCES

Andersson, S.G.E. 1998. Bioenergetics of the obligate intracellular parasite *Rickettsia prowazekii*. Biochimical and . Biophysics Acta **1365**: 105-111.

Axelsen, K.B., and M.G. Palmgren. 1998. Evolution of substrate specificities in the P-type ATPase superfamily. Journal of Molecular Evolution **46:** 84-101.

Barz, W.P., and P. Walter. 1999. Two endoplasmic reticulum (ER) membrane proteins that facilitate ER-to-Golgi transport of glycosylphosphatidylinositol-anchored proteins. Molecular Biology Cell 1999, **10:** 1043-1059.

Bays, N.W., and R.Y. Hampton. 2002. Cdc48-Ufd1-Npl4: stuck in the middle with Ub. Current Biology **12:** R366-R371.

Beckers, C.J., J.F. Dubremetz, O. Mercereau-Puijalon, and K.A. Joiner. 1994. The *Toxoplasma gondii* rhoptry protein ROP 2 is inserted into the parasitophorous vacuole membrane, surrounding the intracellular parasite, and is exposed to the host cell cytoplasm. Journal of Cell Biology **127:** 947-961.

Bell, W., W. Sun, S. Hohmann, S. Wera, A. Reinders, C. De Virgilio, A. Wiemken, and J.M. Thevelein. 1998. Composition and functional analysis of the *Saccharomyces cerevisiae* trehalose synthase complex. Journal of Biological Chemistry **273:** 33311-33319.

Biderre, C., F. Babin, and C.P. Vivarès. 2000. Fatty acid composition of four microsporidian species compared to that of their host fishes. Journal of Eukaryotic Microbiology **47:** 7-10.

_____, M. Pagès, G. Méténier, E.U. Canning, and C.P. Vivarès. 1995. Evidence for the smallest nuclear genome (2.9 Mb) in the microsporidium *Encephalitozoon cuniculi*. Molecular and Biochemical Parasitology **74:** 229-231.

Bigliardi, E., M.G. Riparbelli,.M.G. Selmi, L. Bini, S. Liberatori, V. Pallini, A. Bernuzzi, S. Gatti, M. Scaglia, and L. Sacchi. 1999. Evidence of actin in the cytoskeleton of microsporidia Journal of Eukaryotic Microbiology **46:** 410-415.

_____, and L. Sacchi. 2001. Cell biology and invasion of the microsporidia. Microbes and Infection **3:** 373-379.

Boehm, M., and J.S. Bonifacino. 2002. Genetic analyses of adaptin function from yeast to mammals. Gene **286:** 175-186.

Bohne, W., D.J. Ferguson, K. Kohler, and U. Gross. 2000. Developmental expression of a tandemly repeated, glycine- and serine-rich spore wall protein in the microsporidian pathogen *Encephalitozoon cuniculi*. Infection and Immunity **68:** 2268-2275.

Bonafonte, M.T.., J. Stewart, and J.R. Mead. 2001. Identification of two putative ATP-cassette genes in *Encephalitozoon intestinalis*. International Journal for Parasitology **31:** 1681-1685.

Bonanno, J.B., C. Edo, N. Eswar, U. Pieper, M.J. Romanowski, V. Ilyin, S.E. Gerchman, H. Kycia, F.W. Studier, A. Sali, and S.K. Burley. 2001. Structural genomics of enzymes involved in sterol/isoprenoid biosynthesis. Proceedings of the National Academy of Sciences USA 98: 12896-12901.

Bozdech, Z., and E. Schurr. 1999. Protein transport in the host cell cytoplasm and ATP-binding cassette proteins in *Plasmodium falciparum*-infected erythrocytes. Novartis Found Symposium, **226:** 231-241.

_____, J. VanWye, K. Haldar, and E. Schurr. 1998. The human malaria parasite *Plasmodium falciparum* exports the ATP-binding cassette protein PFGCN20 to membrane structures in the host red blood cell. Molecular and Biochemical Parasitology **97:** 81-95.

Cali, A., L.M. Weiss, and P.M. Takvorian. 2002. *Brachiola algerae* spore membrane systems, their activity during extrusion, and a new structural entity, the multilayered interlaced network, associated with the polar tube and the sporoplasm. Journal of Eukaryotic Microbiology **49:** 164-174.

Chin, L.S., C. Weigel, and L. Li. 2000. Transcriptional regulation of gene expression of sec6, a component of mammalian exocyst complex at the synapse. Molecular Brain Research **79:** 127-137.

Conchon, S., X. Cao, C. Barlowe, and H.R. Pelham. 1999. Got1p and Sft2p: membrane proteins involved in traffic to the Golgi complex. EMBO Journal **18:** 3934-3946.

Coppens, I., A.P. Sinai, and K.A. Joiner. 2000. *Toxoplasma gondii* exploits host low-density lipoprotein receptor-mediated endocytosis for cholesterol acquisition. Journal of Cell Biology **149**: 167-180.

Cornillot, E., G. Méténier, C.P. Vivarès, and E. Dassa. 2002. Comparative analysis of sequences encoding ABC systems in the genome of the microsporidian *Encephalitozoon cuniculi*. FEMS Microbiology Letters **210**: 39-47.

Craighead, M.W., S. Bowden, R. Watson, and J. Armstrong. 1993. Function of the ypt2 gene in the exocytic pathway of *Schizosaccharomyces* pombe. Molecular Biology Cell **4**: 1069-7106.

Cutler, J.E. 2001. N-glycosylation of yeast, with emphasis on *Candida albicans*. Medical Mycology **39**: Suppl 1:75-86.

Daleke, D.L., and J.V. Lyles. 2000. Identification and purification of aminophospholipid flippases. Biochimical and Biophysics Acta **1486**: 108-127.

Dassa, E., and P. Bouige. 2001. The ABC of ABCS: a phylogenetic and functional classification of ABC systems in living organisms. Research in Microbiology **152**: 211-229.

Dean, M., A. Rzhetsky, and R. Allikmets. 2001. The human ATP-binding cassette (ABC) transporter superfamily. Genome Research **11**: 1156-66.

Douglas, S., S. Zauner, M. Fraunholz, M. Beaton, S. Penny, L. T. Deng, X.N. Wu, M. Reith, T. Cavalier-Smith, and U.G. Maier. 2001. The highly reduced genome of an enslaved algal nucleus. Nature **410**: 1091-1096.

El Alaoui, H., J. Bata, D. Bauchart, J.C. Dore, and C.P. Vivares. 2001. Lipids of three microsporidian species and multivariate analysis of the host-parasite relationship. Journal of Parasitology **87**: 554-559.

El Alaoui, H., J. Bata, P. Peyret, and C.P. Vivarès. 2001. *Encephalitozoon cuniculi* (Microspora): characterization of a phospholipid metabolic pathway potentially linked to therapeutics. Experimental Parasitology **98**: 171-179.

Epple, U.D., I. Suriapranata, E.L. Eskelinen, and M. Thumm. 2001. Aut5/Cvt17p, a putative lipase essential for disintegration of autophagic bodies inside the vacuole. Journal of Bacteriology **183**: 5942-5955.

Ernst, J.F., and S.K. Prill. 2001. O-glycosylation. Medical Mycology **39**: Suppl 1: 67-74.

Fast, N.M., and J.P. Keeling. 2001. Alpha and beta subunits of pyruvate dehydrogenase E1 from the microsporidian *Nosema locustae*: mitochondrion-derived carbon metabolism in microsporidia. Molecular and Biochemical Parasitology **117**: 201-209.

Finger, F.P., T.E. Hughes, and P. Novick.1998. Sec3p is a spatial landmark for polarized secretion in budding yeast. Cell **92**: 559-571.

Fischer von Mollard, G. and T.H. Stevens. 1999. The *Saccharomyces cerevisiae* v-SNARE Vti1p is required for multiple membrane transport pathways to the vacuole. Molecular Biology Cell **10**: 1719-1732.

Fotiadis, D., L. Hasler, D.J. Muller, H. Stahlberg, J. Kistler, and A. Engel. 2000. Surface tongue-and-groove contours on lens MIP facilitate cell-to-cell adherence. Journal of Molecular Biology **300**: 779-789.

Frixione, E., L. Ruiz, J. Cerbon, and A.H. Undeen. 1997. Germination of *Nosema algerae* (Microspora) spores: conditional inhibition by D2O, ethanol and Hg2+ suggests dependence of water influx upon membrane hydration and specific transmembrane pathways. Journal of Eukaryotic Microbiology **44**: 109-116.

Gardner,M.J., N. Hall, E. Fung, O. White, M. Berriman et al. 2002. Genome sequence of the human malaria parasite *Plasmodium falciparum*. Nature **419**: 498-511.

Hashimoto, H., M. Abe, A. Hirata, Y. Noda, H. Adachi, and K. Yoda. 2002. Progression of the stacked Golgi compartments in the yeast *Saccharomyces cerevisiae* by overproduction of GDP-mannose transporter. Yeast **19**: 1413-1424.

Hauser, M., V. Narita, A.M. Donhardt, F. Naider, and J.M. Becker. 2001. Multiplicity and regulation of genes encoding peptide transporters in *Saccharomyces cerevisiae*: Molecular Membrane Biology **18**: 105-112.

Hayman, J.R., S.F. Hayes, J. Amon, and T.E. Nash. 2001. Developmental expression of two spore wall proteins during maturation of the microsporidian *Encephalitozoon intestinalis*. Infection and Immunity **69:** 7057-7066.

Henrissat, B., E. Deleury, and P.M. Coutinho. 2002. Glycogen metabolism loss: a common marker of parasitic behaviour in bacteria? Trends in Genetics **18:** 437-440.

Hsu, S.C., C.D. Hazuka, D.L. Foletti, and R.H. Scheller. 1999. Targeting vesicles to specific sites on the plasma membrane: the role of the sec6/8 complex. Trends in Cell Biology **9:** 150-153.

Ikezawa, H. 2002. Glycosylphosphatidylinositol (GPI)-anchored proteins. Biological Pharmacology Bulletin **25:** 409-417.

Jantti, J., M.K. Aalto, M. Oyen, L. Sundqvist, S. Keranen, and H. Ronne. 2002. Characterization of temperature-sensitive mutations in the yeast syntaxin 1 homologues Sso1p and Sso2p, and evidence of a distinct function for Sso1p in sporulation Journal of Cell Science **115:** 409-420.

Jazwinski, S.M., and A. Conzelmann. 2002. LAG1 puts the focus on ceramide signaling. Intternaional Journal for Biochemistry and Cell Biology **34:** 1491-1495.

Johnson, A.E., and M.A. van Waes. 1999. The translocon: a dynamic gateway at the ER membrane. Annual Reviews in Cell Developmental Biology **15:** 799-842.

Katinka, M.D., S. Duprat, E. Cornillot, G. Méténier, F. Thomarat, G. Prensier, V. Barbe, E. Peyretaillade, P. Brottier, P. Wincker, F. Delbac, H. El Alaoui, P. Peyret, W. Saurin, M. Gouy, J. Weissenbach, and C. P. Vivares. 2001. Genome sequence and gene compaction of the eukaryote parasite *Encephalitozoon cuniculi*. Nature **414:** 450-453.

Kinclova, O., S. Potier, and H. Sychrova. 2002. Difference in substrate specificity divides the yeast alkali-metal-cation/H(+) antiporters into two subfamilies. Microbiology **48:** 1225-1232.

Knop, M., A. Finger, T. Braun, K. Hellmuth, and D.H. Wolf. 1996. Der1, a novel protein specifically required for endoplasmic reticulum degradation in yeast. EMBO Journal **15:** 753-763.

LaGier, M.J., J.S. Keithly, and G. Zhu. 2002. Characterisation of a novel transporter from *Cryptosporidium parvum*. International Journal for Parasitolology **32:** 877-887.

Lagorce, A., V. Le Berre-Anton, B. Aguilar-Uscanga, H. Martin-Yken, A. Dagkessamanskaia, and J. Francois. 2002. Involvement of GFA1, which encodes glutamine-fructose-6-phosphate amidotransferase, in the activation of the chitin synthesis pathway in response to cell-wall defects in *Saccharomyces cerevisiae*. European Journal of Biochemistry **269:** 1697-1707;

Lander, E.S., L.M. Linton, B. Birren, C. Nusbaum, M.C. Zody et al. Initial sequencing and analysis of the human genome. Nature **409:** 860-921.

Lauer, S., J. VanWye, T. Harrison, H. McManus, B.U. Samuel, N.L. Hiller, N. Mohandas, and K. Haldar. 2000. Vacuolar uptake of host components, and a role for cholesterol and sphingomyelin in malarial infection. EMBO Journal **19:** 3556-3564.

Leitch, G.J., M. Scanlon, A. Shaw, and G.S. Visvesvara. 2001. Role of P glycoprotein in the course and treatment of *Encephalitozoon* microsporidiosis. Ant microbial Agents and Chemotherapy **45:** 73-78.

Leitch, G.J., M. Scanlon, G.S. Visvesvara, and S. Wallace. 1995. Calcium and hydrogen ion concentrations in the parasitophorous vacuoles of epithelial cells infected with the microsporidian *Encephalitozoon hellem*. Journal of Eukaryotic Microbiology **42:** 445-451.

Lill, R., and G. Kispal. 2000. Maturation of cellular Fe-S proteins: an essential function of mitochondria. Trends in Biochemical Science **25:** 352-236.

Lipschutz, J.H., and K.E. Mostov. 2002. Exocytosis: the many masters of the exocyst. Current Biology **12:** R212-R214.

Luberto, C., D.L. Toffaletti, E.A. Wills, S.C. Tucker, A. Casadevall et al. 2001. Roles for inositol-phosphoryl ceramide synthase 1 (IPC1) in pathogenesis of C. neoformans. Genes Development **15:** 201-212.

Luo, Z., aand D. Gallwitz. 2003. Biochemical and genetic evidence for the involvement of yeast Ypt6-GTPase in protein retrieval to different golgi compartments. Journal of Biological Chemistry **278**: 791-799.

Lussier, M., A.M. Sdicu, and H. Bussey. 1999. The KTR and MNN1 mannosyltransferase families of *Saccharomyces cerevisiae*. Biochemical and Biophysics Acta **1426**: 323-334.

Maia, J.C. 1994. Hexosamine and cell wall biogenesis in the aquatic fungus *Blastocladiella emersonii*. FASEB Journal **8**: 848-853.

Méténier, Guy, Vivarès CP. "Genomics of unicellular eukaryotic parasite: Microsporidial paradigm" In *Organelles, genomes and eukaryotic phylogeny: an evolutionary synthesis in the age of* genomics, Robert P. Hirt, Horner DS eds. Taylor and Francis Books Ltd, London. In press.

Mittleider, D., L.C. Green, V.H. Mann, S.F. Michael, E.S. Didier, and J.P. Brindley. 2002. Sequence survey of the genome of the opportunistic microsporidian pathogen, *Vittaforma corneae*. Journal of Eukaryotic Microbiology. **49**: 393-401.

Monod, M., S. Capoccia, B. Lechenne, C. Zaugg, M. Holdom, and O. Jousson. 2002. Secreted proteases from pathogenic fungi. International Journal of Medical Microbiology **292**: 405-419.

Munro, S. 2001. What can yeast tell us about N-linked glycosylation in the Golgi apparatus? FEBS Letters **498**: 223-227.

Nickel, W., B. Brugger, and F.T. Wieland. 2002. Vesicular transport: the core machinery of COPI recruitment and budding. Journal of Cell Science **115**: 3235-3240.

Nishikawa, A., J.B. Poster, Y. Jigami, and N. Dean. 2002. Molecular and phenotypic analysis of CaVRG4, encoding an essential Golgi apparatus GDP-mannose transporter. Journal of Bacteriology **184**: 29-42.

Nwaka, S., and H. Holzer. 1998. Molecular biology of trehalose and the trehalases in the yeast *Saccharomyces cerevisiae*. Progress in Nucleic Acid Research and Molecular Biology **58**: 197-237.

Orlean, P. 1992. Enzymes that recognize dolichols participate in three glycosylation pathways and are required for protein secretion. Biochemical Cell Biology **70**: 438-447.

Pal, A., B.S. Hall, D.N. Nesbeth, Hi Field, and M.C. Field. 2002. Differential endocytic functions of *Trypanosoma brucei* Rab5 isoforms reveal a glycosylphosphatidylinositol-specific endosomal pathway. Journal of Biological Chemistry **277**: 9529-9539.

Parlati, F., O. Varlamov, K. Paz, J.A. McNew, D. Hurtado T.H. Sollner, and J. E. Rothman. 2002. Distinct SNARE complexes mediating membrane fusion in Golgi transport based on combinatorial specificity. Proceedings of the National Academy of Science USA **99**: 5424-5429.

Pelham, H.R. 1998. Getting through the Golgi complex. Trends in Cell Biology **8**: 45-49.

_____. 2002. Insights from yeast endosomes. Current Opinions in Cell Biology **14**: 454-462.

Peng R., and D. Gallwitz. 2002. Sly1 protein bound to Golgi syntaxin Sed5p allows assembly and contributes to specificity of SNARE fusion complexes Journal of Cell Biology **157**: 645-655.

Rehling, P., T. Darsow, D. J. Katzmann, and S.D. Emr. 1999. Formation of AP-3 transport intermediates requires Vps41 function Nature Cell Biology **1**: 346-353.

Richard, M., P. De Groot, O. Courtin, D. Poulain, F. Klis, and C. Gaillardin. 2002. GPI7 affects cell-wall protein anchorage in *Saccharomyces cerevisiae* and *Candida albicans*. Microbiology **148**: 2125-2133.

Robibaro, B., H.C. Hoppe, M. Yang, I. Coppens, H.M. Ngo, T.T. Stedman, K. Paprotka, and Joiner KA . 2001. Endocytosis in different lifestyles of protozoan parasitism: role in nutrient uptake with special reference to *Toxoplasma gondii*. International Journal for Parasitology **31**: 1343-1353.

_____, T.T. Stedman, I. Coppens, H.M. Ngo, M. Pypaert, T. Bivona, H.W. Nam, and K.A. Joiner. 2002. *Toxoplasma gondii* Rab5 enhances cholesterol acquisition from host cells. Cellular Microbiology **4**: 139-152.

Rodman, J.S., and A. Wandinger-Ness. 2000. Rab GTPases coordinate endocytosis. Journal of Cell Science **113**: 183-192.

Rozmajzl, P.J., M. Kimura, C.J. Woodrow, S. Krishna, and J.C. Meade. 2001. Characterization of P-type ATPase 3 in *Plasmodium falciparum*. Molecular and Biochemical Parasitology **116**: 117-126.

Rudge, S.A., C. Zhou, and J. Engebrecht. 2002. Differential regulation of *Saccharomyces cerevisiae* phospholipase D in sporulation and Sec14-independent secretion. Genetics **160**: 1353-1361.

Sabharanjak, S., P. Sharma, R.G. Parton, and S. Mayor. 2002. GPI-anchored proteins are delivered to recycling endosomes via a distinct cdc42-regulated, clathrin-independent pinocytic pathway. Developmental Cell **2**: 411-423.

Saliba, K.J., and K. Kirk. 2001. Nutrient acquisition by intracellular apicomplexan parasites: staying in for dinner. International Journal for Parasitology **31**: 1321-30.

Sato, K., M. Sato, and A. Nakano. 2001. Rer1p, a retrieval receptor for endoplasmic reticulum membrane proteins, is dynamically localized to the Golgi apparatus by coatomer. Journal of Cell Biology **152**: 935-944.

Schmid, S.L., M.A. Mc Niven, and P. De Camilli. 1998. Dynamin and its partners: a progress report. Current Opinions in Cell Biology **10**: 504-512.

Sever, S. 2002. Dynamin and endocytosis. Current Opinions in Cell Biology **14**: 463-467.

Sokolova, Y., E. Snigirevskaya, E. Morzhina, S. Skarlato, A. Mironov, and Y. Komissarchik. 2001. Visualization of early golgi compartments at proliferate and sporogenic stages of a microsporidian *Nosema grylli*. Journal of Eukaryotic Microbiology **41** Suppl: 86S-87S.

Stamnes, M. 2002. Regulating the actin cytoskeleton during vesicular transport. Current Opiniona in Cell Biology **14**: 428-433.

Stedman, T.T., A.R. Sussmann, and K.A. Joiner. 2003. *Toxoplasma gondii* Rab6 mediates a retrograde pathway for sorting of constitutively secreted proteins to the Golgi. Journal of Biological Chemistry **278**: 5433-5443

Steiner, H.Y., F. Naider, and J.M. Beckers. 1995. The PTR family: a new group of peptide transporters. Molecular Microbiology **16**: 825-834.

Takvorian, P.M., and A. Cali. 1994. Enzyme histochemical identification of the Golgi apparatus in the microsporidian, *Glugea stephani*. Journal of Eukaryotic Microbiology **41**: 63S-64S.

Toh-E, A, and T. Oguchi. 2002. Genetic characterization of genes encoding enzymes catalyzing addition of phospho-ethanolamine to the glycosylphosphatidylinositol anchor in *Saccharomyces cerevisiae*. Genes Genetic Systmatics **77**: 309-322.

Tolner, B., K. Roy, and F.M. Sirotnak. 1997. Organization, structure and alternate splicing of the murine RFC-1 gene encoding a folate transporter. Gene **89**: 1-7.

Tsui, M.M., W.C. Tai, and D.K. Banfield. 2001. Selective formation of Sed5p-containing SNARE complexes is mediated by combinatorial binding interactions. Molecular Biology Cell **12**: 521-538.

Undeen, A.H., and E. Frixione. 1991. Structural alteration of the plasma membrane in spores of the microsporidium Nosema algerae on germination. Journal of Protozoology **38**: 511-518.

_____, and R.K. Vander Meer. 1999. Microsporidian intrasporal sugars and their role in germination. Journal of Invertebrate Pathology **73**: 294-302.

Vavra, J., and J.I.R. Larsson. 1999. "Structure of the microsporidia". In *The Microsporidia and Microsporidiosis*, Wittner Murray , Weiss Louis M eds. ASM, Washington, 1999.

Venter, J.C., M.D. Adams, E.W. Myers, P.W. Li, Mural RJ et al. 2001. The sequence of the human genome. Science **291**: 1304-1351.

Vivarès, C.P, 2001 : Introduction: the microsporidial world, a paragon for analyzing intracellular parasitism. Microbes and Infection **3**: 371-372.

_____, B.J. Martin, and H.J. Ceccaldi. 1980. Acides gras de trois microsporidies (Protozoa) et de leur hôte *Carcinus mediterraneus* (Crustacea) sain et parasité par *Thelohania maenadis*. Z Parasitenkunde **61** : 99-107.

Vivarès, C.P., M. Gouy, F. Thomarat, and G. Méténier. 2002. Functional and evolutionary analysis of a eukaryotic parasitic genome. Current Opinions in Microbiology **5:** 499-505.

_____, and G. Méténier. 2000. Towards the minimal eukaryotic parasitic genome. Current Opinions in Microbiology **3:** 463-467.

Walmsley, A.R., M.P. Barrett, F. Bringaud, and G.W. Gould. 1998. Sugar transporters from bacteria, parasites and mammals: structure-activity relationships. Trends in Biochemical Science **23:** 476-481.

Weidner, E., and A. Findley. 2002. Peroxisomal catalase in extrusion apparatus posterior vacuole of microsporidian spores. Biological Bulletin **203:** 212.

_____, A.M. Findley, V. Dolgikh, and J. Sokolova. 1999. "Microsporidian biochemistry and physiology". In *The Microsporidia and Microsporidiosis*, Wittner Murray , Weiss Louis M eds. ASM, Washington, 1999.

Williams, B.A., R.P.Hirt, J.M. Lucocq, and T.M. Embley. 2002. A mitochondrial remnant in the microsporidian *Trachipleistophora hominis*. Nature 2002, **418:** 865-869.

Wolf, Y.I.., L. Aravind, and E.V. Koonin. 1999. Rickettsiae and Chlamydiae: evidence of horizontal gene transfer and gene exchange. Trends in Genetics **15:** 173-175.

# INDEX

## A

Actinomyxida 191
albendazole 159, 161-164, 180-190
antibiotics 41, 44-47
apicoplast 39-47
astrocytes 57, 59, 60, 68, 72, 75-80, 82, 84

## B

B cells 72, 78, 79
Bcl-3 51, 54, 61
benzimidazoles 163, 183, 185
*Brachiola algerae* 112, 115, 119-121, 128, 134, 137, 139, 162, 167, 209
bradyzoite 1, 11, 67-69, 89-105
bradyzoite differentiation 89, 92-94, 96, 98, 100, 103-105

## C

CD40 51, 61
CD4$^+$ T cells 69, 70, 72, 74, 76, 80-82, 84, 85, 147, 148
CD8$^+$ T cells 71-74, 76, 79-83, 85, 146-148, 155
cat 1, 4-6, 8, 9, 11, 14, 15, 17, 39, 68, 129, 133
chemotherapy 39, 40
CNS 67-69, 72, 76
congenital 1, 5, 11, 13
cyst 56, 63, 65, 111, 114, 117, 120, 121

## D

dendritic cells 69-72, 74, 75, 78, 80, 85
dense granules 26, 27
drug development 21, 22

## E

encephalitis 67, 68, 81, 82
*Encephalitozoon cuniculi* 128, 131-133, 137-159, 161-164, 166, 170, 180, 182, 183, 185, 187-190 190, 194, 196, 197, 200, 223-242
*Encephalitozoon hellem* 131, 132, 138, 140, 137, 138, 149, 151, 156, 194-197
*Encephalitozoon intestinalis* 128, 130, 138, 140, 141, 146, 161, 163, 166, 184, 185, 186
*Enterocytozoon bieneusi* 128-130, 137-141, 145, 150-152, 154, 155, 160, 161, 179, 180, 182, 186, 210
endocytosis 223, 224, 228, 229
endospore 231
endosymbiosis 40, 41, 224

## F

fatty acid 39, 41-43, 45, 46
fatty acid biosynthesis 41, 43, 46, 233
fumagillin 159, 162, 163, 165-167,
fungi 189, 192-202, 211, 231, 232, 234, 238, 240

## G

gliding motility 22, 38

## H

heat shock protein 95, 101, 102

## I

IDO 55, 57, 59, 60
IFN-γ 51-61, 64, 68, 70-84, 140, 146, 148

IgA 142
IgG 142-145
IgM 142, 144, 145
IGTP 51, 55, 56, 60
IL-1 72, 74, 76, 77, 81
IL-4 81- 84, 147, 148
IL-6 76, 77, 81, 83
IL-10 51, 55, 61, 147, 153
IL-12 51, 54, 61, 65, 146, 148, 153
IL-10 76, 78, 81, 82, 84
IL-12 70, 71, 74-76, 80, 81, 83, 84
IL-18 51, 54, 61
iNOS 53-58, 61, 76, 78

# M

macrogamete 112
macrophage 54, 58, 69
meronts 224-226, 135, 138
merozoite 90, 113
metrocyte 112
microgamete 112
microglia 55, 56, 72, 76
microneme 24, 26, 27, 29-34, 34, 91
microtubule 24, 28
Microsporidia 127, 131, 134-137, 139-
    155, 159-163, 165-167, 170, 176, 179,
    189-211, 223-233, 236, 239, 241
moving junction 23, 24, 25, 224
Myxosporidia 190, 191

# N

NF-κB 51, 54
NK cells 53, 54, 70, 71, 74, 75, 80, 81
neutrophils 69, 70
nitric oxide 51, 53-57, 76, 77, 79, 81, 82
Nosema locustae 195, 196, 198, 199, 202

# O

octospore 209
oligoamine 159, 176
oocyst 1, 4, 11, 15, 17, 68, 69, 90, 113

# P

pansporoblast 203, 211
parasitophorous vacuole 3, 23, 25, 27,
    36, 40, 45, 90, 94-96, 138, 224, 225,
    227, 238, 242
Plasmodium berghei 44
Plasmodium falciparum 41-46, 195, 224,
    227, 229-231
pentamine 159, 176, 177
polar filament 190-192, 200, 203, 205,
    206, 211
polar tube 224, 232, 235, 238
polyamine 159, 167-176, 178-179
protease 21, 25, 27-29, 31-36
protein synthesis 39, 41, 44, 46

# R

ROI 55, 58
resistance 51, 53, 54, 56-63, 65
rhoptry 25, 29, 34, 35, 91, 225

# S

sarcocyst 111, 112, 114-117, 119, 120
Sarcocystis species 111, 112, 114-121
Sarcocystis cruzi 115-117
Sarcocystis fusiformis 112
Sarcocystis hirsuta 115, 116
Sarcocystis hominis 114
Sarcocystis meischeriana 111
Sarcocystis suihominis 114
sarcocystosis 111, 116, 117, 119-121
secretory protein 21, 26, 28-30, 32, 33,
    35, 36
subtilisin 27, 28, 34
spore 129, 130, 132, 135, 139, 144, 162,
    164, 165, 170, 172, 173, 177-179, 225,
    231, 234, 238, 239, 241, 242
spore coat 144, 190, 200, 205
sporoblast 203
sporocysts 4, 5, 111, 113-116, 118, 119,
    121
sporogonial plasmodium 203, 210
sporogony 112, 203, 205, 231, 235, 236,
    238

sporoplasm 224, 228, 235, 242
sporozoite 2, 4, 5, 69, 89-91, 98, 100, 101
stress conditions 92, 99, 102

---

# T

T cells 70-76, 78-85, 140, 142, 146-148
TGF-β 51, 55, 61, 65, 80
TNF-α 54, 56- 62, 70, 72, 74, 76, 77-81, 84
tachyzoite 1-3, 23, 25, 27, 33, 34, 40, 51-54, 56, 57, 59, 60, 62, 64, 67, 68, 71, 76, 77, 83, 89, 90, 93, 98-106
tetramine 176, 177, 179

thalidomide 188
tissue cyst 1, 3, 4, 7, 9, 11, 16, 67-69, 73, 78, 82, 83, 89-91, 99
*Toxoplasma gondii* 1-7, 9-13, 15-17, 21, 22, 25, 29, 36, 39, 40, 42-47, 51-65, 67-85, 89-106, 225, 228, 229, 237
toxoplasmic encephalitis 51, 53, 56-58, 62-64
toxoplasmosis 5, 7-11, 13, 14, 17, 21, 35, 36, 47, 61, 63-65
tryptophan 51, 56, 59